International Scientific and Theoretical Conference "Modeling and Computer Engineering in Mechanical Engineering: Theory, Practice, and Innovation" (MCEME-2022)

IOP Conference Series: Materials Science and Engineering
Volume 1277

Lviv, Ukraine
28 September – 1 October 2022

ISBN: 978-1-7138-7141-5
ISSN: 1757-8981

Printed from e-media with permission by:

Curran Associates, Inc.
57 Morehouse Lane
Red Hook, NY 12571

Some format issues inherent in the e-media version may also appear in this print version.

Institute of Physics
Dirac House, Temple Back
Bristol BS1 6BE UK

Phone: 44 1 17 929 7481
Fax: 44 1 17 920 0979

techtracking@iop.org

Additional copies of this publication are available from:

Curran Associates, Inc.
57 Morehouse Lane
Red Hook, NY 12571 USA
Phone: 845-758-0400
Fax: 845-758-2633
Email: curran@proceedings.com
Web: www.proceedings.com

TABLE OF CONTENTS

PREFACE

International scientific and theoretical conference "Modeling and Computer Engineering in Mechanical Engineering: Theory, Practice, and Innovation" - MCEME-2022 dedicated to the topics of modeling and computer engineering in mechanical engineering, took place over three days, respectively from Thursday, September 29 to Saturday, October 1 in 2022. The plenary session was held on Friday, September 30, 2022. Considering that Ukraine was attacked by Russia and there is a war, the program and organizational committees decided to hold the conference in a hybrid mode: both virtually, through Google Meet, and face to face, a plenary meeting in the conference hall, a meeting by sections in three designated auditoriums of the Department of Machine Design and Automotive Engineering, Lviv Polytechnic National University, Ukraine.

Despite periodic alarm sirens, also during the meetings, and the need to hide from rocket fire, the conference took place, and some of the conference participants from Ukraine and the Republic of Poland were also present in person.

The participants in the conference were informed, constantly on the news about the progress of the peer review process (there was a team of 19 academic reviewers), through general messages, posted on the conference website, but also individually, through the platform www.morressier.com.

The initial difficulties and compliance with the Morressier platform were overcome relatively quickly, and the positive results were not long in coming. Conference participants were informed, in advance, about the program of events, presentations, in general, all components of the MCEME-2022 scientific event. The conference program was posted on the conference website: https://science.lpnu.ua/mceme-2022

The conference was opened with a welcome speech by the Vice-Rector for Scientific Research of the Lviv Polytechnic National University, Dr.Sc.Tec., professor Ivan Demydov. Also, the Director of the Institute of Mechanical Engineering and Transport, Dr.Sc.Tec., professor Oleksii Lanets, addressed the participants and guests with a welcome speech Vice-Rector for Science and Innovation Vilnius Gediminas Technical University Prof., Dr. Dalius Navakauskas gave a welcome speech. Director of the Research Institute of Mechanics and Automation of Agro-Industrial Production of the National Academy of Agrarian Sciences of Ukraine, professor, Dr.Sc.Tec., Valerii Adamchuk. A member of the program committee who took part, a professor, Dr.Sc.Tec. Michal Bembenek (Akademia Górniczo-hutnicza im. Stanisława Staszica w Krakowie) made a welcome speech.

IOP Conf. Series: Materials Science and Engineering 1277 (2023) 011001 doi:10.1088/1757-899X/1277/1/011001

Next, it became necessary to announce a break in connection with the announcement of a rocket bombardment of the city of Lviv. The break lasted almost 2 hours.

Then the plenary session continued. The report "THE method of experimental determination of tangential stresses on the surface of flow around with a gaseous medium" was presented (authors: T. V. Dmytriv, O. S. Lanets, V. T. Dmytriv, I. V. Dmytriv, I. M. Horodetskyy, and N. G. Horodetska), reported by Vasyl Dmytriv, prof., Dr.Sc.Tec., Department of Machine Design and Automotive Engineering, Lviv Polytechnic National University, Ukraine. Then the partners of the conference made a welcome speech.

After the lunch break, the work of the MCEME-2022 conference was resumed, in three sections in three different halls by the decision of the editorial committee:

- SECTION I - Latest Technologies and Materials in Mechanical Engineering - Chairman: V. Stupnytskyy, Dr.Sc.Tec., professor
- SECTION II – Modeling of Technical Systems and Processes; - Experimental Mechanics - Chairman: M. Bembenek, DScTec, Professor / P. Yatsunskyi, PhD
- SECTION III – Dynamics and Strength of Machines; - New Concepts in Vibration Technologies and Automated Systems in MechanicaL Engineering - Chairman: Z. Stotsko, Dr.Sc.Tec., Professor / I. Dmytriv, PhD, associate professor

The conference program was controlled by the organizers and coordinated by the organizing committee, each section began its work with the presentation of the articles presented at the conference in LPNU (Lviv, Ukraine), after which the online presentation was continued by the authors who could not be physically present.

After the presentation of each participant's report, comments, and clarifications requested by the participants, discussions took place. In total, 15 minutes were allotted for each article, 5 minutes of which were for discussion. It was rare that these intervals were not observed.

The two independent reviewers assigned to each article had the mission to generate a Review Report in which each work was evaluated, based on a Scoring Review System, in five categories: [1] Novelty; [2] Scope, topic, and structure; [3] Relevance; [4] Quality and citations / - displaying results, interpretation and consistently; [5] Scientific feasibility.

According to the reviewers' comments, the Editor was the one who had the mission to make the decision 1) -immediate acceptance; 2) - the return of the article for review by the authors; or 3) - the rejection of the article. Thus, two independent reviewers reviewed each paper, in part, in a Single anonymous system, the authors not knowing the identity of the reviewers, the editor-in-chief validating, or not, and the decisions of the two reviewers for each article. The acceptance rate of the works was 72%. Were rejected 28% of the works. The large number of rejected works in the review process is explained by the seriousness and professionalism of the reviewers, as well as a significant increase in demandingness. When evaluating plagiarism or self-plagiarism, literature was not taken into account, which was also evaluated together with the text of the article.

The program committee and the organizing committee sincerely thank all the participants of the MCEME-2022 conference, because they were able to participate in a difficult time for the scientific community of Ukraine and with their presentations added scientific value to the event organized and coordinated by us.

With deep consideration,
Program Committee and Organizing Committee
International Scientific and Theoretical Conference
"Modeling and Computer Engineering in Mechanical Engineering: Theory, Practice, and Innovation" – MECME-2022
September 29, 2022

IOP Conf. Series: Materials Science and Engineering 1277 (2023) 011002 doi:10.1088/1757-899X/1277/1/011002

Peer Review Statement

All papers published in this volume have been reviewed through processes administered by the Editors. Reviews were conducted by expert referees to the professional and scientific standards expected of a proceedings journal published by IOP Publishing Publishing.

- **Type of peer review:** Double Anonymous
- **Conference submission management system:** Morressier
- **Number of submissions received:** 47
- **Number of submissions sent for review:** 41
- **Number of submissions accepted:** 38
- **Acceptance Rate (Submissions Accepted / Submissions Received × 100):** 80.9
- **Average number of reviews per paper:** 2.05
- **Total number of reviewers involved:** 20
- **Contact person for queries:**
  Name: Vasyl Dmytriv
  Email: vasyl.t.dmytriv@lpnu.ua
  Affiliation: Lviv Polytechnic National University - Machine Desing and Automotive Engineering

The use of machine learning methods to predict the processes and results of high-voltage electric discharge processing of titanium powder in kerosene with the implementation of volume-distributed multi-spark discharge

M S Prystash, S F Prystash, A S Torpakov, Ye V Lypian, O M Syzonenko

Institute of Pulse Processes and Technologies of NAS of Ukraine, Bohoyavlenskyi ave., 43-A, 54018, Mykolaiv, Ukraine

Abstract: The possibility of using machine learning methods to predict the results of high-voltage electric discharge treatment of titanium powder in a hydrocarbon liquid is studied. Distribution surfaces for the average particle diameter of Titanium powder. The amount of Titanium carbide formed during processing, and the number of spherical particles of titanium powder depending on the interelectrode gap and the number of pulses, when using volume-distributed multi-spark discharge and with Titanium powder concentration in kerosene of 0.07 kg / dm^3, pulse repetition frequency 0.3 Hz and the energy of single discharge of 1 kJ, were obtained.

Keywords: MACHINE LEARNING, RANDOM FOREST, LOGISTIC REGRESSION, HIGH VOLTAGE ELECTRIC DISCHARGE, TITANIUM.

1. Introduction

Traditional methods of analyzing the composition and properties of materials are time-consuming and expensive. Modern laboratories, on average, need up to 10 years to develop new materials and up to 20 years to begin their mass production. However, the use of modeling technologies in the development of new materials allows to reduce the time of development and introduction of new materials by almost two times due to the high speed of obtaining results, repeatability, safety, quick detection of errors and low cost. There are two large groups of modern modeling and forecasting methods [1]:

- modeling based on physical models;
- modeling based on large databases (machine learning).

The approach to modeling based on physical models consists in observing the physical phenomenon of interest, forming a correct representation of it, formulating this representation in the form of mathematical equations, and finally solving these equations. Accurate engineering calculations give this method physical realism.

The advantages of this method include:

- it is based on physics and logic;
- it is easy to generalize to similar physical problems.

The main disadvantages are as follows:
- it can take a lot of time and cost a lot;
- it has a large error in problems with nonlinearity;
- limit conditions are taken into account;
- with increasing complexity of calculations, it is difficult to reconcile the results with real tasks.

Modeling based on large databases has emerged due to the increase in digital data we receive about the world around us, and high-quality and constant access to storage and prediction resources makes this method increasingly popular. Contrary to the physical model-based approach, this method is based on the assumption that as data comes from both well-studied and not-so-well-studied branches of physics relevant to the question under study, the development of a model based on large databases data can take into account all the physics of the phenomenon together.

The advantages of this method include:
- it takes into account data and experience for the previous long period of time;
- resulting models are stable and quickly make predictions.

The disadvantages:
- poor generalization of undetected problems;
- data deviations are reflected in forecasts created by the model;
- most modern algorithms work like a "black box".

High-voltage electric discharge treatment (HVED) of powder mixtures is one of the modern methods of grinding and changing the phase composition of powders, which is distinguished not only by its efficiency, but also by the low cost of the process in comparison with classical grinding methods. However, the physical mechanism of dispersion processes when using HVED in a liquid has not yet been sufficiently studied. In works [2, 3], it is suggested that all the impact factors of HVED can be divided into two groups – mechanical, which include the compression wave, hydroflows during the pulsation of the vapor-gas cavity, and cavitation, and thermal, which include the low-temperature plasma in the discharge channel and microplasma formations between particles, as well as a discharge current flowing through the particles of the processed metal and leading to electric erosion destruction of the powder particles.

The main role in the destruction of material particles located in the discharge zone is played by the shock wave generated by the spark discharge channel [2, 3]. In addition, volumetric microcavitation, destruction of particles when they hit the walls of the discharge chamber, and mutual abrasion of particles during mixing in the entire volume of the chamber are involved in the dispersion process to varying degrees. The choice of titanium powder for this study is due to the relevance and wide possibilities of its use as a filler, modifier, or component of metal matrix composites.

Taking this into account, the use of the modeling method on large databases can provide more accurate prediction results of the processes and results of HVED processing. Therefore, the **goal of this work** is to investigate the possibility of using machine learning methods to predict the processes and results of high-voltage electric discharge processing of titanium powder in a hydrocarbon liquid.

2. Methodology

In general, the process of machine learning can be represented by the diagram shown in figure 1. It consists of the input data, the algorithm, and the results that were obtained. As an addition, the data of the previous "training" of the model can be added to the process, which allows you to assess the accuracy of the forecast. This model is called "controlled".

Logistic regression and the Random forest method were chosen as modeling algorithms. The implementation language was the Python programming language and the Collaboratory application, or Colab for short, which is a product of Google Research. Colab allows writing and executing arbitrary Python code through the browser, and is particularly well-suited for machine learning and data analysis.

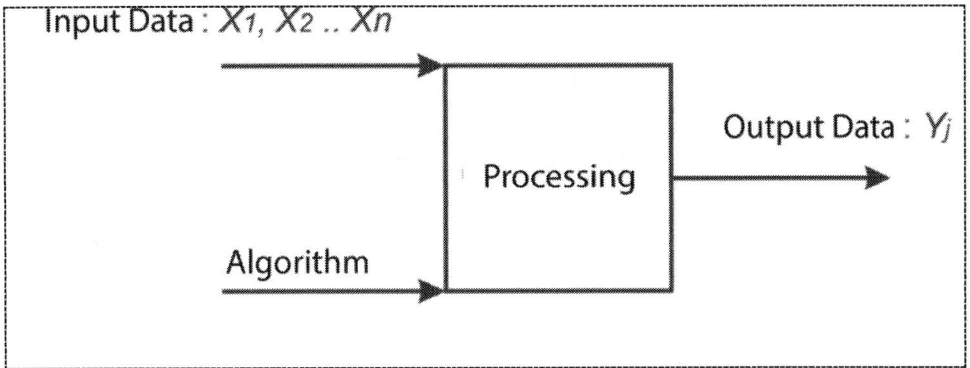

Figure 1. General scheme of modeling using machine learning

Logistic regression is a classification algorithm used to relate data to a discrete set of classes. Unlike linear regression, which outputs continuous numerical values, logistic transforms the raw data using a sigmoid (see figure 2). This is one of the simplest algorithms that can be used to solve classification problems [4]. Sigmoid formula:

$$f(x) = \frac{1}{1 + e^{-(x)}},$$

where x – processed value.

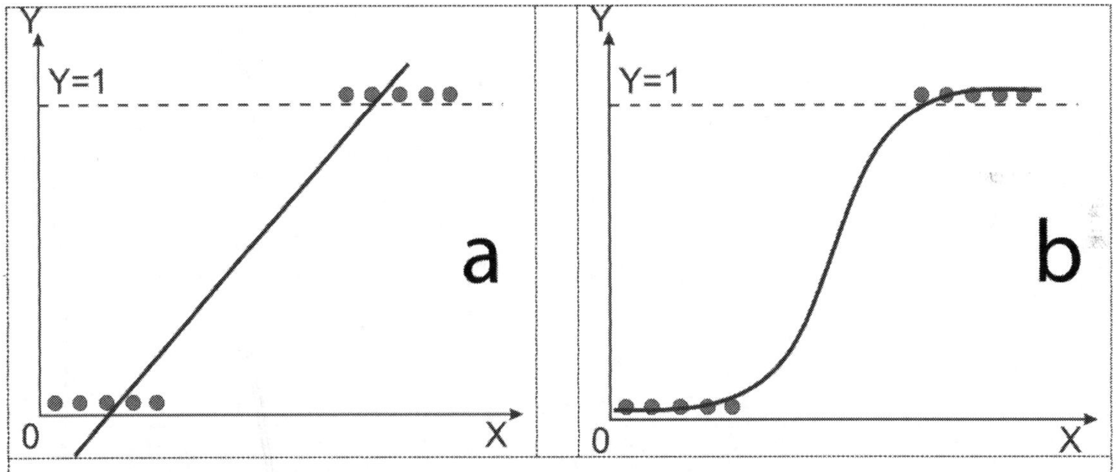

Figure 2. Illustration of linear (a) and logistic (b) regression methods

Random forest is a machine learning method for classification, regression, and other tasks that works by building numerous decision trees during model training. Random fores produces an averaged forecast (regression) of the built trees (see figure 3) [5, 6].

Advantages of the Random forest method:

- The ability to efficiently process data with a large number of attributes and classes.
- Insensitivity to scaling (and in general to any monotonic transformations) of feature values.
- Both continuous and discrete features are handled equally well. There are methods for building trees from data with missing feature values.
- There are methods of assessing the significance of individual features in the model.
- There is an internal assessment of the model's ability to generalize.

Disadvantages of the method:
- The algorithm is prone to overtraining on some tasks, which creates an error.
- The large size of the received models and, accordingly, high equipment requirements.

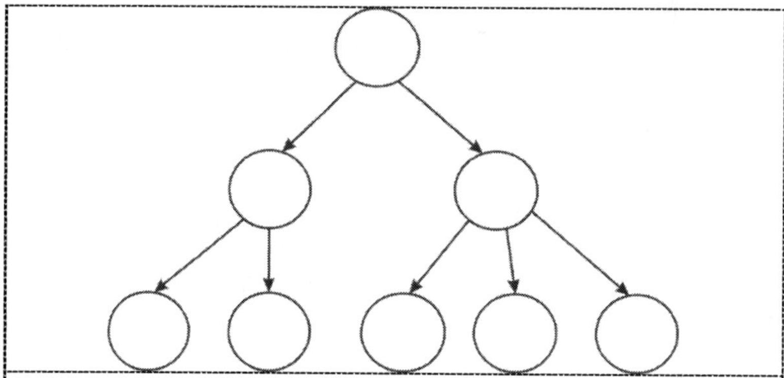

Figure 3. Illustration of Random forest method

The results of titanium powder high voltage electric discharge treatment in kerosene with various electrode systems obtained in the period from 2013 to 2021 and presented in works [7-9] were used as data for modeling. The use of different electrode systems makes it possible to implement a different distribution of plasma in the metal powder – liquid system, which affects both the hydrodynamic properties of the liquid and the process of grinding and phase transformations of titanium powder. The use of a single-point system allows you to implement a spark discharge (SD), and the use of a three-point system leads to the realization of a volume-distributed multi-spark discharge (VMD) [7-9] (see figure 4). As the results of SD implementation are already sufficiently presented in the cited works, the VMD needs a more detailed consideration.

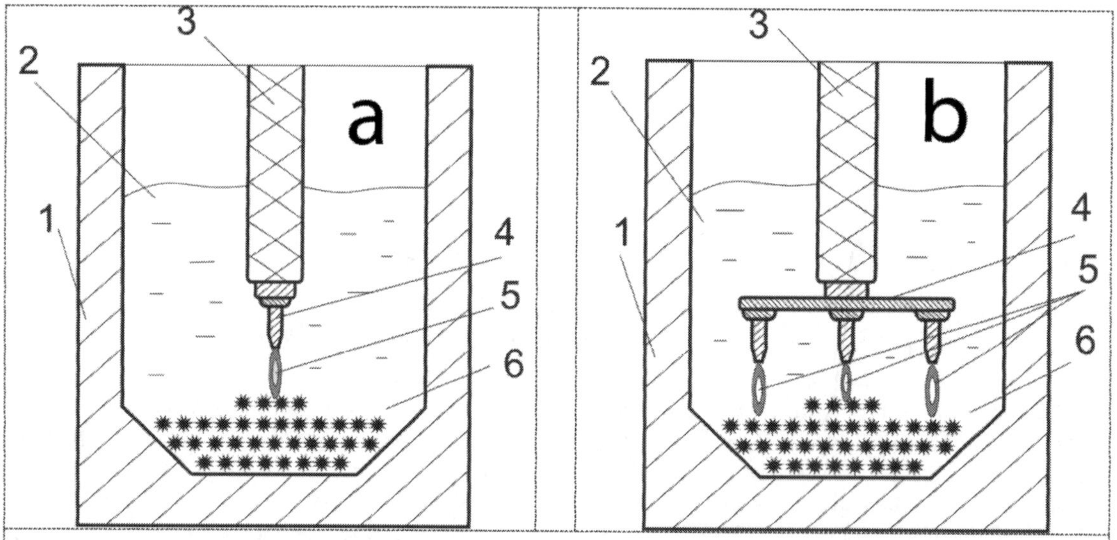

Figure 4. General scheme of implementation of SD (a) and VMD (b). 1 — processing chamber; 2 – hydrocarbon liquid; 3 – insulator; 4 – electrode; 5 – plasma channel(s); 6 – processed metal powder

The data used included information on the number of processing pulses, the interelectrode gap, the pressure in the discharge channel and the pressure on the chamber walls, which is discussed in detail in works [2, 3], as well as optical microscopy data of titanium powder particles and the amount of synthesized titanium carbide calculated using X-ray diffraction patterns of powders processed with the RIR method [10].

One of the modeling tasks was to determine the basic elementary characteristics of HVED processing, which can be directly controlled, and which would influence the results of the processing itself. For this purpose, pre-processing of the data was carried out. This made it possible to exclude fixed values for all selected treatments, namely:

- The concentration of titanium powder in the hydrocarbon liquid, which was 0.07 g/dm^3;
- The energy of a single discharge, which was equal to 1 kJ;
- The discharge passage frequency, which was 0.3 Hz.

This made it possible to determine the main variables that will become input data, namely the interelectrode gap and the number of impact pulses that affect the formation of the discharge channel and the characteristics of the medium. Plasma temperature, pressure in the discharge channel, pressure on the chamber wall, average particle diameter of the powder after treatment, the number of spherical particles and the amount of titanium carbide formed after HVED treatment were used as the output data.

A controlled machine learning model was used to predict the output data. For this, the data was divided in the ratio of 80 / 20 percent to determine the accuracy of the forecast, for which the Accuracy metric was used. Accuracy is an indicator that determines the overall accuracy of the model's prediction across all classes. This is especially useful when each class is equally important. Accuracy is calculated as the ratio of the number of correct predictions to their total number [11].

3. Results and discussion

According to the Accuracy metric, the accuracy of forecasts by logistic regression and Random forest methods is as follows:

- for plasma temperature – 40% and 50%, respectively;
- for pressure values in the discharge channel – 40% and 50%, respectively;
- for pressure values on the chamber wall – 50% and 70%, respectively;
- for the average size of the obtained titanium particles, the number of spherical particles, and the amount of titanium carbide formed – 60% and 70%, respectively.

As the result of the use of the logistic regression model, the following dependences of the plasma channel temperature on the size of the interelectrode gap and the number of processing pulses were obtained when using VMD in kerosene (see figure 5, a). It was found that the distribution is quite uneven, and the obtained values are about 28,000 K for the number of pulses from 200 to 1,000 in almost all values of the interelectrode gap, and up to 35,000 K when the number of pulses increases from 1,000 to 1,400. It should be noted separately that the plasma temperatures at 35,000K are characteristic for a gap from 2 mm to 14 mm, and are also observed with a linear increase in the interelectrode gap and the number of impact pulses, and in some cases can be up to 37,000 K. Zones of a decrease in the plasma temperature to ~ 28,000 K are also observed on the distribution, which can be associated with the changes occurring in the kerosene – titanium powder system.

The distributions obtained using Random forest differ from those obtained using the logistic regression method (see figure 5, b). According to Random forest modelling results, a plasma channel is not formed at all when using VMD in kerosene with the interelectrode gap from 2 mm to 8 mm. Furthermore, the plasma temperature in almost all possible cases is about 35,000 K over the entire range of the number of processing pulses.

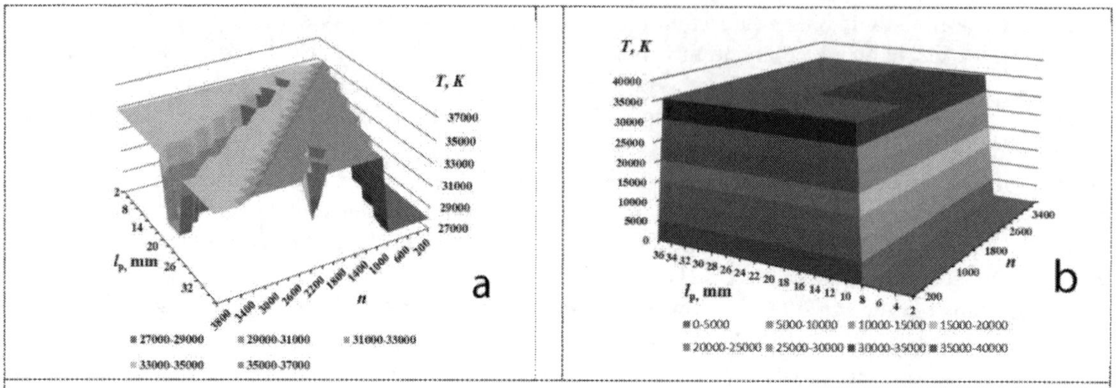

Figure 5. The distribution of the plasma temperature values in the discharge channel in case of VMD depending on the interelectrode gap and the number of pulses, obtained according to logistic regression (a) and Random forest (b) models

The distribution of pressure values in the discharge channel depending on the interelectrode gap and the number of pulses when using VMD according to logistic regression model is shown in figure 6,a. It can be seen that pressure values from 800 MPa to 1,300 MPa are characteristic to the zone of linear increase in both the interelectrode gap and the number of impact discharges. It should also be noted that the highest pressure value (up to 1,800 MPa) in the discharge channel is characteristic to the area with an inter-electrode gap of 2 mm to 6 mm starting with 1,000 processing pulses.

According to Random forest modelling results (see figure 6,b), the pressure in the discharge channel is zero when the interelectrode gap if from 2 mm to 8 mm. When the gap is from 8 to 18 mm, the pressure has a value of 800 MPa with the number of pulses from 200 to 3000, and its maximum values are up to 1.4 GPa, which are observed when the number of processing pulses if from 3,000 to 4,000. In the interval of interelectrode gap values from 20 mm to 36 mm when the number of pulses is from 2,000 to 4,000 the pressure is equal to 600 MPa.

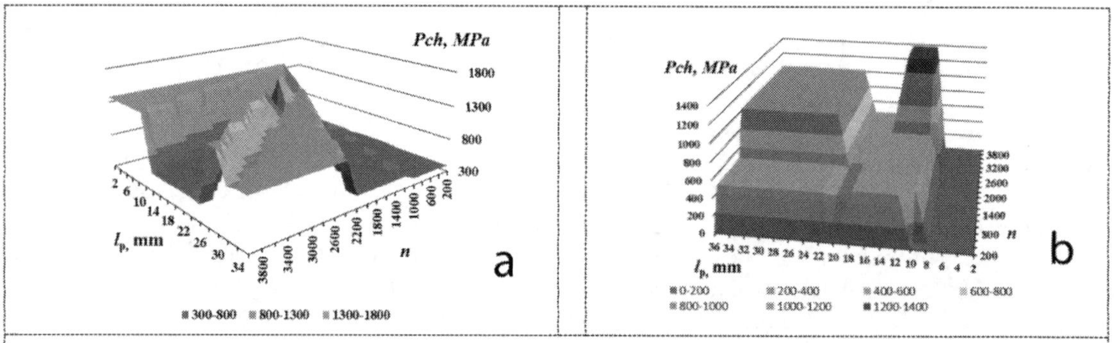

Figure 6. The distribution of pressure values in the discharge channel in case of VMD depending on the interelectrode gap and the number of pulses, obtained according to logistic regression (a) and Random forest (b) models

Consistently high values of pressure on the chamber wall (up to 130 MPa), according to logistic regression models, are characteristic to zones belonging to the linear dependence of the increase in the interelectrode gap and the number of processing discharges (see figure 7, a). The highest pressure values on the chamber wall of 150 MPa are characteristic to zones corresponding to the highest pressure values in the discharge channel – these are areas with an interelectrode gap from 2 mm to 6 mm and the number of processing pulses higher than 1,000.

According to Random forest modelling results, the distribution of pressure values on the wall of the processing chamber with VMD in kerosene is similar to the distribution of pressure in the discharge channel. Thus, the pressure on the chamber wall (see figure 7,b) is zero when the interelectrode gap is from 2 mm to 8 mm, for the gap from 8 to 18 mm it has a value of 40 MPa with the number of pulses from 200 to 3,000, and the maximum values up to 140 MPa are observed when the number of processing pulses is from 3,000 to 4,000. In the interval of interelectrode gap values from 20 mm to 36 mm and the number of pulses from 2,000 to 4,000, the pressure on the wall of the processing chamber is equal to 120 MPa.

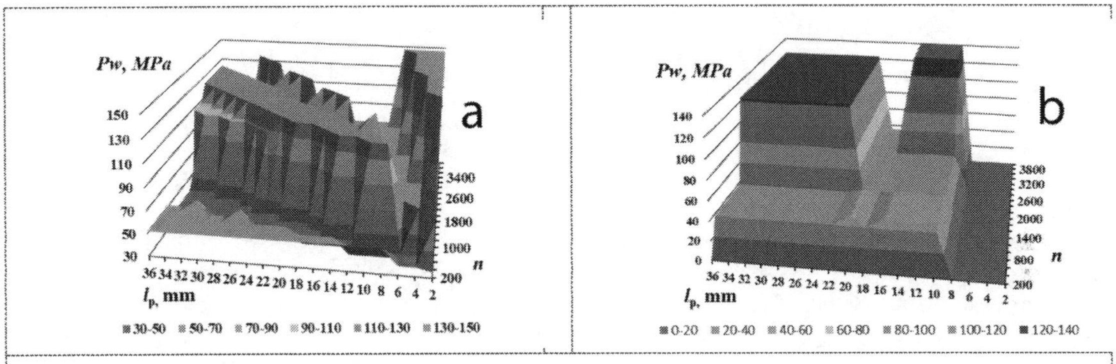

Figure 7. The distribution of pressure values on the wall of the processing chamber in case of VMD depending on the interelectrode gap and the number of pulses, obtained according to logistic regression (a) and Random forest (b) models

As a result of HVED treatment, due to the hydrodynamic impact and the impactof the plasma of the discharge channel on titanium powder in kerosene, its grinding, spheroidization of particles and chemical transformations with the formation of titanium carbide take place [7-9].

From figure 8,a, it can be seen that according to the logistic regression model, the grinding process begins already with 200 processing pulses at almost all values of the interelectrode gap and allows reducing the average particle diameter of Ti powder to 9 μm. Such a decrease in the average diameter is both due to the grinding and due to the formation of free solid phase carbon in the process of kerosene destruction. A further increase in the number of impact pulses allows reducing the value of the average diameter of Ti powder particles to 5 μm, and in the zone with an interelectrode gap from 2 mm to 14 mm and the number of pulses from 2,000 to 4,000 it allows obtaining particles with a diameter of ~ 4 μm.

The average diameter of Ti powder particles according to the Random forest model when using VMD in kerosene at an interelectrode gap from 2 mm to 8 mm remains unchanged and is 60 μm (see figure 8, b). With an increase in the interelectrode gap from 8 mm to 36 mm, the average diameter of Ti powder particles is approximately 9 μm starting from treatment modes with at least 200 processing pulses.

According to logistic regression models, HVED processing with VMD in almost all possible modes leads to an increase in the number of spherical powder particles due to the influence of the plasma of the discharge channel (see figure 9,a). Thus, at values of the interelectrode gap from 10 mm to 36 mm, and the number of processing pulses from 200 to 1,000, the number of spherical particles increases to 27%. Increasing the number of pulses from 1,000 to 4,000 allows increasing the number of spherical particles to 30% at all values of the interelectrode gap. The use of an interelectrode gap from 10 mm to 36 mm and the number of processing pulses from 1,000 to 2,000 will allow obtaining up to 33% of spherical particles, which is due to higher temperature values in the discharge channel.

According to Random forest model, the number of spherical particles (see figure 9mb) when using VMD in kerosene with an interelectrode gap in the range from 8 to 36 mm increases from 20% to 34% when the number of processing pulses increases from 200 to 1000. However, when increasing number of pulses from 1,000 to 4,000, the number of spherical particles drops to 30%.

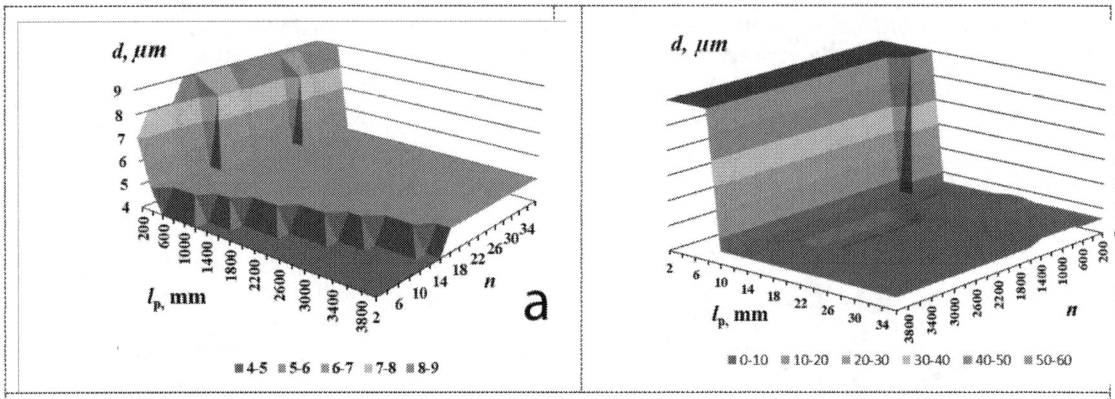

Figure 8. The distribution of Ti powder particles average diameter after VMD treatment depending on the interelectrode gap and the number of pulses, obtained according to logistic regression (a) and Random forest (b) models

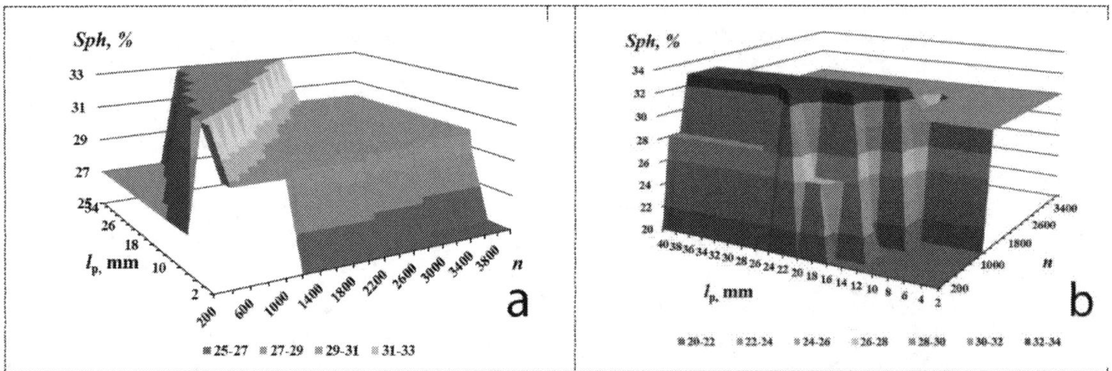

Figure 9. The distribution of the number of spherical particles of Ti powder after VMD treatment depending on the interelectrode gap and the number of pulses, obtained according to logistic regression (a) and Random forest (b) models

Fig. 10 a shows the data on the amount of titanium carbide formed during HVED processing with VMD according to logistic regression models. According to the obtained data, the amount of synthesized carbide gradually increases from 0 to 40% with an increase in the number of processing pulses from 200 to 4,000. However, for an interelectrode gap from 14 mm to 36 mm, this process actively starts from 1,000 processing pulses, and at 4,000 pulses the amount of formed carbide does not exceed 30%.

The predicted amount of titanium carbide that will be formed after HVED treatment using VMD in kerosene according to Random forest is shown on figure 10,b. Accordig to this model, carbidization does not occur in the range of interelectrode gap values from 2 mm to 8 mm. When the gap is from 8 to 36 mm, and the number of processing pulses is from 200 to 1000, about 15% of titanium carbide is formed. When the interelectrode gap is from 8 to 36 mm, and the number of processing pulses is from 1,000 to 4,000, the quantity of synthesized titanium carbide is about 25%.

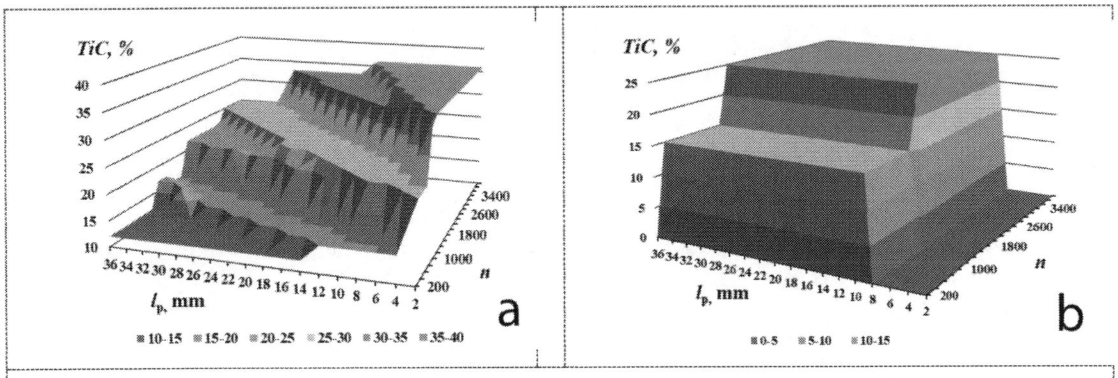

Figure 10. The distribution of the amount of titanium carbide formed during HVED processing with VMD depending on the interelectrode gap and the number of pulses, obtained according to logistic regression (a) and Random forest (b) models

4. Conclusions

The possibility of using machine learning methods to predict the values of the plasma temperature in the discharge channel, the pressure in the discharge channel, the pressure on the chamber walls, the average value of the particle size of titanium powder, the number of spherical particles and the amount of formed titanium carbide as a result of high-voltage electric discharge processing of titanium powder in kerosene with the implementation of VMD depending on the interelectrode gap and the number of processing pulses is shown.

It was shown that, according to logistic regression forecasts, the highest values of plasma temperature and pressure during processing can be obtained by proportionally increasing the value of the number of processing pulses in accordance with the value of the interelectrode gap, and grinding of titanium powder occurs in almost all of the studied modes. However, the greatest spheroidization of powder particles is characteristic to an interelectrode gap of more than 18 mm, while carbidization of titanium occurs better with a gap of up to 18 mm. An increase in the number of processing pulses for the area with the interelectrode gap up to 18 mm leads to the increase in the amount of synthesized titanium carbide.

Prediction results, obtained using Random forest, are more accurate if compared to experimental data and indicate that when using an interelectrode gap from 8 mm to 36 mm, plasma temperature values reach 35,000 K, the pressure in the discharge channel is 1.5 GPa and at the same interelectrode gap range and the number of pulses from 3,000 to 4,000 the pressure on the chamber walls is 140 MPa. The average size of titanium powder particles with an interelectrode gap from 10 mm to 36 mm is almost independent of the number of processing pulses, and the number of spherical particles is the largest with a number of processing pulses from 200 to 1,500. When the value of the interelectrode gap is from 8 mm to 36 mm, and the number of processing pulses is from 1,500 to 4,000, carbidization of titanium at the level of 25% can be achieved.

5. References

[1] Samarskii A A, Mikhailov AP 2001 *Principles of Mathematical Modelling: Ideas, Methods, Examples* 1st ed. (London: CRC Press).

[2] Gavrilov G N, Gorovenko G G, Malyushevskii P P, Ryabin A G 1979 *Pulsed discharge technology of mineral mediums treatment* (Kyiv: Naukova Dumka)
[3] Kurets V I, Usov A F, Tsukerman V A 2002 *Electric pulse disintegration of materials* (Apatity: Publishing house of the Kola Scientific Center of the Russian Academy of Sciences)
[4] Hassti T, Tibshirani R, Friedman J 2009 *The Elements of Statistical Learning, 2nd edition* (Berlin: Springer)
[5] Breiman L 2001 *Machine Learning* **45** 5–32
[6] Raschka S, Mirjalili V 2019 Machine learning and deep learning with Python, scikit-learn, and TensorFlow2 (Birmingham: Packt Publishing Ltd)
[7] Syzonenko O M, Loboda P I, Zaichenko A D, Solodkyi Ye V, Torpakov A S, Prystash M S, Trehub V O 2017 *Journal of Superhard Materials* **39, 4** 243–50
[8] Syzonenko O M, Prystash M S, Taftai E I, Torpakov A S, Lypian Ye V 2021 *Materials Science* **57** 209–14
[9] Sizonenko O, Prokhorenko S, Torpakov A, Żak A, Lypian Y, Wojnarowska-Nowak R, Polit J, Sheregii E M 2018 *AIP Advances* **8, 8** 085317
[10] Chipera S J, Bish D L 1995 *Powder Diffraction* **10** 47–55
[11] Powers D M W 2011 *Journal of Machine Learning Technologies* **2, 1** 37–63

Acknowledgments
Presented work was performed with partial financial support from Research Council of Lithuania and Ministry of Education and Science of Ukraine in the framefork of "Application of high-concentrated energy flows for producing nanostructured polyfunctional composite materials" joint project according to results of joint Ukrainian-Lithuanian R&D projects for the period of 2022 – 2023 contest.
Also, authors would like to express gratitude to the Armed Forces of Ukraine for their bravery which made this work possible even in the dark times of war.

Theoretical principles and practical aspects of the theory of transformation of interaction measures with external objects and the environment for ground robot complexes

V Zalypka[1]
[1]National Academy of Ground Forces, Lviv, Ukraine

zalypka_w@ukr.net

Abstract. The guiding documents regarding ground robotic complexes (GRC) of the USA were analyzed and it was found that the basis of the development of modern GRC is artificial intelligence. It was found out that GRC is the preservation of life and health of servicemen in battle. It was established that the transformation of the means of interaction with external objects and the environment (MIEOE) is due to the need to improve the operational properties of the GRC. A new idea is proposed to provide a dual function of the MIEOE for the GRC due to their transformation. Individual cases of the transformation of the wheel drive and the design of the manipulators were analyzed, and a mathematical description of the transformation of the circle into figures for the construction of the wheel drives of the GRC of increased cross-country ability was carried out. It has been established that the promising MIEOE GRC can be manufactured in a combination of different materials and combine the latest technologies for the use of gallium nitride transistors, soft crystals and graphene thread (films).

1. Introduction

At the current stage of the development of the armed forces of the leading countries of the world and Ukraine, and taking into account the intensity of hostilities in various theaters of war [1-5], in particular during the Russian-Ukrainian war, the main trend is the gradual displacement of servicemen from those areas of official activity that are the most dangerous for their life and health, while they are being replaced by automated and robotic systems that completely take over the functions previously performed by military personnel. The results of processing the available information sources indicate that a similar trend is occurring in the development of ground-based robotic military complexes (GRC). Today, it is quite obvious to understand that only the proposals of new ideas, principles and technologies in the creation of new samples of GRC are able to ensure a significant improvement of their operational properties, such as passability, stability, maneuverability and, as a result, ensure superiority on the battlefield over the enemy and the survivability of such means in general In this regard, research related to solving the problems of improving the operational properties of GRC acquire special importance. One of the ways to improve operational properties is the transformation of means of interaction with external objects and environment (MIEOE) for GRC. The main scientific achievements regarding the creation and application of GRC in Ukraine were reflected in the works of such authors as V. Horbulin [6], O. Guslyakov, A. Dovgopoliy, I. Chepkov [7], O. Kuprinenko [8], O. Shershakov, V. Shemaev, V. Begma, N. Sklyar [9], however, this field is new for Ukrainian specialists and many essential aspects of the development and application of GRC require research. Thus, research that will improve the operational properties of the GRC due to the transformation of the MIEOE is relevant and important in the scientific and technical aspect.

The object of research: means of interaction with external objects and the environment for ground robotic complexes.

The subject of research: the process of transformation of the means of interaction with external objects and the environment for ground robotic complexes.

The purpose of the work: to determine the theoretical foundations and practical aspects of the transformation of the means of interaction with external objects and the environment for ground robotic complexes.

2. Materials and methods

The analysis of the Strategy of Robotic and Autonomous Systems of the US Army, the Integrated Road Map of the US, the Concept of Development and Application of GRC (Platforms) in the units of the Armed Forces of Ukraine was carried out. According to the results of the analysis, the obtained results are synthesized and the classification of GRC is given. The basic principles of theoretical mechanics, the theory of car motion, methods of the theory of differential equations are used to develop mathematical models that allow studying the kinematics and dynamics of the GRC movement. The basis of the methods of researching the theory of transformation of MIEOE GRC are: new principles of drive mechanisms; artificial intelligence technologies; fractal geometry for describing object transformations, finite element method; homology; affine transformations; functions of complex variables; Riemann sphere.

3. Results

Analyzing the US Army Robotic and Autonomous Systems Strategy (as a leader in this segment) (RAS) until 2035, the US Army Operational Concept until 2040 and the Integrated Unmanned Systems Roadmap for 2017-2042 [1-4] can be found how their military-political leadership will integrate new technologies into future projects to secure an advantage over an increasingly technologically prepared adversary: these are the development of fully autonomous robotic systems with extensive use of artificial intelligence that must interact with each other and with humans regardless of purpose and type. From the provisions of the Concept of the development and use of GRC (platforms) in units of the Ground Forces of the Armed Forces of Ukraine [5], we state that the GRC during use as part of units of the Armed Forces of Ukraine complement (replace) traditional types of weapons and military equipment in almost all forms and methods application of subdivisions. Although there are many problematic issues in Ukraine that hold back the development of GRC for the needs of the Armed Forces of Ukraine, there is an understanding that in the future they will be an integral part of units that perform combat tasks according to their purpose. As a result, relevant scientific research is still being conducted even under conditions of aggression by the Russian Federation and with limited financial and logistical support.

So, the GRC is an unmanned robotic platform, remote control points associated with it, the necessary control and control lines and other elements that allow automating the performance of assigned tasks, including without the participation of the operator [5] 'figure 1'.

Figure 1. GRC of the world's leading countries for various purposes

During the application of the GRC by military units [1-5]:

- complement (replace) traditional weapons and military equipment, ensuring the achievement of the set goal with minimal losses of personnel and reducing the impact of the human factor;
- can be used to solve a wide range of tasks (reconnaissance, shock, special) in various environmental conditions;
- can be used in all types of combat and during stabilization (specific) actions and special operations.

During the application of the GRC, combat tasks will be performed, as well as the tasks of combat, rear, technical and medical support. Thus, given the nature of the assigned tasks, it is most expedient to classify GRC according to their functional purpose and divide them into three groups [5]:

- combat GRC;
- special GRC, which are divided into GRC of combat support and GRC of combat operations support;

multi-purpose robotic platforms (MRP), which, depending on the purpose, have appropriate equipment (replaceable specialized modules) and are divided into (MRP of rear support; MRP of technical support; MRP of medical support).

GRC consists of three systems: transport, special, control. The transport system is a means of transport intended for the delivery of special and technological equipment to the place of performance of the assigned task. The vehicle consists of a chassis, body and power plant. Depending on the type of operating environment, the undercarriage can be: tracked, wheeled, wheeled-tracked, semi-tracked, walking, wheel-walking, rotary, with loop, screw, water jet and jet engines. The control system includes: a motion control system, a system for coordinating the functioning of systems, and a special equipment control system. A special system is: special technological equipment, in particular, executive bodies (manipulators) and the technological equipment control subsystem. So, by the MIEOE for the GRC, we will understand their movers and manipulators. However, there is a significant functional difference between them: the former provide the mobility of the GRC, and the latter are specific in nature and contain various structures for different purposes of manipulating the object 'figure 2'.

Figure 2. GRC manipulator with end effector (capture)

Based on the above, the author is considering the possibility of providing them with new functions due to the transformation of MIEOE, in particular, the manipulator function of the driver and vice versa, or increasing the passability, stability, maneuverability of the GRC by transitioning, due to the transformation, from one driver to another: from wheeled to tracked or walking, from tracked to wheeled or walking, from walking to wheeled or tracked. This article proposes to analyze partial cases of the transformation of the most common drives (wheeled, tracked) of some currently known approaches in existing concepts, as well as to consider the first stage of research on the kinematic description of the transformation of wheels and the design features of manipulators and the perspective of their transformation.

To begin with, it is necessary to find out what a transformation is and what effect it has on objects. Transformation is a change, transformation of the appearance, form, essential properties of something. It includes the so-called rigid (displacement, reflection, rotation), non-rigid (expansion, shift) transformations. It should be noted that a special case of rigid transformations is affine (self-similar) transformations, as well as the fact that, unlike rigid transformations, the size or shape of a solid (figure) does not change with these transformations. Of course, when we talk about non-rigid transformation, it is necessary to understand that some properties of objects can change and they can have both a positive and a negative effect on them as a whole, so this should be taken into account when developing new technological solutions. Some well-known examples of successful, but not yet widely implemented, are proposed to be considered below.

The means of transportation of tomorrow can simultaneously play the role of a car and a water vehicle. The creators of the Amphibious Hybrid [10] – a symbiosis of a car and a motorboat 'figure 3'can be counted among the latter.

Figure 3. Concept Amphibious Hybrid

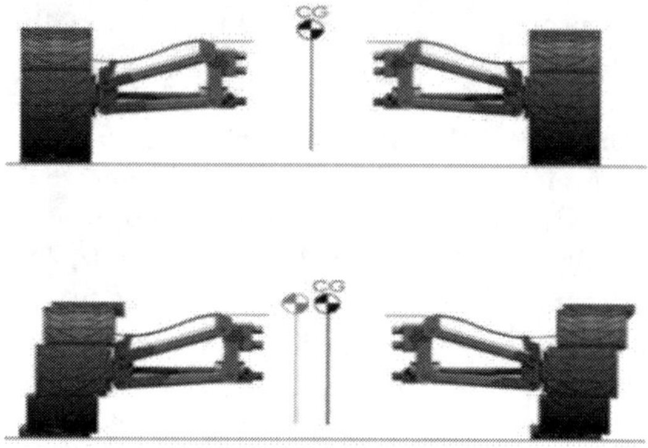

Figure 4. Dynamicaly Augmentig Wheel System

For example, if the car is driving on the highway, the wheels are no different from ordinary ones. In the snow, the wheels of the Amphibious Hybrid change shape, thereby increasing the grip on the road surface. If the vehicle is used for moving on water, the wheels take the form of propellers, that is, they "open" completely.

The next example of transformation, the development of DAWS (Dynamically Augmentig Wheel System) 'figure 4' is the transformation of an ordinary car wheel into a caterpillar [11]. Of course, a decent amount of time will pass before the development of a real model of a car with such a track wheel.

Automatic wheel conversion. According to road conditions, the automatic conversion of wheels and tracks can be implemented 'figure 5' to meet the requirements of passability in different road conditions [12].

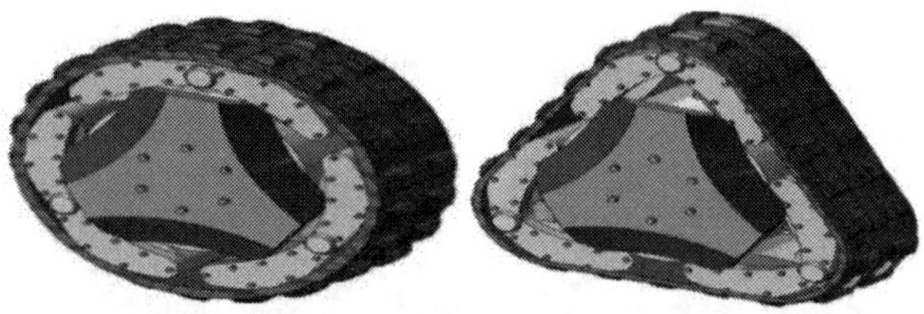

Figure 5. Automatic conversion: wheel-track

The above examples of wheel drive transformations increase cross-country ability, but do not have a dual function other than providing vehicle mobility and transforming only into propeller and crawler respectively.

It is obvious that the prerequisite for the GRC, which determines the need for the transition from a wheeled to an alternative (tracked, walking) drive, is primarily road conditions (terrain), because in roadless conditions or as a result of missile and artillery destruction, such extreme conditions occur when the movement on a wheel drive it becomes impossible and it is advisable to use a crawler or a walking one, and also taking into account the possibilities of transformation, a manipulator can also be involved, in general, various combinations are possible.

So, let's consider the first mathematical (kinematic) description of some possible wheel transformations. For description, we make the most simplified model, when we describe the wheel rim with one flat line. The mathematical description of curves in a plane can be implemented in various ways. One option is to use graphs of functions like:

$$y = f(x) \tag{1}$$

where x and y have the sense of rectangular Cartesian coordinates of the plane, and the function f sets the connection between them, that is, it makes it possible to single out the necessary ones among them, or rather to match each one x one y. For example $y=\sqrt{1-x^2}; y=-\sqrt{1-x^2}$ – the graphs of these functions define, respectively, the upper and lower halves of a circle of unit radius centered at the origin. As you can see, this way of describing the points of the plane that make up the curve is not optimal, because with the explicit specification of the function, each value of the argument corresponds to exactly one value of the function, and therefore objects such as a circle require several analyticals expressions.

Another option is to implicitly specify a set of points, for example:

$$(x-x_0)^2 + (y-y_0)^2 = R^2. \tag{2}$$

Thus, equation (2) defines the set of points that is a circle of radius R with the center in $(x_0; y_0)$. Implicitly specifying the relationship between the abscissa and the ordinate of a point in the shape $F(x;y)=0$ in general, it is quite general, but difficult to apply and requires additional analysis during research and use.

The optimal option is the parametric setting of the curve in the form:

$$\begin{cases} x = x(s), \\ y = y(s), \end{cases} s_0 \le s \le s_1. \tag{3}$$

So, equation (2) can be implemented as:

$$\begin{cases} x = x_0 + R\cos(s), \\ y = y_0 + R\sin(s), \end{cases} 0 \le s \le 2\pi, \tag{4}$$

closed circle curve (initial and final points coincide, there are no self-intersection points).

Such a task makes it possible both to study the shape of the curve and to conveniently visualize it. Note that the task is usually ambiguous and several different parameterizations can describe one curve; yes, the previous curve can also be specified as follows:

$$\begin{cases} x = x_0 + R\cos(s), \\ y = y_0 + R\sin(s), \end{cases} -\pi \le s \le \pi \text{ and } \begin{cases} x = x_0 + R\sin(s), \\ y = y_0 + R\cos(s), \end{cases} 0 \le s \le 2\pi,$$

as well as in other equivalent ways. Without reducing the generality (by properly choosing the frame of reference), it is possible to set in equation (4) $R=1$, $x_0 = y_0 = 0$, having received:

$$\begin{cases} x = \cos(s), \\ y = \sin(s), \end{cases} 0 \le s \le 2\pi. \tag{5}$$

Let's consider several options for the implementation of a continuous transformation of a circle into a number of geometric shapes, as a basis for the development of transformer wheels for GRC: an ellipse (two separate points of support), a star-shaped element, an S-shaped element (two isolated points of support), a multi-hinged element. We denote the parameter responsible for the transformation t and ask $0 \leq t \leq 1$.

Ellipse. Correlation:

$$\begin{cases} x = \cos(s) + \frac{t}{2}\cos(s), \\ y = \sin(s) \ \ \frac{t}{2}\sin(s), \end{cases} \quad 0 \leq s \leq 2\pi, \tag{6}$$

t = 0;0.25;0.5;0.75;1, sets the transformation shown in ‘figure 6’. If t increases to 2, the ellipse is transformed into a segment, if we remove 2 we get ‘figure 7’.

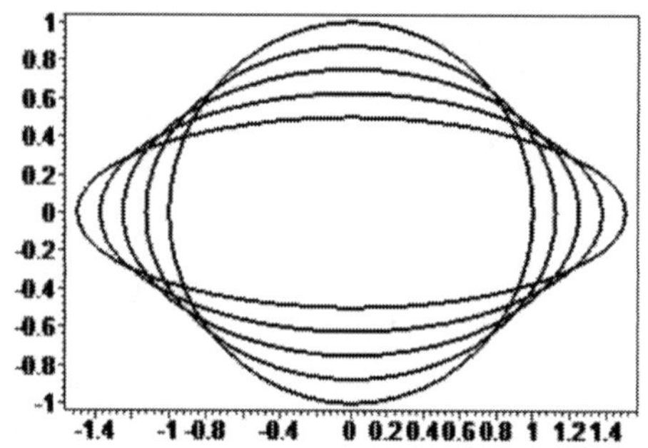

Figure 6. Implementation of continuous transformation of a circle into an ellipse

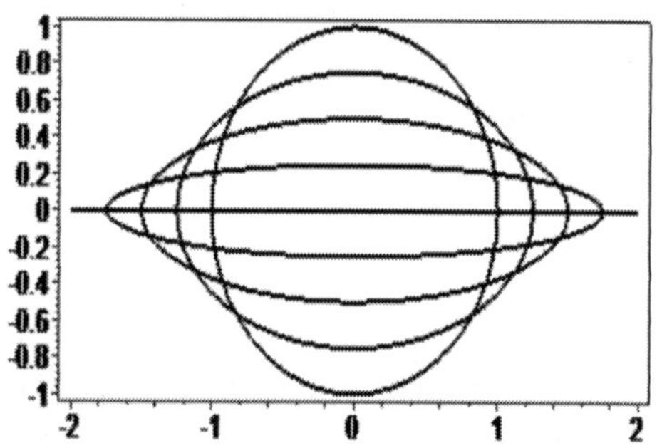

Figure 7. Implementation of a continuous transformation of an ellipse into a segment

Star-3 corners. Correlation:

$$\begin{cases} x = \cos(s) + \frac{t}{2}\cos(2s), \\ y = \sin(s) \ \ \frac{t}{2}\sin(2s), \end{cases} \quad 0 \leq s \leq 2\pi, \tag{7}$$

t = 0;0.25;0.5;0.75;1, sets the transformation shown in ‘figure 8’. If t increases to 2, we get shapes with self-intersection points ‘figure 9’.

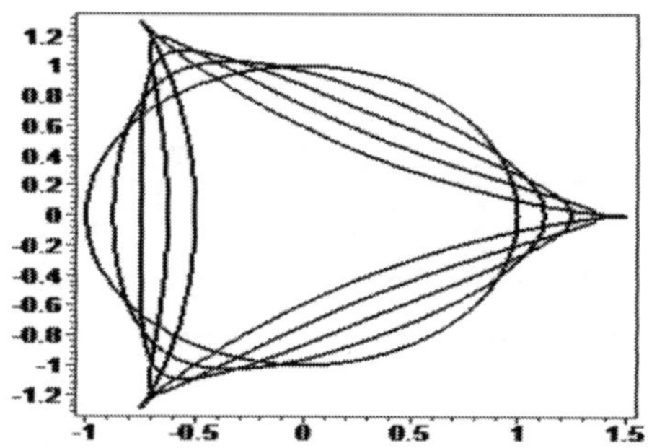

Figure 8. Implementation of continuous transformation of a circle into a triangular star

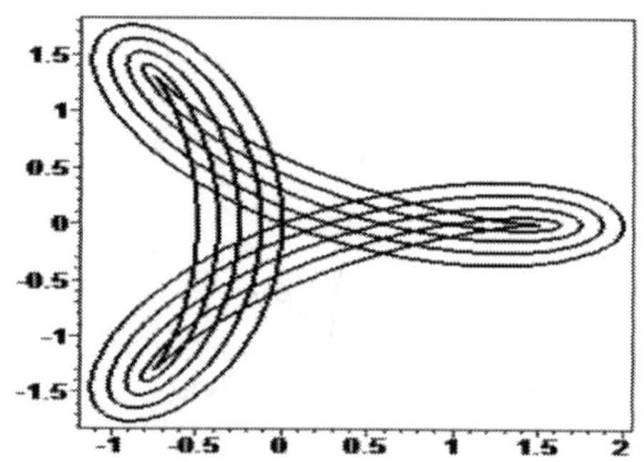

Figure 9. Implementation of a continuous transformation of a circle into a triangular star with self-intersection points

Star-4 corners. Correlation:

$$\begin{cases} x = \cos(s) + \dfrac{t}{4}\cos(2s), \\ y = \sin(s) \quad \dfrac{t}{4}\sin(2s), \end{cases} \quad 0 \leq s \leq 2\pi, \tag{8}$$

t = 0;0.25;0.5;0.75;1, specifies the transformation shown in 'figure 10'. If t increases to 2, we get figures with self-intersection points 'figure 11'.

S-shaped element. Correlation:

$$\begin{cases} x = \cos(s) - (-1)^{[s/\pi]}t, \\ y = \sin(s), \end{cases} \quad 0 \leq s \leq 2\pi, \tag{9}$$

t = 0;0.25;0.5;0.75;1, sets the transformation shown in 'figure 12'.

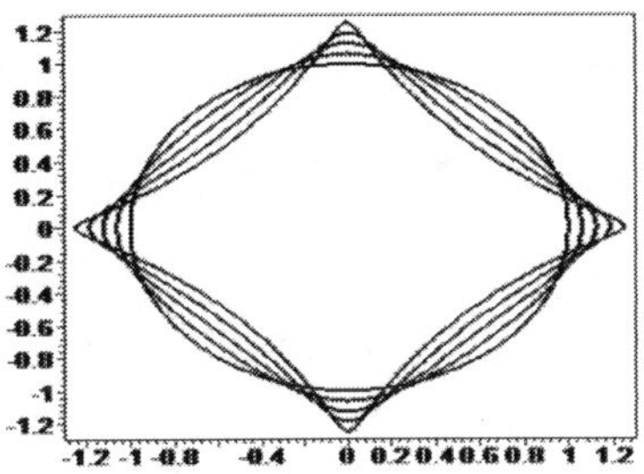

Figure 10. Implementation of a continuous transformation of a circle into a four-pointed star

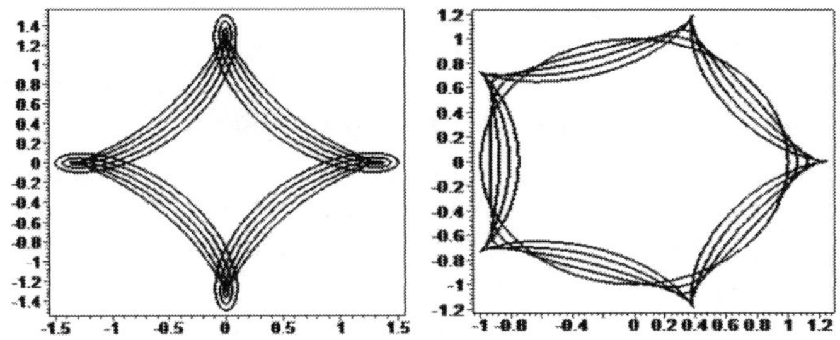

Figure 11. Implementation of a continuous transformation of a circle into a four- and five-pointed star with points of self-intersection

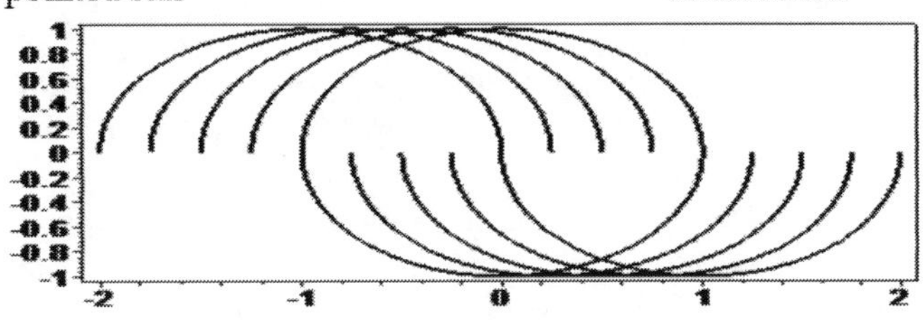

Figure 12. Implementation of a continuous transformation of a circle into an S-shaped element

Thus, based on the known options for changing the shape of the wheels and the figures I proposed, which are elements of the initial implementation of the transformation of the wheel drive into other forms than the rounded one, it can be predicted that they will improve the possibility and stability of the GRC in off-road conditions. This statement can be verified with the help of computer simulations, as well as by directly executing such engines, in particular, on a 3D printer. The obtained results will be published in the following articles.

If cases of transformation of thrusters are found in today's engineering solutions, then everything is not so simple with manipulators and in general for the scientific community this is a new area for research.

So, the manipulator is an electronically controlled mechanism consisting of several segments and performs a task interacting with the environment [13]. They are also usually called robots, or robot-manipulators, they are widely used in industrial production and in dangerous areas of human life, which of course is the military. They consist of a set of links and connections 'figure 13' and usually use hydraulic, pneumatic and electric drives. Links are defined as rigid sections that form a mechanism, and joints are defined as connections between two links. The device attached to the manipulator that interacts with the environment to perform tasks is called an end effector (capture) 'figure 14'.

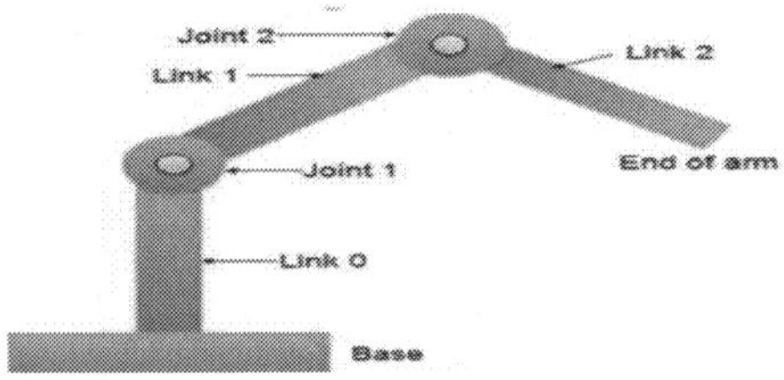

Figure 13. An example of a manipulator

Figure 14. An example of the end effector of the manipulator

If we talk about the transformation of GRC manipulators to give them a double function, that is, in addition to the traditional one, also the function of a mover (wheeled, tracked, walking), it should be noted that the perspective of such developments primarily depends on the material of execution, it is clear that the existing ones will not be able to properly provide the necessary change in form, but fundamentally new ones that are gradually beginning to be used in the technical field would be quite suitable. I suggest you consider them.

There are some potentially disruptive technologies in corporate and university labs. One of them is the use of gallium nitride [14] for the production of transistors. This material has a wider band gap than silicon, from which classic transistors are made. Thanks to this, gallium nitride transistors work at higher temperatures. The operating temperature of GaN transistors is 150-200 °C and can theoretically reach 500°C. Also, GaN transistors withstand higher currents than silicon transistors.

In general, the advantages of such transistors in comparison with traditional ones are related to the minimum intrinsic resistance and, as a result: higher limiting current density; increased electrical strength; high thermal conductivity; a wide range of operating temperatures; high frequency characteristics; minimum noise level. As a result, gallium nitride transistors reduce switching losses and conduction losses several times compared to traditional transistors. They can reduce the heating of engine control units in GRC. Also, gallium nitride transistors are smaller in size, which allows to reduce the mass and dimensions of final devices and reduce their cost.

Japanese researchers from Tokyo's Waseda University have obtained a new type of crystals [15] that change shape as a result of a cyclical process of changing temperature effects. Under the influence of one or another temperature regime, the crystals bend and straighten, repeating the movement of a worm. This opens up opportunities for creating a new type of soft robotics, more flexible and adaptable to the physical environment, in particular for GRC. In the future, this may give impetus to the production and industrial application of MIEOE for a new type of GRC. Moving away from the transformation of the rotation of the electric motor to use the change of shape will allow us to think about the creation of structures similar to the muscle tissue of living organisms. Animal muscle tissue is highly efficient, but it uses chemical energy sources. Such promising materials and MIEOE designed on their basis will make possible the appearance of the GRC with a minimum number of traditional mechanical elements.

Combination of graphene production technologies on an industrial scale with existing microelectronics technologies. Many developments are underway, which will lead to the creation of new materials - graphene films and threads [16]. Graphene thread (film) will be characterized by high strength combined with thinness. Such qualities will make it possible to develop executive actuators of manipulators for GRC, which have more compact dimensions while simultaneously increasing accuracy and increasing the drive force.

It is obvious that engineering thought generates ideas for the creation of new materials, which will fundamentally influence new construction solutions in the future. Considering this, as well as the possibilities of transformation, the perspective of my further research consists in the development of appropriate mathematical models that will describe the transformation process of MIEOE for GRC with the aim of acquiring a dual function, in their verification by means of computer simulation, design in the Solid Works program, and carrying out natural experiments.

So, based on the results of the initial research in this new idea regarding the combination of functions of MIEOE for GRC due to their transformation, it is possible to formulate such a scientific novelty and practical significance of the obtained results.

The scientific novelty of the obtained research results is that: the structural solutions for the transformation of MIEOE for GRC received further development, for the first time it was proposed to apply the transformation in order to provide MIEOE with dual functions.

The practical significance of the research results: the research data can be used as a basis for the development of fundamentally new GRC with dual-function MIEOE, which will increase their passability, stability and, as a result, survivability in general.

Conclusion

The Strategy of the Army Robotics and Autonomous Systems, the Operational Concept of the US Army until 2040 and the Roadmap for the development and implementation of robotics, in particular ground systems of the US, were analyzed and it was clarified how their military-political leadership will integrate new technologies into future projects in order to ensure their superiority over all more technologically prepared opponent.

It was found that the basis of the development of modern GRC of the leading countries of the world is the use of innovative technologies, in particular artificial intelligence.

It has been established that during hostilities, and in particular during the Russian-Ukrainian war, GRC play an important role in preserving the life and health of servicemen. Therefore, for the high-quality performance of tasks according to their purpose, and therefore an advantage over the enemy, they must be in sufficient quantity and they must have improved operational properties.

It was found out that the transformation of MIEOE is due to the need to increase the operational properties of GRC, such as: passability, stability, maneuverability and survivability in general.

A new idea has been proposed to provide a dual function of MIEOE for GRC due to their transformation.

Individual cases of the transformation of the wheel drive and the design features of the manipulators were analyzed, as well as a mathematical description of the transformation of the circle into figures, which can be used for the construction of high-passability GRC drives, was carried out.

It has been established that the transformation of MIEOE for GRC takes place due to the movement of their constituent parts to acquire the appropriate form, which is provided for the performance of the specified task without losing the primary amount of material (except in some cases). In order to acquire the most optimal form, MIEOE for GRC can be manufactured in a combination of different materials (hard, flexible, plastic) and combine the latest technologies for the use of gallium nitride transistors, soft crystals and graphene thread.

References

[1] U.S. Army Robotic and Autonomous Systems Strategy.2020.

[2] U.S. Department of the Army. The U.S. Army Operating Concept – Win a Complex World 2020 - 2040. TRADOC Pamphlet 525-3-1, October 7, 2014.

[3] U.S. Department of the Army. U.S. Army Capabilities Integration Center's System Threat Assessment Report (STAR) for Soldier as a System (SaaS) for 2022.

[4] Unmanned Systems Integrated Roadmap FY2017-2042.

[5] The concept of the development and use of ground robotic complexes (platforms) in units of the Ground Forces of the Armed Forces of Ukraine// The Command of the Ground Forces of the Armed Forces of Ukraine together with the Scientific Center of the Ground Forces of the National Academy of Ground Forces named after Hetman Petro Sahaidachny - December 2021.

[6] Gorbulin V. How to defeat Russia in the war of the future. Kyiv: Bright Books. 2020. 256 p.

[7] Guslyakov O, Dovgopoliy A and Chepkov I Critical technologies for the creation of ground robotic complexes of heavy and medium classes. Armament and military equipment. DOI: https://doi.org/1034169/2414-0651.2020.1(25).24-34.

[8] Kuprinenko O Problems of creating terrestrial robotics complexes for the needs of the Armed Forces of Ukraine. Armament and military equipment. DOI: https://doi.org/1034169/2414-0651.2021.4(32).26-34.

[9] Regarding the development of the production of unmanned robotic systems based on public-private partnership: analytical note dated 12/17/2020. / Shershakov O, Shemaev V, Begma V and Sklyar N Kyiv: Nats. Institute of Strategic Studies. 5 p.

[10] Future Amphibious Hybrid Concept Vehicle with Intelligent Wheel System / Available: https://www.tuvie.com/future-amphibious-hybrid-concept-vehicle-with-intelligent-wheel-system/.

[11] DAWS / Available: https://pyottdesign.com/product/daws.

[12] Yuchuan Yang, Tong Cai and Xiaojun Xu. Track Design of Track-Wheel Transport Vehicle and Strength Analysis of Important Components. IOP Conf. Series: Materials Science and Engineering 782 (2020) 022058. doi:10.1088/1757-899X/782/2/022058.

[13] Engineering Fundamentals / Available: https://www.maplesoft.com/content/ /13/MapleDocument_13/Position,%20Orientation%20and%20Coordinate%20Transformations.pdf.

[14] How GaN is changing the future of semiconductors / Available: https://www.techradar.com/news/how-gan-is-changing-the-future-of-semiconductors.

[15] Shodai Hasebe, Yuki Hagiwara, Jun Komiya, Meguya Ryu, Hiroki Fujisawa, Junko Morikawa, Tetsuro Katayama, Daiki Yamanaka, Akihiro Furube, Hiroyasu Sato, Toru Asahi and Hideko Koshima / Photothermally Driven High-Speed Crystal Actuation and Its Simulation / Journal of the American Chemical Society / DOI: https://doi.org/10.1021/jacs.1c03588.

[16] Jin Xinga, Feng Chunfang, Deepalekshmi Ponnamma, Yi Zhifeng, Jyotishkumar Parameswaranpillai, SabuThomas and Nisa V Salim / Review on exploration of graphene in the design and engineering of smart sensors, actuators and soft robotics / Chemical Engineering Journal Advances https://doi.org/10.1016/j.ceja.2020.100034.

IOP Conf. Series: Materials Science and Engineering 1277 (2023) 012003 doi:10.1088/1757-899X/1277/1/012003

THE INFLUENCE OF THE RELATIVE MOTION DYNAMICS OF THE WHEELED VEHICLE SPRUNG AND UNSPRUNG PART ON THE MOTION STABILITY ALONG CURVED ROAD SECTIONS

R Nanivskyi

PhD in Engineering, associate professor, Head of the Scientific and Organizational Department of the Hetman Petro Sahaidachnyi National Army Academy, Lviv, Ukraine, ORCID – 0000-0001-6504-1178

Abstract. The research method of the influence of the main suspension parameters on the motion stability of a wheeled vehicle was developed on the basis of a flat model of the dynamics of its relative motion. The mathematical model is a system of two nonlinear differential equations of the second order. The basic parameters of transverse angular oscillations of the sprung and unsprung part, which determine the dynamics of relative motion, are described using the basic principles of the perturbation method in combination with Ateb-functions. Kinetostatic equations and conditions of motion stability in the aggregate made it possible to determine the critical value of the steady motion speed as a function of the main parameters that characterize the suspension and the oscillations amplitudes of the sprung and unsprung part.

It was established that an increase in the oscillations amplitude of the unsprung part from 0.01 rad to 0.015 rad causes a decrease in the critical speed of steady motion by up to 35%.

The main results can serve as a basis for the suspension modernization of wheeled vehicle, and their reliability is confirmed by obtaining, in the extreme case, known results concerning suspensions with linear characteristics of elastic elements and damping devices

1. Introduction

The main operational characteristics of a wheeled vehicle are motion smoothness and stability, controllability and passability [1-3]. They are determined by the dynamics of the sprung and unsprung part, which are influenced by external (road surface characteristics) and internal factors (vehicle mass distribution, power characteristics of the suspension system, tires, etc.). The dynamic wheel interaction with the road surface is determined by the oscillations of the sprung and unsprung part, the tire condition, etc. Most wheeled vehicles use suspensions in which the relationship between restoring forces and deformations or deformations and their speeds is described by nonlinear dependencies. However, the main analytical studies of the impact on the wheeled vehicle operational characteristics were carried out according to the linear law of the change in the restoring force of elastic elements and

IOP Conf. Series: Materials Science and Engineering 1277 (2023) 012003 doi:10.1088/1757-899X/1277/1/012003

tires [4-6]. As for analytical studies of the influence of nonlinear restorative characteristics of elastic elements and tires on the main operational characteristics, the problem remains open due to the lack of a mathematical apparatus for constructing analytical solutions of nonlinear differential equations that describe the dynamics of the selected wheeled vehicle model. Therefore, there is a problem of choosing the minimum possible calculation models, which, on the one hand, would maximally reflect the influence of the main properties of the wheeled vehicle suspension and tires on the dynamics of sprung and unsprung part, and on the other hand, would have an analytical solution.

2. Material and method

Most wheeled vehicles are designed in such a way [7] that they are symmetrical with respect to the vertical plane that passes through the mass center of the sprung and unsprung part, and therefore the transverse angular oscillations for them do not depend on the longitudinal-angular ones and vice versa. The above, to some extent, serves to simplify the calculation and so the mathematical models of the wheeled vehicle dynamics depending on the problem under consideration. Thus, to study the motion controllability it is sufficient to consider the relative longitudinal-angular oscillations of the sprung and unsprung part system, to the motion smoothness and passability – longitudinal-angular and vertical, to the motion stability – transverse angular [8, 9].

An effective way to solve this problem is the method of mathematical modeling of a dynamic process with obtaining an analytical solution of differential equations. However, analytical studies are related to taking into account the dynamics of the sprung and unsprung part, related to the solutions construction to the system of ordinary nonlinear differential equations. Obtaining, in the general case, analytical relations for them, which would be the basis for influence evaluating the amplitude-frequency characteristics of the sprung and unsprung part oscillations on the main operational characteristics, is a problem that cannot be solved mathematically. Therefore, in a number of works [8-10], in order to partially solve the existing problems, the calculation models were simplified as much as possible, with limit taking into account only the sprung part oscillations. At the same time, even from the analysis of linear mathematical models [10, 11], it follows that the wheeled vehicle unsprung part oscillations significantly affect the main operational characteristics during movement along a path with irregularities.

Another approach based on experimental studies [6] of the influence of certain factors on the dynamics of a sprung part with a suspension non-linear elastic characteristic cannot respond to a number of important questions, which concern, in particular, resonance processes during the wheeled vehicle movement along a path with an ordered system of irregularities; the influence of individual irregularities on the oscillations frequency, because the last one depends on the amplitude, etc. In addition, experimental studies require significant material and time resources. Thus, the most effective methods of studying the influence of the dynamics of sprung and unsprung part on operational characteristics are those that are based on the combination of analytical results of specified parts ocsillations studies with the basic positions of motion stability, controllability and smoothness. Exactly such approach is being developed in the study of the wheeled vehicle motion stability along a curved section of the road with individual irregularities. Therefore, the purpose of the work is to obtain analytical ratios that determine the influence on the steady motion critical speed of the main parameters that describe the characteristics of the suspension, tires and the sprung and unsprung part dynamics for a two-axle wheeled vehicle with dependent suspension with a continuous bridge moving along a curvilinear section of the road with fixed speed V and curvature radius of the road section ρ.

As for the elastic properties of tires (F_{1i}) and elastic elements of the suspension system (F_{2i}),they depend on their deformations, in the following we will accept these deformations as (Δ_{1i}, Δ_{2i})and they are described by the dependences $F_{1i} = c_1(\Delta_{1i})^{\nu_1+1}$ and $F_{2i} = c_2(\Delta_{2i})^{\nu_2+1}$, respectively, and the viscous friction forces of tires and damping devices are proportional to the speed of their deformations $\dot{\Delta}_{1i}$ or $\dot{\Delta}_{2i}$ in the degree $2s_i+1: R_{1i} = \alpha_1(\dot{\Delta}_{1i})^{2s_1+1}$, $R_{2i} = \alpha_2(\dot{\Delta}_{2i})^{2s_2+1}$ (i = 1 for tires, elastic elements and damping devices of the starboard side and, accordingly, i = 2 - of the left side, $c_1, c_2, \alpha_1, \alpha_2$ - known constants). As for the parameters ν_1 and ν_2, which indicate the deviation of the elastic properties of tires and elastic elements from the linear law, they must additionally satisfy the conditions for the oscillating process existence in the sprung and unsprung part. This will be true if $\nu_s + 1 = \frac{2m_s + 1}{2n_s + 2}, m_s, n_s = 0,1,2,...$. With the help of these functions, it is possible to describe the characteristics of tires and elastic elements of a wide range of wheeled vehicles.

Results

Under the above assumptions, the mathematical model of relative motion has two degrees of freedom. The angles of rotation of the corresponding parts around the longitudinal axes passing through the mass centers of the corresponding parts in their relative static position are taken as generalized coordinates for it: $\varphi(t)$- unsprung part, $\psi(t)$- sprung part. The specified parameters for the proposed model of the wheeled vehicle dynamics are connected by differential equations:

$$\begin{aligned} I_{c1}\ddot{\varphi} &= -c_2 a\left\{(\Delta_{st.} - \psi a - \Delta_{1st.} + \varphi\frac{L}{2})^{\nu_2+1} + (\Delta_{st.} + \psi a - \Delta_{1st.} - \varphi\frac{L}{2})^{\nu_2+1}\right\} - c_1\frac{L}{2}\left\{\left(\Delta_{1st.} + \varphi\frac{L}{2}\right)^{\nu_1+1} - \left(\Delta_{1st.} - \varphi\frac{L}{2}\right)^{\nu_1+1}\right\} + R_{\varphi}(\dot{\psi}, \dot{\varphi}), \\ I_c\ddot{\psi} &= c_2 a(\Delta_{st.} - \psi a - \Delta_{1st.} + \varphi\frac{L}{2})^{\nu_2+1} - c_2 a(\Delta_{st.} + \psi a - \Delta_{1st.} - \varphi\frac{L}{2})^{\nu_2+1} + R_{\psi}(\dot{\psi}, \dot{\varphi}), \end{aligned} \quad (1)$$

in which I_c, I_{c1} the inertia moments of the wheeled vehicle sprung and unsprung part relative to the longitudinal axes passing through their mass centers, $R_{\psi}(\dot{\psi}, \dot{\varphi})$ and $R_{\psi}(\dot{\psi}, \dot{\varphi})$ - respectively, resistance force moments of damping devices and tire viscous forces relative to the axes specified above.

To find the tire dynamic pressure on the support surface, first of all, it is necessary to find the amplitude-frequency characteristic of the two-mass system under consideration and this is connected with the construction of the analytical solution of the differential equation system (1). The basis for constructing an approximate analytical solution of the differential equation system is the following physically justified provisions that apply to the wheeled vehicle motion along arbitrary sections of the path: a) the maximum values of the resistance forces of the suspension system damping devices are much smaller than the maximum values of the elastic element forces; b) the coefficients of tire "stiffness" are significantly greater than the corresponding coefficients of elastic elements, i.e $c_1 >> c_2$. This follows from the fact that for all wheeled vehicles, the tire static deformations are much smaller than the static deformations of the suspension system. The above allows us to present of the differential equation system (2) without changing its accuracy in the form:

$$\begin{aligned} I_{c_1}\ddot{\varphi} + c_1 L\left(\varphi\frac{L}{2}\right)^{\nu_1+1} &= c_2 a\left\{(\Delta_{st.} - \psi a - \Delta_{1st.} + \varphi\frac{L}{2})^{\nu_2+1} + (\Delta_{st.} + \psi a - \Delta_{1st.} - \varphi\frac{L}{2})^{\nu_2+1}\right\}^{\nu_2+1} + R_{\varphi}(\dot{\varphi}, \dot{\psi}, \dot{z}), \\ I_c\ddot{\psi} + 2c_2 a(a\psi)^{\nu_2+1} &= 2c_2 a(\nu_2+1)(a\psi)^{\nu_2}\varphi L + R_{\psi}(\dot{\varphi}, \dot{\psi}, \dot{z}). \end{aligned} \quad (2)$$

Based on the given comparative assessment of the main characteristics of the suspension and tires, it follows that the maximum values of the right parts of the differential equation system (3) are significantly smaller than the maximum values of the corresponding second terms of the left parts. The latter is a justification for the general ideas application of perturbation methods [12, 13] when constructing the solution of the nonlinear equation system (2). The solutions of the unperturbed equations corresponding to system (2) are expressed [8] through periodic Ateb-functions [14, 15] in the form of system (3):

$$\varphi = a_{\varphi}\begin{Bmatrix} sa\left(\nu_1+1,1,\omega_{\varphi}(a_{\varphi})t+\gamma_0\right) \\ ca\left(\nu_1+1,1,\omega_{\varphi}(a_{\varphi})t+\bar{\gamma}_0\right)\end{Bmatrix}, \vartheta_{\varphi} = \omega_{\varphi}(a_{\varphi})t+\gamma, \quad \psi = a_{\psi}\begin{Bmatrix} sa\left(\nu_2+1,1,\omega_{\psi}(a_{\psi})t+\phi_0\right) \\ ca\left(\nu_2+1,1,\omega_{\psi}(a_{\psi})t+\bar{\phi}_0\right)\end{Bmatrix}, \vartheta_{\psi} = \omega_{\psi}(a_{\psi})t+\phi \quad (3)$$

in which a_ψ, a_φ, ϑ_ψ, ϑ_φ - oscillation amplitudes and phases of the sprung and unsprung part unperturbed motion, ϕ, γ - initial phases of unperturbed motion. As for the natural frequencies $\omega_\psi(a_\psi)$, $\omega_\varphi(a_\varphi)$, they depend on the amplitudes of the corresponding parts, as for most nonlinear oscillating systems, and take the value:

$$\omega_\psi(a_\psi) = \sqrt{\frac{(v_2+2)c_2 a^{v_2+2}}{I_C}} a_\psi^{\frac{v_2}{2}}, \quad \omega_\varphi(a_\varphi) = \sqrt{\frac{2(v_1+2)c_1\left(\frac{L}{2}\right)^{v_2+2}}{I_{C1}}} a_\varphi^{\frac{v_1}{2}} \tag{4}$$

The presented dependencies show that for the regressive characteristics of the suspension and tires, for larger amplitudes of oscillations, the natural frequencies of the corresponding parts are smaller and, at the same time, for progressive ones, they are larger, while for larger values of static deformations, the natural frequencies are smaller.

To find the laws of time change of functions $\varphi(t)$ and $\psi(t)$ we generalize the main ideas of Van der Pol's method to the differential equation system (1). This allows to find the solution of the considered nonlinear differential equation system in the form

$$\varphi = a_\varphi(t)\,\mathrm{sa}\,(v_1+1,1,\vartheta_\varphi(t)), \cdots \vartheta_\varphi(t) = \omega_\varphi(a_\varphi)t + \gamma(t), \quad \psi = a_\psi(t)\,\mathrm{sa}\,(v_2+1,1,\vartheta_\psi(t)), \quad \vartheta_\psi(t) = \omega_\varphi(a_\varphi)t + \phi(t), \tag{5}$$

where the unknown time functions $a_\varphi(t)$, $a_\psi(t)$, $\vartheta_\varphi(t)$, $\vartheta_\psi(t)$ are found so that the relation (5) satisfies the original differential equation system (3).

Without delving into mathematical calculations for the first approximation, the specified parameters are determined by ordinary differential equations

$$\frac{da_\psi}{dt} = -\frac{(\alpha_2)a_\psi}{\Pi_2 I_c}\left(\frac{2a_\psi\omega_\psi(a_\psi)}{(v_2+2)}\right)^{2s_2}\frac{\Gamma\left(\frac{1}{v_2+2}\right)\Gamma\left(\frac{2s_2+2}{2}\right)}{2\Gamma\left(\frac{1}{v_2+2}+\frac{2s_2+2}{2}\right)}, \quad \frac{d\vartheta_\psi}{dt} = \sqrt{\frac{(v_2+2)c_2 a^{v_2+2}}{I_C}} a_\psi^{\frac{v_2}{2}}, \tag{6}$$

$$\frac{da_\varphi}{dt} = -\frac{(\alpha_2)a_\varphi}{\Pi_1 I_{c1}}\left(\frac{2a_\varphi\omega_\varphi(a_\varphi)}{(v_1+2)}\right)^{2s_1}\frac{\Gamma\left(\frac{1}{v_1+2}\right)\Gamma\left(\frac{2s_1+2}{2}\right)}{2\Gamma\left(\frac{1}{v_1+2}+\frac{2s_1+2}{2}\right)}, \quad \frac{d\vartheta_\varphi}{dt} = \sqrt{\frac{(v_1+2)c_1 a^{v_1+2}}{I_{C_1}}} a_\varphi^{\frac{v_1}{2}}. \tag{7}$$

Thus, for the first approximation, the oscillatory process of the sprung and unsprung part system is described by dependencies (3), in which the parameters a_ψ, a_φ, ϑ_ψ, ϑ_φ are connected by the first-order differential equation system (6), (7). For an unambiguous description of the specified parameters, it is necessary to additionally specify the initial conditions for them. It is known that path irregularities are the cause of the sprung and unsprung part oscillations. It is believed that the distance between them is such that after the wheeled vehicle overcomes one of them, thanks to the available damping devices and the viscous friction forces of the tires, at the contact moment of the tires with the following bump, the oscillations of both parts stop. This allows to choose such initial conditions for the differential equation system that are consistent with the "reaction" of the tire at the moment of its contact with the bump. Taking into account that during the wheeled vehicle motion: the tire is in constant contact with the road irregularity; the road irregularity has small length; the vehicle moves with a constant speed; with a sufficient accuracy, it can be assumed that the relative component of the speed of the wheeled vehicle unsprung part at the moment when the wheel hits the bump is equal to the projection of the wheeled vehicle speed on the vertical.

Thus, the basis for determining the initial amplitude of transverse angular oscillations is the ratio: $V\sin\beta = \frac{L}{2}\frac{d\varphi}{dt}_{|t-0}$, β - the inclination angle to the horizontal section of the road tangent to the bump at the contact point between the tire and the bump. Accepting the above, we get

$$\frac{L}{2}\left\{\frac{da_\varphi}{dt} sa(v_1+1{,}1,\omega_\varphi(a_\varphi)t) + \frac{2a_\varphi}{v_1+2}\omega_\varphi(a_\varphi)ca(1,v_1+1,\omega_\varphi(a_\varphi)t)\right\}_{|t=0} = V\sin\beta. \quad (8)$$

Algebraic equation (8) allows to determine the initial value of the oscillation amplitude of the wheeled vehicle unsprung part caused by the road irregularities

$$a_{\varphi|t=0} = \left\{\frac{1}{L}V\sin\beta\sqrt{\frac{2(v_1+2)I_{C1}}{L(P+Q)}\left(\frac{2\Delta_{1st}}{L}\right)^{v_1+1}}\right\}^{\frac{2}{v_1+2}} \quad (9)$$

In a similar way, the initial value of the transverse angular oscillation amplitude the sprung part is found. The condition for its finding is the speed equality of the elastic elements attachment points to the sprung and unsprung part at the initial moment of time. We receive:

$$a_{\psi|t=0} = \left\{\frac{V\sin\beta}{L}\sqrt{\frac{2(v_2+2)I_C}{Q}\left(\frac{\Delta_{st}}{a}\right)^{v_2+1}}\right\}^{\frac{2}{v_2+2}} \quad (10)$$

Thus, the amplitude-frequency characteristic of the sprung and unsprung part oscillations is described by ordinary differential equations (6), (7) under the initial conditions (9) and (10).

As for the critical speed values of steady motion in view of overturning, they are determined, as mentioned above, by the amount of dynamic interaction of the tires with the road surface. The basis for the analytical assessment of the indicated interaction can be the kinetostatic equations of the sprung and unsprung part system and the analytical relations obtained above, which describe the kinematics of the relative motion of the indicated parts. As for the kinetostatic equations of the system under consideration, they take the following form for the case of wheeled vehicle motion along a curved section with a constant speed:

$$N_1+N_2-P-Q=0,\quad F_1+F_2-\Phi_P-\Phi_Q=0,\quad N_1L-(P+Q)\frac{L}{2}+M^{\Phi}_{K.Q}+M^{\Phi}_{K.P}+\Phi_P H_{C_1}+\Phi_Q H_C=0, \quad (11)$$

in which Φ_Q, Φ_P - the main vectors of the inertial forces of the sprung and unsprung part translational motion; $M^{\Phi}_{K.Q}, M^{\Phi}_{K.P}$ - the main inertia force moments of the specified parts relative to the contact point of the left side tire and the road surface; N_i and F_i - respectively, the vertical and horizontal components of the road surface action on the wheel of the right $(i=1)$or left $(i=2)$ side; H_{C_1}, H_C - distances between the road surface and the sprung and unsprung part mass centers.

According to the definition, the main inertia force vectors of the sprung and unsprung part translational motion are equal $\Phi_Q = \frac{Q}{g}\frac{V^2}{\rho}, \Phi_P = \frac{P}{g}\frac{V^2}{\rho}$ - the main inertia force moments of the sprung and unsprung part are determined by dependencies

$$M^{\Phi}_{K.Q} = I_{K.Q}\frac{d^2\psi}{dt^2} = I_{K.Q}\frac{2\omega_\psi^2(a_\varphi)}{v_2+2}a_\psi sa^{v_2+1}(v_2+1{,}1,\vartheta_\psi),\quad M^{\Phi}_{K.P} = I_{K.P}\frac{d^2\varphi}{dt^2} = I_{K.P}\frac{2\omega_\varphi^2(a_\varphi)}{v_1+2}a_\varphi sa^{v_1+1}(v_1+1{,}1,\vartheta_\varphi), \quad (12)$$

in which $I_{K.Q}$, $I_{K.P}$ - the inertia moments of the sprung and unsprung part relative to the contact point of the left side tire and the road surface: $I_{K.Q} = I_C + \frac{Q}{g}\left(H_C^2 + \frac{L^2}{4}\right)$, $I_{K.P} = I_{C_1} + \frac{P}{g}\left(H_{C_1}^2 + \frac{L^2}{4}\right)$.

The relations obtained above are basic for determining the critical speed values of steady motion with regard to overturning and shearing. The condition for determining the critical speed in view of overturning transformed for the considered calculation model is $N_1 = 0$. It's obtained

$$V = \sqrt{\frac{(P+Q)\frac{L}{2} - I_{K.Q}\frac{2\omega_{\psi}^2(a_{\varphi})}{\nu_2+2}a_{\psi} - I_{K.P}\frac{2\omega_{\varphi}^2(a_{\varphi})}{\nu_1+2}a_{\varphi}}{PH_{C_1}+QH_C}\rho g} \qquad (13)$$

A special case of the obtained dependence when $a_{\varphi} = 0$ s the critical speed value of steady motion in view of overturning without taking into account the unsprung part oscillations. In Fig. 1. is presented, according to dependence (13), the dependence of the critical speed of steady motion with regard to overturning at the parameter values specified below.

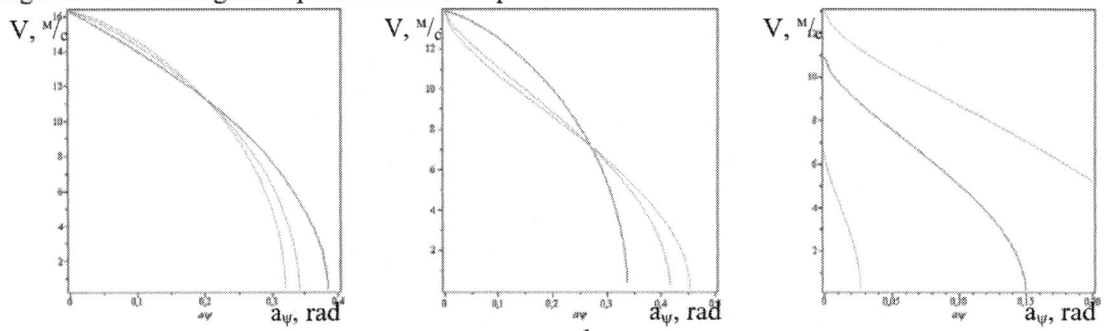

a *b* *c*

Fig. 1. Dependence of the critical speed value on the amplitude of the sprung part oscillations at different characteristics of the suspension and tires

a) $\nu_2 = 0$ *red*, $\nu_2 = \frac{2}{9}$ *green*, $\nu_2 = \frac{2}{3}$ *brown*, $\nu_1 = \frac{2}{9}$, $\Delta = 0.15м$, $a_{\varphi} = 0$, $\Delta_1 = 0.03м$;

b)

$\nu_2 = \frac{2}{3}$ *red*, $\nu_2 = \frac{-2}{7}$ *green*, $\nu_2 = \frac{-2}{5}$ *brown*, $\nu_1 = \frac{2}{7}$, $\Delta = 0.2м$, $a_{\varphi} = 0{,}01$, $\Delta_1 = 0.04\ м$;

c)

$\nu_2 = \frac{-2}{5}$, $\nu_1 = 0$ *red*, $\nu_1 = \frac{2}{11}$ *green*, $\nu_1 = \frac{-2}{11}$ *brown*, $\Delta = 0.15м$, $a_{\varphi} = 0{,}01$, $\Delta_1 = 0.06\ м$.

The presented graphics show that the critical speed value with regard to overturning: without taking into account the unsprung part oscillations is greater for the suspension system with a greater value of elastic element static deformation. An increase in the elastic element static deformation from 0.15 m to 0.25 m causes an increase in the critical speed by 27% at the oscillation amplitude $a_{\psi} = 0.2$ rad and $\nu_2 = \frac{2}{9}$, and at $a_{\psi} = 0.3$ rad - by 35% ; with a larger value of the nonlinearity parameter ν_2 the critical speed value for a small oscillation amplitude is smaller and for a "large" oscillation amplitude of the sprung part is larger; for suspension system with a progressive law of change in the restoring force for small amplitudes the oscillations are greater than with a linear one at the same static deformation values, and for large ones the opposite is true. At the sprung part oscillation amplitude $a_{\psi} = 0.1$ rad and the elastic element static deformation of 0.25 m the critical speed value of steady motion at $\nu_2 = \frac{2}{3}$ is greater by 5% than for a suspension with a linear law, and at $a_{\psi} = 0.4$ rad - is smaller by 11%.

As for the influence of the unsprung part oscillations and static deformation of the tire, an increase in the static deformation of tires from 0.03 m to 0.05 m causes an increase in the critical speed of motion at $a_\psi = 0.1$ rad, $a_\varphi = 0.01$ rad by 9% ($\nu_2 = \frac{2}{9}, \Delta = 0.3м$), and at $a_\psi = 0.2$ rad and $a_\varphi = 0.01$ rad by 29%; an increase in the unsprung part oscillation amplitude causes a decrease in the critical speed of steady motion. An increase in a_φ from 0.01 rad to 0.015 rad causes a decrease in the critical speed by 35% at $a_\psi = 0.1$ rad, $\nu_2 = \frac{2}{9}, \Delta = 0.3м, \Delta_1 = 0.05м$, and at $a_\psi = 0.2$ rad - by 45%.

In the case of the regressive characteristic of elastic elements and progressive characteristic of tires, the critical speed value of steady motion is a little larger for the small sprung part oscillation amplitudes at the larger values of the parameter ν_1 and, on the contrary, is smaller - for the large sprung part oscillation amplitudes. At the sprung part oscillation amplitude of 0.1 rad an increase in the parameter ν_2 from $\frac{-2}{5}$ to $\frac{-2}{5}$ causes an increase in the critical speed of steady motion by 3%, and at the oscillation amplitude of 0.4 rad – the critical speed value of steady motion increases by 33% at following values of the system parameters $\nu_1 = \frac{2}{7}, \Delta = 0.2м, a_\varphi = 0{,}01, \Delta_1 = 0.04$ м; or the regressive characteristic of elastic elements and tires, the critical speed value of steady motion is greater than the critical speed value for the regressive law of change in the suspension system elastic properties and linear elastic tire properties.

Conclusions

The results obtained in the work show that the critical speed values of steady motion taking into account the transverse angular oscillations of the unsprung part, even with small oscillation amplitudes of up to 30%, are smaller compared to the case when this factor is neglected. At the same time, the transverse angular oscillations of the wheeled vehicle sprung and unsprung part significantly reduce the critical speed value of steady motion in view of the overturning, so the increase in the oscillation amplitude of the unsprung part a_φ from 0.01 rad to 0.015 rad causes a decrease in the critical speed of steady motion by 35% at $a_\psi = 0.1$ rad, and at $a_\psi = 0.2$ rad - by 45% at all other specified values of suspension parameters.

It was determined that for wheeled vehicles with a progressive law of change in the elastic element characteristics, larger values of their static deformation and tires correspond to a larger critical speed value of steady motion. An increase in the static deformation of tires from 0.03 m to 0.05 m causes an increase in the critical speed of motion at $a_\psi = 0.1$ rad, $a_\varphi = 0.01$ rad by 9%, and at $a_\psi = 0.2$ rad and $a_\varphi = 0.01$ rad by 29%.

Thus, the use of a suspension with a progressive law of change in the elastic element characteristics on wheeled vehicles increases the critical speed value of steady motion in comparison with elastic elements with a linear or regressive law of change in the restoring force. At the same time, neglecting of the unsprung part oscillations even at small amplitudes of its oscillations leads to overestimated results of the critical speed values of steady motion

References

[1] Misaghi, S., Tirado, C., Nazarian, S., Carrasco, C. (2021), Impact of pavement roughness and suspension systems on vehicle dynamic loads on flexible pavements. *Transp. Eng.* 3, P.100045.
[2] Vasyl'yev A.YU., Hrubel' M.H., Shatalov O.YE., Nanivs'kyy R.A., Dudar YE.YE. (2015), "Vplyv kharakterystyky pidvisky na stiykist' bron'ovanykh armiys'kykh avtomobiliv bahatotsil'ovoho pryznachennya" [Effect of Suspension Characteristics on Stability of Armored Army Multi-Purpose

Vehicles], *Mechanics and mechanical engineering* No. 2, (Kharkiv: Technical collection of NTU KhPI), p. 68-75.

[3] Leitner, B.; Decký, M.; Ková˘c, M. Road (2019), Pavement longitudinal evenness quantification as stationary stochastic process. Transport 34, P.195–203.

[4] Park S., Popov A., Cole D. (2004), Influence of soil deformation in off-road heavy vehicle suspension vibration. *Journal of Terramechanics* 41, P. 41–68.

[5] Ružinskas, A., Giessler, M., Gauterin, F., Wiese, K.;, Bogdevi˘cius, M. (2021), Experimental investigation of tire performance on slush. *Eksploat. I Niezawodn. Maint. Reliab.* 23, P.103–109.

[6] Hussan S., Kamal M.A., Hafeez I., Ahmad N., Khanzada S., Ahmed, S. (2020), Modelling asphalt pavement analyzer rut depth using different statistical techniques. *Pavement Des.*,21, P.117–142.

[7] Vasyl'yev A.YU., Hrubel' M.H., Shatalov O.YE., Nanivs'kyy R.A., Dudar YE.YE. (2015), "Vplyv kharakterystyky pidvisky na stiykist' bron'ovanykh armiys'kykh avtomobiliv bahatotsil'ovoho pryznachennya" [Effect of Suspension Characteristics on Stability of Armored Army Multi-Purpose Vehicles], *Mechanics and mechanical engineering* No. 2, (Kharkiv, Technical collection of NTU KhPI), p. 68-75.

[8] Bogdan Sokil, Oleg Lyashuk, Mariya Sokil, Yuriy Vovk, Volodymyr Dzyura, Victor Aulin, Roman Khoroshun (2021), Interpreting the main power characteristics choice of the wheel vehicles guided cushioning system *Mechanical Engineering in Transport.*, 23(2), P. B139-B149

[9] Sokil B.I., Nanivs'kyy R.A., Hrubel' M. H., Yal'nyts'kyy O.D (2015), "Vplyv kharakterystyk pidvisky avtomobiliv spetsial'noho pryznachennya na poperechno-kutovi kolyvannya pidresorenoyi masy" [Sokil B.I., Nanivskyi R.A., Grubel M.G., Yalnytskyi O.D. (2015), Influence of the characteristics of the suspension of special-purpose vehicles on the transverse angular oscillations of the sprung mass]. *Highway of Ukraine*, No. 1-2(243-244), Kyiv, p. 18-20.

[10] Hrubel' M.H., Kraynyk L.V., Kuprinenko O.M. (2019), "Metodolohiya otsinky opornoyi prokhidnosti kolisnoyi viys'kovoyi avtomobil'noyi tekhniky" [Methodology for assessing the reference cross-country ability of wheeled military vehicles], *Scientific and technical magazine Armaments and military equipment*, No. 4(24), K., c. 22-35.

[11] Tyan F. Hong Y., Tu S. (2009), Generation of Random Road Profiles. *Journal of Advanced Engineering*, Vol. 4, № 2, P. 151-156.

[12] Kuz'o I. V., Zaluts'kyy YU. V., Zhytenko O. V. (2016), "Modelyuvannya mikroprofilyu dorohy u zadachakh dynamiky kolisnykh mashyn" [Road microprofile modeling in problems of wheeled machine dynamics] Visnyk NULP, *Dynamics, strength and design of machines and devices*, No. 838, pp. 173-179.

[13] Bogolyubov N.N., Mitropol'skiy YU.A (1974), "Asimptoticheskiye metody v teorii nelineynykh kolebaniy" [Asymptotic methods in the theory of nonlinear oscillations] (M.) *Nauka*, 504 P.

[14] Nazarkevych M. (2012), Investigation of Beta- and Ateb-function dependencies. *Bulletin of the National University.*(Lviv Polytechnic), P.207–216 (in Ukrainian).

[15] Olshanskiy V.P., Olshanskiy S.V. & Tishchenko L.N. (2015) Kolyvannya dysypatyvnykh ostsylyatoriv. *Dissipative oscillators oscillations.* (Kharkiv, Miskdruk), P 116.

IOP Conf. Series: Materials Science and Engineering 1277 (2023) 012004 doi:10.1088/1757-899X/1277/1/012004

Approximate calculation of natural frequencies of oscillations of the diamond-shaped plates of the discrete-continuous inter-resonance vibrating table

O S Lanets[1], P V Maistruk[1], I A Derevenko[2], R Ya Kachmar[3] and A I Hordieiev[4]

[1]Department of Robotics and Integrated Mechanical Engineering Technologies, Lviv Polytechnic National University, Lviv, Ukraine
[2]Department of Strength of Materials and Structural Mechanics, Lviv Polytechnic National University, Lviv, Ukraine
[3]Department of Motor Vehicle Transport, Lviv Polytechnic National University, Lviv, Ukraine
[4]Department of Mechanical Engineering Technology, Khmelnytskyi National University, Khmelnytskyi, Ukraine

Abstract. Conventional single- and two-mass vibrating machines do not have a sufficient level of energy efficiency. There are vibration machines based on inter-resonance oscillating systems, which make possible the significant reduction in the use of electricity. To implement highly effective inter-resonance operation modes of vibrating machines, the oscillating masses of such a system must precisely calculate values of inertial-rigid parameters, as well as their frequency of oscillations. Since the synthesis of continuous sections into classical discrete models of multi-mass mechanical oscillating systems has recently become widespread, the most optimal option is to use a continuous member as a reactive mass of a vibrating machine. The continuous member combines inertial and stiffness parameters and makes it extremely easy to perform the reactive mass. In particular, rectangular plates are used as continuous members. The rectangular shape of the plate is not the only option for making a continuous member. This paper proposes the construction of a discrete-continuous inter-resonance vibration machine with electromagnets, where a diamond-shaped plate is used as a continuous member. The authors carried out the computation of the first natural frequency of the diamond-shaped plate using the Rayleigh-Ritz method with the equation of the hyperboloid. The determined eigenfrequency was confirmed in the Ansys software.

1. Introduction

Widespread in machine-building, construction, chemical, and food industries, one-mass and two-mass vibration machines do not have a sufficient level of energy efficiency. There are vibrating machines based on inter-resonance oscillating systems, which make it possible to significantly reduce the use of electricity to perform the necessary technological tasks. The most common type of such equipment is inter-resonance vibration machines with electromagnets.

To implement highly effective inter-resonance operation modes of vibrating machines, the oscillating masses of such a system must precisely calculate values of inertial-rigid parameters, as well as certain frequency characteristics. Calculations of the parameters of inter-resonance vibrating machines are given a lot of attention in scientific works [1]. Also recently, the synthesis of continuous members into classical discrete models of multi-mass oscillating systems has become common [2]. The most optimal option is to use a continuous member as a reactive mass of a vibrating machine. The continuous member combines inertial and rigid parameters and allows us to perform reactive mass extremely light. In particular, such a design is considered in the article [3], where the continuous member is made in the form of a simple rectangular plate, that is hinged in the intermediate mass. However, it is obvious [4] that the rectangular shape of the plate is not the only option for the execution of a continuous member.

An important factor for the effective synthesis of a continuous member into vibrating equipment is the correct computation of its eigenvalue frequency of oscillations. Since the plate has many eigenforms and eigenfrequencies of oscillation, in the following we will focus on the first eigenfrequency. Considerable attention is paid to the vibration analysis of simple plates in scientific works [5-7].

Various boundary conditions for attachment of the plates of types SSSS, CCSS, CCCC, CFSS, CSSS, SSSF, SFSS, etc. are analyzed. However, it primarily concerns rectangular functionally graded plates. Plates with a variable cross-section are also being studied. In particular, the bending characteristic of a sandwich plate with a variable cross-section was investigated in [8]. The authors of this paper identified the most adverse regional limit-bearing capacity, the stress-strain ratio, and the three-dimensional degree of stress in the ultimate state. S. K. Jena et al. [9] investigated the free vibration of single-layered graphene nanoribbons of variable cross-section using the Differential Quadrature Method. In [10], using Bessel functions were solved differential equations for free vibrations of the straight beam with variable cross-section. J. Feng and others [11] proposed a general analytical method based on segmentation and iterative calculation to obtain the modal function and natural frequency of a beam with an arbitrary variable cross-section. In the work of R. Vescovini et al. [12], the Ritz method was used to analyze the deflections of composite anisotropic plates. The main attention was paid to the choice of experimental functions regarding the degree and type of anisotropy of the plates. Kozbur et al. [13] proposed a technique for predicting critical values of stresses in the walls of axisymmetric shells. They used Kirchhoff's hypotheses describing the normality of elements concerning the middle surface before loading and after it. The article [14] presents a new theory of hyperbolic shear deformation that can be applied to the analysis of free vibrations of laminated, functionally classified, isotropic, and sandwich composite plates. The authors of the article [15] propose a new way of dimensional discretization of two-dimensional regions to calculate the natural frequencies of curved shells.

The most attention was paid to the calculation of natural frequencies of oscillations of plates with a variable cross-section in [3]. It is proposed to extend this technique to a diamond-shaped plate as a continuous member.

2. Methodology

In this work, the continuous member is synthesized into a discrete system of an inter-resonance vibrating machine, the scheme of which is shown in figure 1. The active m_1, intermediate m_2 and reactive m_3 masses of the vibrating machine are connected in series with the help of systems of elastic nodes c_{12} and c_2. The active mass m_1 is mounted to the foundation with the help of vibration isolators c_{is}.

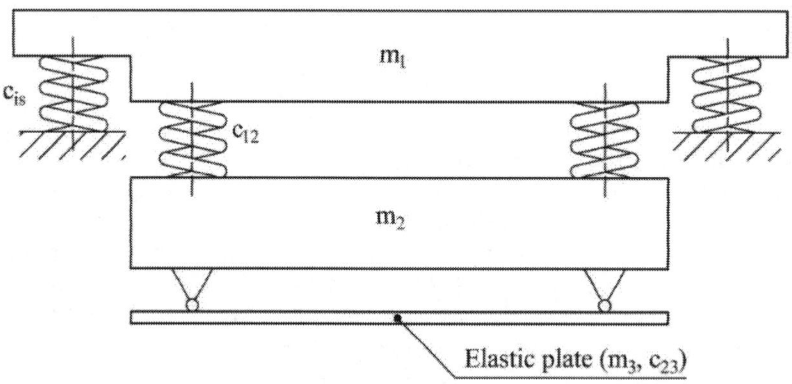

Figure 1. Principal scheme of the inter-resonance vibrating machine.

The reactive mass in the form of a plate is a body with distributed parameters, it blends rigid and inertial parameters (m_3, c_{23}). These two parameters are taken into account only in dynamic processes when they appear as abstract aggregate values.

Analyzing known methods of designing inter-resonance three-mass vibrating machines [1], it is known that the reactive mass needs to be quite light. It is also known, that the reactive mass should have the necessary values of rigid parameters. The design of a vibration machine with electromagnets proposed in the article [3], where elastic plates are used as a reactive mass, is the best option for achieving high work efficiency. The elastic plate is made of structural steel, which is an inexpensive material with magnetic properties. The plate is attached to the intermediate mass with four hinges at points A – D (figure 2).

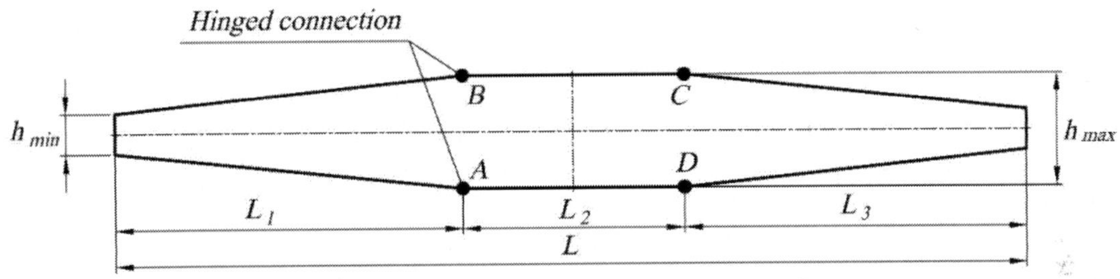

Figure 2. Plate attachment scheme.

The inertial-rigid parameters of the plate are related to the parameter of its eigenfrequency ω. Thus, for the effective synthesis of the continuous member into the discrete system of the vibrating machine, the requisite condition is to accurately calculate the first eigenfrequency of the plate.

It is proposed to use the Rayleigh-Ritz method to compute the first eigenfrequency of a diamond-shaped plate with a hinged attachment at four points. In work [3], when computing the first natural frequency of a rectangular plate, it was found that the deflection occurs along the surface describing the hyperboloid, and the plate is a section of the hyperboloid's surface. The diagram of the deflection of a diamond-shaped plate is shown in figure 3.

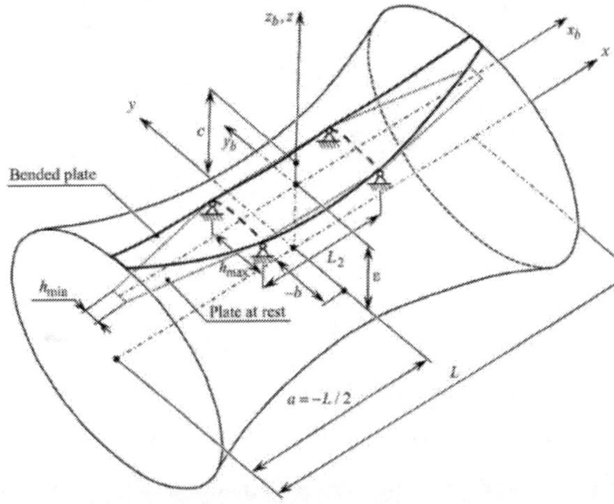

Figure 3. The calculation diagram of the deflection of a diamond-shaped plate on the first eigenfrequency, where x, y, z – coordinates of the absolute reference system; x_b, y_b, z_b – coordinates of the relative reference system; ε – displacement of the relative coordinate system along the axis $0z$; a, b, c – parameters of the hyperboloid.

The coefficient k does not affect the first eigenfrequency of both rectangular plates [3] and plates with a variable cross-section [4], so for further calculations, we take the value $k=1$. We also take the value of the hyperboloid parameter $b=0,6$.

As can be seen from figure 3, the deflection of the proposed plate along the axis $0z$ will be:

$$W(x,y)=z-\varepsilon. \tag{1}$$

Taking into account expression (1), the equation of the deflection of the plate surface will have the following form:

$$W(x,y)=\sqrt{\left(\frac{b}{k}\right)^2\cdot\left[1+\frac{x^2}{(L/2)^2}-\frac{y^2}{b^2}\right]}-\sqrt{\left(\frac{b}{k}\right)^2\cdot\left[1+\frac{(L_2/2)^2}{(L/2)^2}-\frac{(h_{max}/2)^2}{b^2}\right]}. \tag{2}$$

When calculating by the Rayleigh-Ritz method, it should be considered that the kinetic energy of plate vibrations is equal to the potential energy:

$$K=P. \tag{3}$$

To simplify the computations, the plate is divided into three simple sections as shown in figure 4. Sections 1 and 3 are equilateral trapezoids and have the same area. Section 2 is a rectangle with vertices at points A – D.

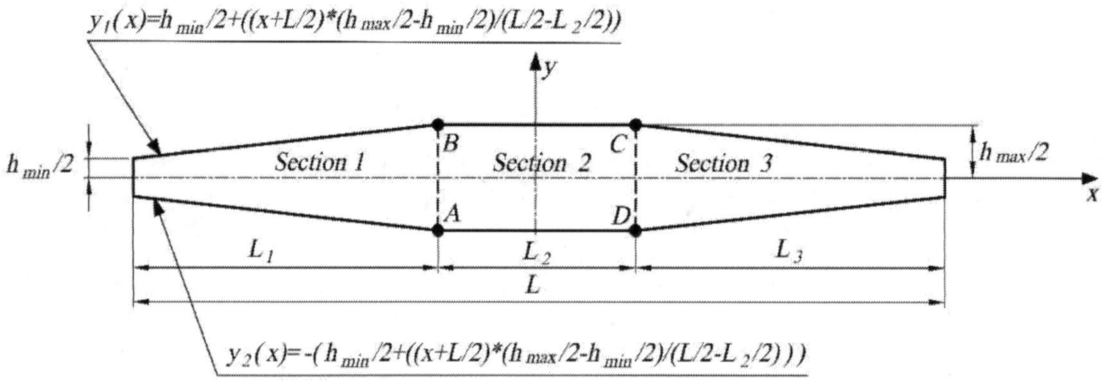

Figure 4. Calculation scheme of a diamond-shaped plate.

As can be seen from figure 4, the sides of section 1 can be described by equations $y_1(x)=\frac{h_{min}}{2}+\frac{\left(x+\frac{L}{2}\right)\cdot\left(\frac{h_{max}}{2}-\frac{h_{min}}{2}\right)}{L/2-L_2/2}$, $y_2(x)=-\left(\frac{h_{min}}{2}+\frac{\left(x+\frac{L}{2}\right)\cdot\left(\frac{h_{max}}{2}-\frac{h_{min}}{2}\right)}{L/2-L_2/2}\right)$. The investigated plate has the following geometric dimensions:

$$L_1=L_3=278\,mm;\ L_2=176\,mm;\ h_{max}=87,5\,mm; h_{min}=0\,mm.$$

For section 1 of a diamond-shaped plate, the ratio of the kinetic energy K_1 to the square of the first circular frequency of oscillations ω^2 during deflection along a hyperboloid's surface at the first eigenfrequency can be determined from the equation [3]:

$$\frac{K_1}{\omega_1^2}=\frac{1}{2}\int_{-L/2}^{-L_2/2}\int_{-\left(\frac{h_{\min}}{2}+\frac{\left(x+\frac{L}{2}\right)\cdot\left(\frac{h_{\max}}{2}-\frac{h_{\min}}{2}\right)}{L/2-L_2/2}\right)}^{\frac{h_{\min}}{2}+\frac{\left(x+\frac{L}{2}\right)\cdot\left(\frac{h_{\max}}{2}-\frac{h_{\min}}{2}\right)}{L/2-L_2/2}}\rho b_n W(x,y)^2\,\mathrm{d}x\mathrm{d}y, \tag{4}$$

where $\rho = 7800\ kg/m^3$ – density of steel.

The ratio K_2/ω^2 for section 2 of the plate will take the form:

$$\frac{K_2}{\omega_1^2}=\frac{1}{2}\int_{-L_2/2}^{L_2/2}\int_{-h_{\max}/2}^{h_{\max}/2}\rho b_n W(x,y)^2\,\mathrm{d}x\mathrm{d}y. \tag{5}$$

Since section 1 and section 3 are symmetrical, they will be valid for the condition $K_1/\omega^2 = K_3/\omega^2$, and therefore the ratio of the total kinetic energy K of the plate to the square of its circular frequency of oscillation ω^2 will be:

$$\frac{K}{\omega^2}=\frac{K_1}{\omega^2}+\frac{K_2}{\omega^2}+\frac{K_3}{\omega^2}=2\cdot\frac{K_1}{\omega^2}+\frac{K_2}{\omega^2}=5{,}55\cdot10^{-3}\,\frac{J\cdot s^2}{rad^2}. \tag{6}$$

The expressions for finding the potential energy of the deflection of section 1 and section 2 of the diamond-shaped plate at the first natural frequency of oscillation will take the form:

$$P_1=\frac{1}{2}\int_{-L/2}^{-L_2/2}\int_{-\left(\frac{h_{\min}}{2}+\frac{\left(x+\frac{L}{2}\right)\cdot\left(\frac{h_{\max}}{2}-\frac{h_{\min}}{2}\right)}{L/2-L_2/2}\right)}^{\frac{h_{\min}}{2}+\frac{\left(x+\frac{L}{2}\right)\cdot\left(\frac{h_{\max}}{2}-\frac{h_{\min}}{2}\right)}{L/2-L_2/2}} \mathrm{D}\left(\begin{array}{l}\left(\frac{\mathrm{d}^2W(x,y)}{\mathrm{d}x^2}+\frac{\mathrm{d}^2W(x,y)}{\mathrm{d}y^2}\right)^2+2(1-\mu)\times\\ \times\left(\left(\frac{\mathrm{d}^2W(x,y)}{\mathrm{d}x\mathrm{d}y}\right)^2-\frac{\mathrm{d}^2W(x,y)}{\mathrm{d}x^2}\cdot\frac{\mathrm{d}^2W(x,y)}{\mathrm{d}y^2}\right)\end{array}\right)\mathrm{d}x\mathrm{d}y, \tag{7}$$

$$P_2=\frac{1}{2}\int_{-L_2/2}^{L_2/2}\int_{-h_{\max}/2}^{h_{\max}/2} \mathrm{D}\left(\begin{array}{l}\left(\frac{\mathrm{d}^2W(x,y)}{\mathrm{d}x^2}+\frac{\mathrm{d}^2W(x,y)}{\mathrm{d}y^2}\right)^2+2(1-\mu)\times\\ \times\left(\left(\frac{\mathrm{d}^2W(x,y)}{\mathrm{d}x\mathrm{d}y}\right)^2-\frac{\mathrm{d}^2W(x,y)}{\mathrm{d}x^2}\cdot\frac{\mathrm{d}^2W(x,y)}{\mathrm{d}y^2}\right)\end{array}\right)\mathrm{d}x\mathrm{d}y, \tag{8}$$

where the parameter D can be determined from the equation:

$$D = E\cdot b_n^3/\left(12\left(1-\mu^2\right)\right), \tag{9}$$

$E = 2{,}1\cdot10^{11}\ Pa$ – Young's modulus, $\mu = 0{,}26$ – Poisson's ratio.

The total potential energy of plate deflection:

$$P = P_1 + P_2 + P_3 = 2\cdot P_1 + P_2 = 543{,}88\ J. \tag{10}$$

The first eigenfrequency of the diamond-shaped plate is[3]:

$$\upsilon=\frac{\omega}{2\pi}=\frac{1}{2\pi}\sqrt{\frac{P}{K}}=\frac{1}{2\pi}\sqrt{\frac{543{,}88}{5{,}55\cdot10^{-3}}}=49{,}83\,Hz. \tag{11}$$

To verify the accuracy of the calculation method, the plate with the specified parameters is investigated in Ansys. It is determined that the first natural frequency of oscillations of the plate is $\upsilon = 49{,}61\,Hz$ (figure 5).

Figure 5. The result of the calculation of the first eigenfrequency of the diamond-shaped plate in Ansys.

3. Results and discussion

According to the numerical calculations in the Ansys software, the first natural frequency of oscillation of the diamond-shaped plate is $\omega = 49{,}61\,Hz$. The value obtained by the finite element method is close to the value determined by the Rayleigh-Ritz method using the hyperboloid equation ($\upsilon = 49{,}83\,Hz$). Therefore, it can be stated that the Rayleigh-Ritz method with the equation of the hyperboloid is appropriate for use in the calculations of the first eigenfrequency of diamond-shaped plates.

4. Conclusions

Achieving better energy efficiency of vibration machines with electromagnets is associated with the use of the inter-resonance operation regimen of this type of equipment. This encourages the introduction of new multi-mass designs of vibrating machines, and therefore, new methods of their calculation and design. One of the ways for the further development of vibration technologies is the improvement of discrete-continuous inter-resonance vibration machines, where a continuous member is synthesized into a discrete model of the oscillating system. In particular, this work presents a vibrating machine where a diamond-shaped flexible plate is used as a reactive mass. An important factor in the implementation of this type of construction is the correct selection of system parameters. One of the most important parameters in vibration technology is the natural frequency of oscillations. To compute the first eigenfrequency of a diamond-shaped plate, it is proposed to apply the Rayleigh-Ritz method using the equation of the hyperboloid. The value of the first natural frequency of oscillations of the diamond-shaped plate established using this method is confirmed in the Ansys software, which allows the use of the Rayleigh-Ritz method with the equation of the hyperboloid for this type of task.

References

[1] Lanets O S 2018 *Osnovy rozrakhunku ta konstruyuvannya vibratsiynykh mashyn. Knyha 1. Teoriya ta praktyka stvorennya vibratsiynykh mashyn z harmoniynym rukhom robochoho orhana [Fundamentals of analysis and design of vibrating machines. Book 1. Theory and*

Practice of Development of Vibratory Machines with Harmonic Motion of the Working Element Body] Lviv: Lviv Polytechnic Publishing House [in Ukrainian].

[2] Lanets O, Kachur O, Borovets V, Dmyterko P, Derevenko I, Zvarich A 2020 Vstanovlennya vlasnoyi chastoty kontynual'noyi dilyanky mizhrezonansnoyi vibromashyny z vykorystannyam nablyzhenoho metodu Releya-Rittsa [Establishment of the original frequency of the continual section of the interreson research machine Rayleigh–Ritz method] *Industrial Process Automation in Engineering and Instrumentation* **54** 5-15 [in Ukrainian].

[3] Maistruk P, Lanets O, Stupnytskyy V 2021 Approximate Calculation of the Natural Oscillation Frequency of the Vibrating Table in Inter-Resonance Operation Mode *Strojnícky časopis - Journal of Mechanical Engineering* **71(2)** 151-166

[4] Lanets O, Maistruk P, Maistruk V, Derevenko I 2022 Approximate calculation of natural frequencies of oscillations of the plate with variable cross-section of the discrete-continuous inter-resonance vibrating table *Ukrainian Journal of Mechanical Engineering and Materials Science* **8 (2)** 41-50

[5] Sharma A, Sharma P, Chauhan P, Bhadoria S 2018 Study on Harmonic Analysis of Functionally Graded Plates Using Fem *International Journal of Applied Mechanics and Engineering* **23(4)** 941-961

[6] Taehyun K, Usik L 2018 Vibration Analysis of Thin Plate Structures Subjected to a Moving Force Using Frequency-Domain Spectral Element Method *Shock and Vibration* **2018** 1-27

[7] Saeed A, Hassan H, Wael E 2013 Vibration attenuation using functionally graded material *World Academy of Science, Engineering and Technology* **7(6)** 1111-20

[8] Ece M, Aydogdu M, Taskin V 2007 Vibration of a variable cross-section beam *Mechanics Research Communications* **34** 78-84

[9] Jena S, Chakraverty S 2018 Free Vibration Analysis of Variable Cross-Section Single-Layered Graphene Nano-Ribbons (SLGNRs) Using Differential Quadrature Method *Frontiers in Built Environment* **4** A63

[10] Boiangiu M, Ceausu V, Untaroiu C 2016 A transfer matrix method for free vibration analysis of Euler-Bernoulli beams with variable cross section *Journal of Vibration and Control* **22(11)** 2591-02

[11] Feng J, Chen Z, Hao S, Zhang K 2020 An Improved Analytical Method for Vibration Analysis of Variable Section Beam *Mathematical Problems in Engineering* **2020** ID 3658146

[12] Vescovini R, Dozio L, D'Ottavio M, Polit O 2018 On the application of the Ritz method to free vibration and buckling analysis of highly anisotropic plates *Composite Structures* **192** 460-74

[13] Kozbur H, Shkodzinsky O, Kozbur I, Gashchyn N 2020 Prediction of the boundary states for thin-walled axisymmetric shells under internal pressure and tension loads *Strojnícky časopis – Journal of Mechanical Engineering* **70(1)** 57 – 68

[14] Mahi A., Adda Bedia E, Tounsi A 2015 A new hyperbolic shear deformation theory for bending and free vibration analysis of isotropic, functionally graded, sandwich and laminated composite plates *Applied Mathematical Modelling* **39 (9)** 2489–08

[15] Borković A, Radenković G, Majstorović D, Milovanović S, Milašinović D, Cvijić R 2020 Free vibration analysis of singly curved shells using the isogeometric finite strip method *Thin-Walled Structures* **157** 107-125

Methodology of investigation the dynamics of longitudinally moving systems under the action of impulse perturbations

A ANDRUKHIV[1], N HUZYK[2,3], B SOKIL[2] and M SOKIL[1]

[1]Lviv Polytechnic National University, 12 Bandera street, 79013, Lviv, Ukraine
[2]Hetman Petro Sahaidachnyi National Army Academy, 32 Heroes of Maidan street, 79026, Lviv, Ukraine

Abstract. A methodology for researching oscillations of one-dimensional elastic systems under the influence of impulse disturbances has been developed. The mathematical model of the considered process is boundary value problems for hyperbolic equations of the second order with an irregular right-hand side and a mixed derivative with respect to linear and time variables. As for the irregularity of the right part, it takes into account the impulse effect on the research object and has the most important theoretical and practical periodic character; and with the help of the mixed derivative in the Euler variables, the longitudinal movement of the studied systems is partially taken into account. Using the basic ideas of asymptotic integration of equations with partial derivatives, the basic propositions of the wave theory of motion adapted to the considered class of problems, the principle of single-frequency oscillations in nonlinear systems with concentrated masses and distributed parameters, the method of regularization of impulse disturbances, equations in the standard form were obtained that describe the determining parameters of nonlinear oscillations for resonant and non-resonant cases. It is shown: that the longitudinal movement of the systems has a significant effect on the natural frequency of oscillations, and therefore on the period of impulse disturbances during which the resonance process takes place; the amplitude of the passage of resonance depends significantly on the speed of movement. The results obtained in the work can serve as a basis for choosing the operating modes of some classes of systems characterized by longitudinal movement in order to avoid resonant processes in them.

Key words: impulse perturbations, longitudinally moving systems, resonance, amplitude, frequency, mathematical model.

INTRODUCTION

Systems that are characterized by longitudinal movement (rope hoists, belt transmissions, elastic bodies) are widely used in various industries, including mechanical engineering, the construction industry, military equipment, etc. The peculiarity of calculation models is that they do not allow using classical methods of integration of partial differential equations for their research. This creates significant problems when analyzing the oscillations of such systems. Their numerical simulation does not give generalized results regarding the influence of the speed of longitudinal movement on the main parameters of their dynamics. Therefore, the development of analytical methods for the study of such systems, even with continuous perturbation of their oscillations, is a complex mathematical problem.

The work deals with a more complex and important problem – the problem of the dynamics of such systems under the action of a periodic impulse perturbations, which is modeled using the Dirac delta function. For the analytical study of the considered class of systems, the main idea of the works [1]-[5] is generalized to a new class of boundary value problems. This made it possible, using the properties of completeness and orthogonality of certain functions and mathematical transformations that do not change the accuracy of the mathematical model, to transform impulse actions on longitudinally moving systems to a form that allows spreading the basic ideas of asymptotic integration. As a result, for the case of the periodic action of the impulse perturbations, equations in the standard form were obtained, which describe the amplitude and frequency of oscillations of the object under study, both for resonant and non-resonant cases.

MATERIAL AND METHOD

The mathematical model of oscillations of one-dimensional elements, which is characterized by a constant speed V of longitudinal movement, is a differential equation

$$w_{tt} + 2Vw_{xt} - \left(\alpha^2 - V^2\right)w_{xx} = \varepsilon F\left(w, w_x, w_t\right), \qquad (1)$$

where $w(x,t)$ denotes movement of the cross-section of a flexible element with the Euler coordinate x at an arbitrary moment of time t, the function $F(w, w_x, w_t)$ is an analytical approximation of the entire set of nonlinear forces, and a small parameter ε indicates a small magnitude of nonlinear forces compared to the magnitude of the tension force ($\alpha^2 = T\rho^{-1}, T$ – tension force, ρ – linear mass of the investigated element for transverse oscillations and $\alpha^2 = E\rho^{-1}, E$ – modulus of elasticity for longitudinal oscillations) under certain boundary conditions.

In the case when a periodic impulse force with a period τ of small magnitude is additionally applied to the element at the point with coordinate x_0, equation (1) is transformed into the form

$$w_{tt} + 2Vw_{xt} - \left(\alpha^2 - V^2\right)w_{xx} = \varepsilon F\left(w, w_x, w_t\right) + \varepsilon\delta(x - x_0)\sum_{i=1}\delta(t - (i-1)\tau)g_i\left(w, w_x, w_t\right), \qquad (2)$$

where $\varepsilon g_i\left(w(x_0,(i-1)\tau), w_x(x_0,(i-1)\tau), w_t(x_0,(i-1)\tau)\right)$ denotes the magnitude of the impulse forces acting at the moment $i\tau$, $\delta(\ldots)$ is Dirac delta function.

Equation (2) will be considered under boundary conditions of the first kind

$$w(0,t) = w(l,t) = 0, \qquad (3)$$

where l is a known constant.

The purpose of the work is to investigate the dynamic process of the object under consideration, which is equivalent to constructing a solution to equation (2) under boundary conditions (3).

Since the maximum value of the right-hand side of equation (2) is a small value compared to the tension force of the flexible element, to construct the solution of the boundary value problem (2), (3) it is possible to apply the methods of integration of boundary value problems that describe the oscillatory processes of systems with distributed parameters . For this aim, the right-hand side of equation (2), using the properties of Dirac function, will be reduced to the form

$$\sum_{i=1}\delta(t - i\tau)g_i\left(w, w_x, w_t\right) = \cos\theta\sum_{i=1}\delta\left(\frac{\theta}{\mu} - \frac{2(i-1)\pi}{\mu}\right)g_i\left(w, w_x, w_t\right), \qquad (4)$$

where $\theta = \mu t$, $\mu = 2\pi\tau^{-1}$, $\delta(t - (i-1)\tau) = \cos\theta\delta\left(\frac{\theta}{\mu} - \frac{2(i-1)\pi}{\mu}\right)$. This means that the differential equation (2) will be reduced to the form

$$w_{tt} + 2Vw_{xt} - \left(\alpha^2 - V^2\right)w_{xx} = \varepsilon F\left(w, w_x, w_t\right) + \varepsilon\frac{2}{l}\cos\theta\sum_{s=1}\sin\frac{s\pi x_0}{l}\sin\frac{s\pi x}{l}\sum_{i=1}\delta\left(\frac{\theta}{\mu} - \frac{2(i-1)\pi}{\mu}\right)g_i\left(w, w_x, w_t\right). \qquad (5)$$

The second term of the right-hand side of the equation will be considered as a periodic perturbations with frequency Ω, and therefore the first approximation of the single-frequency solution of this equation will be given in the form

$$w(x,t) = aw_0(x,\phi) + \varepsilon W_1(a,x,\phi,\theta), \qquad (6)$$

where $aw_0(x,\phi) = a(\cos(\kappa x - \phi) - \cos(\chi x + \phi))$ is the solution of the unperturbed boundary value problem obtained in [6] with $\phi = \omega t + \phi_0$, $\kappa = \frac{k\pi}{\alpha l}(\alpha + V)$, $\chi = \frac{k\pi}{\alpha l}(\alpha - V)$, $\omega = \frac{k\pi}{\alpha l}\left(\alpha^2 - V^2\right)$, a and ϕ_0 denote the amplitude and the initial phase of unperturbed motion respectively.

The function $W_1(a,x,\phi,\theta)$ should take into account the influence of nonlinear forces and periodic impulse perturbations. This means, it satisfies the boundary conditions that follow from (3), i.e.

$$W_1(a,x,\phi,\theta)\big|_{x=0} = W_1(a,x,\phi,\theta)\big|_{x=l} = 0, \qquad (7)$$

and does not contain terms in its expansions that are proportional to the first mode of the oscillation phase ϕ

$$\int_0^{2\pi} W_1(a,x,\phi,\theta)\begin{Bmatrix}\cos\phi \\ \sin\phi\end{Bmatrix} d\phi = 0. \tag{8}$$

Since the paper considers the first approximation of the asymptotic solution in a form close to the first form of undisturbed motion oscillations, then in the dependences for wave numbers κ and χ and the frequency of oscillations ω, the parameter is equal to one and is lowered below.

It is known that for undisturbed motion, the amplitude and frequency of oscillations are constant values, and for disturbed motion, they are variable. The laws of their change are determined not only by nonlinear forces and impulse actions, but also by the ratio between the frequency (period) of natural oscillations and the frequency (period) of the impulse disturbance. Depending on the relationship between these values, resonant and non-resonant cases are distinguished.

In the non-resonant case, when $m\omega \neq n\Omega$ (m, n are mutually prime numbers), the laws of change of parameters a and ϕ are determined from the system of ordinary differential equations

$$\frac{da}{dt} = \varepsilon A_1(a), \quad \frac{d\phi}{dt} = \omega + \varepsilon B_1(a). \tag{9}$$

The conditions (8) imposed on the function $W_1(a,x,\phi,\theta)$ make it possible to determine the laws of change of the amplitude and frequency of the perturbed motion in the form

$$A_1(a) = \frac{\varepsilon}{\pi l\left[(\omega+\kappa V)^2 + (\omega-\chi V)^2\right]}\int_0^l\int_0^{2\pi}\{f_1(a,x,\phi) + \frac{1}{\pi l\mu}\sum_{i=1}\int_0^{2\pi}\cos\theta\sum_{s=1}\sin\frac{s\pi x_0}{l}\sin\frac{s\pi x}{l}\delta\left(\frac{\theta}{\mu} - \frac{2(i-1)\pi}{\mu}\right)g_i(a,x,\phi)d\theta\}\times$$
$$\{\bar{\rho}(x)\cos\phi + \bar{h}(x)\sin\phi\}d\phi dx,$$

$$B_1(a) = \frac{\varepsilon}{a\pi l\left[(\omega+\kappa V)^2 + (\omega-\chi V)^2\right]}\int_0^l\int_0^{2\pi}\{f_1(a,x,\phi) + \frac{1}{\pi l\mu}\sum_{i=1}\int_0^{2\pi}\cos\theta\sum_{s=1}\sin\frac{s\pi x_0}{l}\sin\frac{s\pi x}{l}\delta\left(\frac{\theta}{\mu} - \frac{2(i-1)\pi}{\mu}\right)g_i(a,x,\phi)d\theta\times$$
$$\{\bar{\rho}(x)\sin\phi - \bar{h}(x)\cos\phi\}d\phi dx,$$

where $\bar{\rho}(x) = (\omega+\kappa V)\sin\kappa x + (\omega-\chi V)\sin\chi x$, $\bar{h}(x) = (\omega+\kappa V)\cos\kappa x - (\omega-\chi V)\cos\chi x$.

A partial case of the considered process is the action on the considered object of a periodic, constant force f_i at a fixed point, i.e. $g_i(w, w_x, w_t) \equiv f_i$. Then the corresponding dependences of change of the amplitude and frequency of the perturbed motion can be transformed into the form

$$A_1(a) = \frac{\varepsilon}{\pi l\left[(\omega+\kappa V)^2 + (\omega-\chi V)^2\right]}\int_0^l\int_0^{2\pi}\{f_1(a,x,\phi)\}\times\{\bar{\rho}(x)\cos\phi + \bar{h}(x)\sin\phi\}d\phi dx,$$

$$B_1(a) = \frac{\varepsilon}{a\pi l\left[(\omega+\kappa V)^2 + (\omega-\chi V)^2\right]}\int_0^l\int_0^{2\pi}\{f_1(a,x,\phi)\}\{\bar{\rho}(x)\sin\phi - \bar{h}(x)cos\phi\}d\phi. \tag{10}$$

From the last relations, we can conclude that in the non-resonant case, the periodic impulse disturbance in the first approximation does not cause changes in either the amplitude or the frequency of the process. Its influence is manifested only in a partial change of the form of the dynamic process, i.e., a periodic disturbance in the non-resonant case affects the form of the function $W_1(a,x,\phi,\theta)$.

Much more interesting from a practical point of view and more difficult to study is the resonant case that occurs when $m\omega \approx n\Omega$. Since in nonlinear systems the dynamic process significantly depends on the phase difference of natural and forced oscillations, i.e. on the parameter $\gamma = \phi - \frac{m}{n}\theta$ than in contrast to the non-resonant case, the laws of change of basic parameters are determined from the system

$$\frac{da}{dt} = \varepsilon A_1(a,\gamma) + \ldots, \quad \frac{d\gamma}{dt} = \omega - \frac{m}{n}\Omega + \varepsilon B_1(a,\gamma)\ldots \tag{11}$$

The task is to determine such functions $A_1(a,\gamma), B_1(a,\gamma), \ldots$ which, together with the asymptotic representation (6), satisfy the basic equation (2), if the functions a and ϕ defined by formula (11) and depended on time are substituted in it. The conditions (8) imposed on the function $W_1(a,x,\phi,\theta)$ allow in this case to obtain a system of equations that connects the unknown functions in the form

$$\bar{\rho}(x)A_1(a,\gamma)+a\bar{\hbar}(x)B_1(a,\gamma)+\left[\overline{\wp}(x)\frac{\partial A_1(a,\gamma)}{\partial\gamma}+a\bar{\hbar}(x)\frac{\partial B_1(a,\gamma)}{\partial\gamma}\right]\left(\omega-\frac{s}{n}\mu\right)=\frac{\varepsilon}{2\pi^2}\sum_q \exp(inq\gamma)\int_0^{2\pi}\int_0^{2\pi}\{f_1(a,x,\phi)$$

$$+\frac{2}{l}\cos\theta\sum_{s=1}\sin\frac{s\pi x_0}{l}\sin\frac{s\pi x}{l}\sum_{i=1}\delta\left(\frac{\theta}{\mu}-\frac{2(i-1)\pi}{\mu}\right)g_1(a,x,\phi)\Big\}\exp(-inq\left(\phi-\frac{s}{n}\theta\right))\cos\phi d\phi d\theta,$$

$$-\bar{\hbar}(x)A_1(a,\gamma)+a\bar{\rho}(x)B_1(a,\gamma)+\left[\bar{\hbar}(x)\frac{\partial A_1(a,\gamma)}{\partial\gamma}-a\,\overline{\wp}(x)\frac{\partial B_1(a,\gamma)}{\partial\gamma}\right]\left(\omega-\frac{s}{n}\mu\right)=\frac{\varepsilon}{2\pi^2}\sum_q \exp(inq\gamma)\int_0^{2\pi}\int_0^{2\pi}\{f_1(a,x,\phi)$$

$$+\frac{2}{l}\cos\theta\sum_{s=1}\sin\frac{s\pi x_0}{l}\sin\frac{s\pi x}{l}\sum_{i=1}\delta\left(\frac{\theta}{\mu}-\frac{2(i-1)\pi}{\mu}\right)g_1(a,x,\phi)\Big\}\exp(-inq\left(\phi-\frac{s}{n}\theta\right))\sin\phi d\phi d\theta.$$

where $\overline{\wp}(x,\phi)=-(\cos(\kappa x+\phi)-\cos(\chi x-\phi))$, $\bar{\hbar}(x,\phi)=\sin(\kappa x+\phi)+\sin(\chi x-\phi)$.

This means that for the general case of a periodic disturbance, it is not possible to find in a closed form the relations that are determined in the first approximation $\frac{da}{dt}$ and $\frac{d\phi}{dt}$. However, the right-hand parts of the specified ratios are periodic functions of the phase difference γ and amplitude a, and therefore the solution for the functions $A_1(a,\gamma)$ and $B_1(a,\gamma)$ must also be sought in the same form. The obtained dependencies provide an opportunity to study the dynamic process both directly in the resonance region and when approaching it.

Conclusions

For one-dimensional systems characterized by a constant speed of longitudinal movement, a method of analytical research of transverse oscillations under the action of a periodic system of impulse disturbances has been developed. The peculiarity of this class of dynamic systems is that it is not possible to apply well-known methods of integration of partial differential equations (Euler or D'Alembert) to describe undisturbed motion. As for disturbed motion, applying the basic ideas of the wave theory of motion adapted to longitudinally moving system; the principle of single-frequency nonlinear oscillations of systems with distributed parameters; basic ideas of asymptotic integration of partial differential equations; properties of delta functions, the asymptotic approximation of the corresponding boundary value problems is constructed in the work. Equations were obtained in the standard form, which describe the main parameters of oscillations for both non-resonant and resonant cases.

It was found that in the non-resonant case, the impulse disturbance partially affects changes in the form of oscillations.

For the resonant case, it is shown that:

– for higher speeds of longitudinal movement of the systems, the natural frequency of oscillations decreases, while the period of the impulse disturbance during which resonance occurs increases;

– the resonant value of the amplitude with the increase in the speed of longitudinal movement first increases, and then decreases;

– as the velocities of the objects under study approach the critical ones ($V=\alpha$), the resonance values of the amplitudes depend significantly on the amplitude of the entry into resonance and the magnitude of the impulse disturbance.

The results obtained in the work can serve as a basis for choosing operational parameters of systems characterized by longitudinal movement in order to avoid resonant phenomena in them or to pass them with the smallest amplitude. The reliability of the method is confirmed by obtaining in the extreme case the known from literary sources, and the method itself can be applied to the study of systems whose mathematical models are similar to the considered ones.

References

[1] Andrukhiv A, Sokil M, Sokil B, Fedushko S, Syerov Yu, Karovic V and Klynina T 2021 Influence of Impulse Disturbances on Oscillations of Nonlinearly Elastic Bodies *Mathematics*. **9 (8)** 819

[2] Chen L Q 2005 Analysis and control of transverse vibrations of axially moving strings *Appl. Mech. Rev.* **58 (2)** 91

[3] Cveticanin L and Pogany T 2012 Oscillator with a sum of non-integer order non-linearities *Journal of Applied Mathematics* **2012** 20

[4] Lyashuk O, Vovk Y, Sokil B, Klendii V, Ivasechko R and Dovbush T 2019 Mathematical model of a dynamic process of transporting a bulk material by means of a tube scraping conveyor *Agricultural Engineering International: CIGR Journal* **21 (1)** 74

[5] Tian Y, Yuan P, Yang F, Gu J, Chen M, Tang J, Su Y, Ding T, Zhang K and Cheng Q 2018 Research on the principle of a new flexible screw conveyor and its power consumption *Applied Sciences* **8(7)** 1038

[6] Mytropolskyy Yu 2012 *Selected works in 2 volumes* (Kyiv, Naukova dumka) p 504

Bending of a plate with circular rigid inclusion and system of cracks on the assumption of strip crack face contact

M S Slobodian[1], O V Bilash[2], R H Seliverstov[3] and L R Kurotchyn[4]

[1]Department of Mechanics, Faculty of Mechanics and Mathematics, Ivan Franko National University of L'viv, 1 Universytetska Street, Lviv, 79000 Ukraine
[2]Department of Engineering Mechanics (Weapons and Equipment of Military Engineering Forces), Faculty Training Specialists Battle (Operational) Software, Hetman Petro Sahaidachnyi National Army Academy, 32 Heroes of Maidan Street, Lviv, 79026, Ukraine
[3]Department of Programming, Faculty of Applied Mathematics and Informatics, Ivan Franko National University of L'viv, 1 Universytetska Street, Lviv, 79000, Ukraine
[4]Department of Mechanics of deformable bodies, Pidstryhach Institute for Applied Problems of Mechanics and Mathematics, Ukrainian National Academy of Sciences, 3b Naukova Street, Lviv, 79060, Ukraine

Abstract. The article presents a solution to the bending problem for an infinite isotropic plate with a circular rigid inclusion and arbitrarily oriented straight through cracks. It is assumed that under the action of an external load at infinity, the faces of all cracks are in smooth contact along a region of constant width near the upper surface of the plate. The solution is obtained using methods of the theory of functions of complex variable and complex potentials and is reduced to a system of singular integral equations, which is numerically solved by the mechanical quadrature method. On the basis of numerical analysis, the graphical dependences of the contact force, forces and moments intensity factors are constructed at various parameters of the problem.

1. Introduction

Plate-shaped structures are widely used in various fields of technology. Their operational characteristics depend on crack-like defects and inclusions, which reduce the range of permissible load. During the operation of plate elements, it is important to know how such defects affect the strength and mechanical characteristics of the structure.

Many authors were engaged in solving plane and bending problems for plates with holes and cracks. Generalizations and methods of solving such problems are presented in monographs [1-5]. But these works do not take into account the crack closure under the action of an external load. The experiments carried out by J. Kalthof, I. Shatsky and A. Burgel [6] testify to the significant influence of the contact of crack faces on stress-strain state of the plate under bending. The model of smooth contact of crack faces along a line along the entire length of cracks (linear contact) was used in works [7-9]. The model of smooth contact of crack faced along a region of constant width (strip contact) is used in works [10-14].

In this work the problem of bending of infinite isotropic plate with circular rigid inclusion and system of arbitrarily oriented straight through cracks is investigated on the assumption of smooth strip contact of crack faces along a region of constant width along the entire length of cracks near the upper surface of plate.

2. Problem statement

Consider an infinite isotropic plate of thickness $2h$ containing rigid cylindrical (circular) inclusion of radius R and system of M arbitrarily oriented straight through cracks of length $2l_k$ $\left(k=\overline{1,M}\right)$. The origin O of a chosen Cartesian coordinate system $Oxy\tilde{z}$ is placed on the middle plane of plate at the center of the inclusion, axis $O\tilde{z}$ is perpendicular to the plate. On the plane Oxy we introduce a polar coordinate system r and θ with the polar axis Ox and the pole at the point . The centers of cracks are located at points (x_{0k}, y_{0k}) and the angle between k -th crack and axis Ox is denoted by α_k $(k=\overline{1,M})$. The centers of the cracks are the origins of local Cartesian coordinate systems $O_k x_k y_k$. The plate is under the action of uniformly distributed bending moments M_x^∞ and M_y^∞ at infinity. In the middle plane S^+ and S^- refer to the areas inside and outside the inclusion, respectively, L_k denotes straight line segment of k -th crack, and L – the interfacial contour at the inclusion boundary (see. Fig. 1).

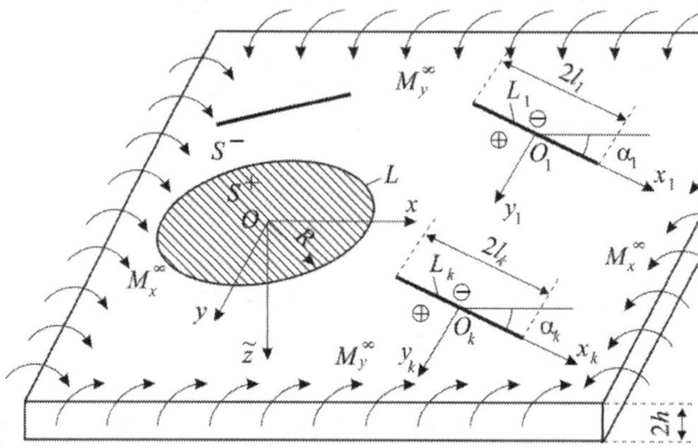

Figure 1. An infinite plate with inclusion and arbitrarily oriented cracks.

Suppose the crack faces smoothly contact along a region of constant width along the entire length of cracks near the upper surface of plate (strip contact). Due to the contact of the crack banks, in addition to bending moments, membrane forces arise. Based on this, we will present the solution of the problem in the form of solutions of two problems: the plate classical bending problem (Kirchhoff-Love theory) and the plane problem of elasticity theory.

The boundary conditions at the interfacial contour of inclusion are

$$w=0,\ \frac{\partial w}{\partial r}=0,\ u_{Pr}=0,\ u_{P\theta}=0,\ x\in L, \tag{1}$$

where w is deflection of the plate, u_{Pr} are $u_{P\theta}$ displacement vector components in the polar coordinate system.

The boundary conditions of strip contact of crack faces are [10-14]

$$P^{\pm}=0,\ M_{y_k}^{\pm}=\beta_k h N_k,\ \tau_{IIx_ky_k}^{\pm}=0,\ \sigma_{Py_ky_k}^{\pm}=-\frac{N_k}{2h},\ \beta_k=1-\frac{\gamma_k}{3},\ x_k\in L_k,\ k=\overline{1,M}, \tag{2}$$

$$\frac{\partial[v_P]}{\partial x_k}+\alpha_k h\left[\frac{\partial^2 w}{\partial x_k \partial y_k}\right]=0,\ \alpha_k=\frac{1}{2}\cdot\left(1+(1-\gamma_k)^2\right),\ \gamma_k=\frac{h_k}{h},\ x_k\in L_k,\ k=\overline{1,M}, \tag{3}$$

where N_k is contact force between faces of k -th crack, h_k – contact zone height of k-th crack, $\sigma_{Py_ky_k}$ i $\tau_{IIx_ky_k}$ – stress tensor components in the Cartesian coordinate system, M_{y_k} – bending moment, P –

generalized Kirchhoff shear force, v_{II} – displacement vector component; $[f]=f^{+}-f^{-}$ (superscripts "+" and "–" stand for limits of the functions as a point of the middle plane approaches the k -th crack, $y_k \to \pm 0$).

Thus, (1)-(3) as interconnected boundary conditions for the plane problem and the plate bending problem describe the crack closure in a two-dimensional setting in the presence of an rigid inslusion in the plate.

3. Solution

Using the complex potentials of the plane problem [4] and the plate bending problem [2], and satisfying the boundary conditions (1)-(3), the solution of the problem is reduced to linear conjugation problems, on the basis of which a following system of singular integral equations in an unknown jump of displacement $G_j(\eta)$ and rotational displacement $Y_j(\eta)$ at crack faces is obtained:

$$\operatorname{Im}\left(\sum_{j=1}^{M}\int_{-1}^{1}\left\{K_{mj}(\eta,\xi)Y_j(\eta)+L_{mj}(\eta,\xi)\overline{Y_j(\eta)}\right\}d\eta\right)==\tilde{c}'_m+\operatorname{Im}\tilde{P}_m(\xi),\ \xi\in[-1,1],\ m=\overline{1,M}, \quad (4)$$

$$\operatorname{Re}\left(\sum_{j=1}^{M}\int_{-1}^{1}\left\{K_{mj}(\eta,\xi)Y_j(\eta)+L_{mj}(\eta,\xi)\overline{Y_j(\eta)}\right\}d\eta\right)+$$
$$+\frac{q}{2}\operatorname{Re}\left(\sum_{j=1}^{M}\int_{-1}^{1}\left\{R_{mj}(\eta,\xi)G_j(\eta)+S_{mj}(\eta,\xi)\overline{G_j(\eta)}\right\}d\eta\right)=\operatorname{Re}\tilde{P}_m(\xi),\ \xi\in[-1,1],\ m=\overline{1,M}, \quad (5)$$

$$\operatorname{Im}\left(\sum_{j=1}^{M}\int_{-1}^{1}\left\{R_{mj}(\eta,\xi)G_j(\eta)+S_{mj}(\eta,\xi)\overline{G_j(\eta)}\right\}d\eta\right)=0,\ \xi\in[-1,1],\ m=\overline{1,M}, \quad (6)$$

where

$$K_{mj}(\eta,\xi)=\begin{cases}\dfrac{\tilde{\kappa}}{\pi i(\eta-\xi)}+\tilde{K}_{mj}(\eta,\xi), m=j,\\ \tilde{K}_{mj}(\eta,\xi)+K^{V}_{mj}(\eta,\xi), m\neq j,\end{cases}\quad L_{mj}(\eta,\xi)=\begin{cases}\tilde{L}_{mj}(\eta,\xi), m=j,\\ \tilde{L}_{mj}(\eta,\xi)+L^{V}_{mj}(\eta,\xi), m\neq j,\end{cases}$$

$$R_{mj}(\eta,\xi)=\begin{cases}1/(\pi(\eta-\xi))+\tilde{R}_{mj}(\eta,\xi), m=j,\\ \tilde{R}_{mj}(\eta,\xi)+R^{V}_{mj}(\eta,\xi), m\neq j,\end{cases}\quad S_{mj}(\eta,\xi)=\begin{cases}\tilde{S}_{mj}(\eta,\xi), m=j,\\ \tilde{S}_{mj}(\eta,\xi)+S^{V}_{mj}(\eta,\xi), m\neq j,\end{cases}$$

$$\tilde{K}_{mj}(\eta,\xi)=\tilde{\kappa}K^{I}_{mj}(\eta,\xi)-\left(1+\frac{e^{-2i\alpha_m}}{\overline{X}_m^2}\right)K^{II}_{mj}(\eta,\xi)-e^{-2i\alpha_m}\left(X_m-\frac{1}{\overline{X}_m}\right)K^{III}_{mj}(\eta,\xi)-\frac{e^{-2i\alpha_m}}{\overline{X}_m^2}K^{IV}_{mj}(\eta,\xi),$$

$$\tilde{L}_{mj}(\eta,\xi)=\tilde{\kappa}L^{I}_{mj}(\eta,\xi)-\left(1+\frac{e^{-2i\alpha_m}}{\overline{X}_m^2}\right)L^{II}_{mj}(\eta,\xi)-e^{-2i\alpha_m}\left(X_m-\frac{1}{\overline{X}_m}\right)L^{III}_{mj}(\eta,\xi)-\frac{e^{-2i\alpha_m}}{\overline{X}_m^2}L^{IV}_{mj}(\eta,\xi),$$

$$K^{I}_{mj}(\eta,\xi)=-\frac{\tilde{\kappa}}{2\pi iX_m}\frac{\lambda_j e^{i\alpha_j}}{\overline{T}_jX_m-1},\ L^{I}_{mj}(\eta,\xi)=-\frac{1}{2\pi i}\left\{\frac{\lambda_j e^{-i\alpha_j}}{\overline{T}_j}+\frac{\lambda_j e^{-i\alpha_j}F_{mj}}{(\overline{T}_jX_m-1)^2}-\frac{\lambda_j e^{-i\alpha_j}}{X_m}\cdot\frac{X_m^2+e^{2i\alpha_j}}{\overline{T}_jX_m-1}\right\},$$

$$K^{II}_{mj}(\eta,\xi)=\frac{1}{2\pi i}\left\{\frac{\lambda_j e^{i\alpha_j}}{T_j}+\frac{\lambda_j e^{i\alpha_j}\overline{F}_{mj}}{(T_j\overline{X}_m-1)^2}-\frac{\lambda_j e^{i\alpha_j}}{\overline{X}_m}\cdot\frac{\overline{X}_m^2+e^{2i\alpha_j}}{T_j\overline{X}_m-1}\right\},\ L^{II}_{mj}(\eta,\xi)=\frac{\tilde{\kappa}}{2\pi i\overline{X}_m}\cdot\frac{\lambda_j e^{-i\alpha_j}}{T_j\overline{X}_m-1},$$

$$K_{mj}^{III}(\eta,\xi)=\frac{1}{\pi i}\frac{\lambda_j}{\left(T_j\bar{X}_m-1\right)^2}\left\{e^{i\alpha_j}+\frac{e^{-i\alpha_j}}{\bar{X}_m^2}+\frac{T_j\bar{X}_m-1}{\bar{X}_m^2}e^{-i\alpha_j}-\frac{e^{i\alpha_j}\bar{F}_{mj}}{\bar{X}_m}-\frac{e^{i\alpha_j}}{\bar{X}_m}\cdot\frac{\bar{F}_{mj}}{T_j\bar{X}_m-1}\right\},$$

$$L_{mj}^{III}(\eta,\xi)=-\frac{\tilde{\kappa}}{\pi i\bar{X}_m^2}\cdot\frac{\lambda_j e^{-i\alpha_j}}{T_j\bar{X}_m-1}\left(1+\frac{1}{2\left(T_j\bar{X}_m-1\right)}\right),\ K_{mj}^{IV}(\eta,\xi)=\frac{\lambda_j\bar{X}_m e^{i\alpha_j}}{2\pi i\left(T_j\bar{X}_m-1\right)},\ L_{mj}^{IV}(\eta,\xi)=-\frac{\lambda_j e^{-i\alpha_j}}{2\pi i\bar{T}_j},$$

$$K_{mj}^{V}(\eta,\xi)=\frac{\tilde{\kappa}\lambda_j e^{i\alpha_j}}{2\pi i}\left(\frac{1}{T_j-X_m}+\frac{e^{-2i\alpha_m}}{\bar{T}_j-\bar{X}_m}\right),\ L_{mj}^{V}(\eta,\xi)=\frac{1}{2\pi i}\cdot\frac{\lambda_j e^{-i\alpha_j}}{\bar{T}_j-\bar{X}_m}\left(1-\frac{T_j-X_m}{\bar{T}_j-\bar{X}_m}e^{-2i\alpha_m}\right),$$

$$\tilde{R}_{mj}(\eta,\xi)=R_{mj}^{I}(\eta,\xi)+\left(1+\frac{e^{-2i\alpha_m}}{\bar{X}_m^2}\right)R_{mj}^{II}(\eta,\xi)+e^{-2i\alpha_m}\left(X_m-\frac{1}{\bar{X}_m}\right)R_{mj}^{III}(\eta,\xi)+\frac{e^{-2i\alpha_m}}{\bar{X}_m^2}R_{mj}^{IV}(\eta,\xi),$$

$$\tilde{S}_{mj}(\eta,\xi)=S_{mj}^{I}(\eta,\xi)+\left(1+\frac{e^{-2i\alpha_m}}{\bar{X}_m^2}\right)S_{mj}^{II}(\eta,\xi)+e^{-2i\alpha_m}\left(X_m-\frac{1}{\bar{X}_m}\right)S_{mj}^{III}(\eta,\xi)+\frac{e^{-2i\alpha_m}}{\bar{X}_m^2}S_{mj}^{IV}(\eta,\xi),$$

$$R_{mj}^{I}(\eta,\xi)=-\frac{1}{2\pi\kappa X_m}\frac{\lambda_j e^{i\alpha_j}}{\bar{T}_jX_m-1},\ S_{mj}^{I}(\eta,\xi)=-\frac{1}{2\pi\kappa}\left\{\frac{\lambda_j e^{-i\alpha_j}}{\bar{T}_j}+\frac{\lambda_j e^{-i\alpha_j}F_{mj}}{\left(\bar{T}_jX_m-1\right)^2}-\frac{\lambda_j e^{-i\alpha_j}}{X_m}\cdot\frac{X_m^2+e^{2i\alpha_j}}{\bar{T}_jX_m-1}\right\},$$

$$R_{mj}^{II}(\eta,\xi)=-\frac{1}{2\pi\kappa}\left\{\frac{\lambda_j e^{i\alpha_j}}{T_j}+\frac{\lambda_j e^{i\alpha_j}\bar{F}_{mj}}{\left(T_j\bar{X}_m-1\right)^2}-\frac{\lambda_j e^{i\alpha_j}}{\bar{X}_m}\cdot\frac{\bar{X}_m^2+e^{2i\alpha_j}}{T_j\bar{X}_m-1}\right\},\ S_{mj}^{II}(\eta,\xi)=-\frac{1}{2\pi\kappa\bar{X}_m}\cdot\frac{\lambda_j e^{-i\alpha_j}}{T_j\bar{X}_m-1},$$

$$R_{mj}^{III}(\eta,\xi)=-\frac{1}{\pi\kappa}\frac{\lambda_j}{\left(T_j\bar{X}_m-1\right)^2}\left\{e^{i\alpha_j}+\frac{e^{-i\alpha_j}}{\bar{X}_m^2}+\frac{T_j\bar{X}_m-1}{\bar{X}_m^2}e^{-i\alpha_j}-\frac{e^{i\alpha_j}\bar{F}_{mj}}{\bar{X}_m}-\frac{e^{i\alpha_j}}{\bar{X}_m}\frac{\bar{F}_{mj}}{T_j\bar{X}_m-1}\right\},$$

$$S_{mj}^{III}(\eta,\xi)=\frac{1}{\pi\kappa\bar{X}_m^2}\frac{\lambda_j e^{-i\alpha_j}}{T_j\bar{X}_m-1}\left(1+\frac{1}{2\left(T_j\bar{X}_m-1\right)}\right),\ R_{mj}^{IV}(\eta,\xi)=-\frac{\kappa\lambda_j\bar{X}_m e^{i\alpha_j}}{2\pi\left(T_j\bar{X}_m-1\right)},\ S_{mj}^{IV}(\eta,\xi)=\frac{\lambda_j e^{-i\alpha_j}}{2\pi\bar{T}_j},$$

$$R_{mj}^{V}(\eta,\xi)=\frac{\lambda_j e^{i\alpha_j}}{2\pi}\left(\frac{1}{T_j-X_m}+\frac{e^{-2i\alpha_m}}{\bar{T}_j-\bar{X}_m}\right),\ L_{mj}^{V}(\eta,\xi)=\frac{1}{2\pi i}\frac{\lambda_j e^{-i\alpha_j}}{\bar{T}_j-\bar{X}_m}\left(1-\frac{T_j-X_m}{\bar{T}_j-\bar{X}_m}e^{-2i\alpha_m}\right),$$

$$\tilde{P}_m(\xi)=-\tilde{\kappa}A-\frac{\tilde{\kappa}B}{X_m^2}+A+\frac{B}{\bar{X}_m^2}-e^{-2i\alpha_m}\left\{-\frac{B}{\bar{X}_m^4}-\frac{A}{\bar{X}_m^2}+2B\frac{X_m\bar{X}_m-1}{\bar{X}_m^4}-B-\frac{A}{\bar{X}_m^2}\right\},$$

$$Z_{0j}=\frac{z_{0j}}{R}=\frac{x_{0j}}{R}+i\frac{y_{0j}}{R},\ x_j=l_j\xi,\ t=l_j\eta,\ \lambda_j=\frac{l_j}{R},\ q=-\frac{3}{2}(1+\upsilon),\ T_j=Z_{0j}+\lambda_j\eta e^{i\alpha_j},$$

$$X_j=Z_{0j}+\lambda_j\xi e^{i\alpha_j},\ F_{mj}=X_m-Z_{0j}-\frac{e^{2i\alpha_j}}{X_m}+\bar{Z}_{0j}e^{2i\alpha_j},\ A=-0.375(1+\rho)(1-\upsilon),$$

$$B=0.75(1-\rho)(1+\upsilon),\ \rho=\frac{M_x^\infty}{M_y^\infty},\ \mu=\frac{E}{2(1+\upsilon)},\ \tilde{\kappa}=\frac{3+\upsilon}{1-\upsilon},\ \kappa=\frac{3-\upsilon}{1+\upsilon},\ i^2=-1,$$

$$Y_m(\eta)=Y_{m1}(\eta)+iY_{m2}(\eta),\ G_m(\eta)=G_{m1}(\eta)+iG_{m2}(\eta),\ y_m(t)=\frac{M_y^\infty Y_m(l_m\eta)}{Eh^3},\ g_m'(t)=\frac{M_y^\infty G_m(l_m\eta)}{h^2},$$

$Y_{m1}(\eta)$, $Y_{m2}(\eta)$, $G_{m1}(\eta)$, $G_{m2}(\eta)$ – real functions, E – Young's modulus; υ – Poisson's ratio.

Satisfying boundary condition (3), we obtain

$$G_{m1}(\eta)+\frac{1+\tilde{\kappa}}{(1+\kappa)(1+\upsilon)}Y_{m2}(\eta)=0,\ \eta\in[-1,1],\ m=\overline{1,M}. \tag{7}$$

To the system (4)-(7) we add the equations

$$\int_{-1}^{1}Y_m(\eta)d\eta=0,\ \int_{-1}^{1}\eta Y_{m1}(\eta)d\eta=0,\ m=\overline{1,M}, \tag{8}$$

which ensures that rotational displacements and deflection of the plate are single-valued when bypassing the contour of each crack.

The displacements also must be single-valued, therefore

$$\int_{-1}^{1}G_m(\eta)d\eta=0,\ m=\overline{1,M}. \tag{9}$$

Thus, to determine the unknown functions $Y_m(\eta)$, $G_m(\eta)$, $m=\overline{1,M}$, we have a system of singular integral equations (4)-(9), which will be solved numerically by mechanical quadrature method [3, 5].

The dimensionless contact force $N_m^*(\xi)$ between the crack faces is determined by the formulas

$$N_m^*(\xi)=\frac{hN_m(\xi)}{2M_y^\infty}=-\mathrm{Re}\left(\sum_{j=1}^{M}\int_{-1}^{1}\left\{R_{mj}(\eta,\xi)G_j(\eta)+S_{mj}(\eta,\xi)\overline{G_j(\eta)}\right\}d\eta\right),\ m=\overline{1,M}. \tag{10}$$

4. Numerical analysis and conclusions

Numerical analysis is carried out at $\upsilon=0.3$, $\gamma_1=h_1/h=0.13$, 200 nodal values of mechanical quadrature method, one crack; $z_0=1+\varepsilon$, $\varepsilon=d/R$.

Figures 2-3 illustrate graphical dependence of the reduced contact force $N^*=hN/M_y^\infty$ between crack faces on dimensionless coordinate $\xi=x_1/l$. Plot lines labeled by 1, 2, 3, 4, 5 are obtained at $\alpha=0^\circ$, $\alpha=30^\circ$, $\alpha=45^\circ$, $\alpha=60^\circ$ and $\alpha=90^\circ$, respectively.

Figure 2 corresponds to the $\varepsilon=d/R=1$, $\lambda=l/R=0.5$, $\rho=M_x^\infty/M_y^\infty=1$. Numerical analysis has shown that for such load at $\alpha<28^\circ$ the contact force N^* is less at the crack tip a $(\xi=-1)$ closest to the inclusion than the far tip b, but at $28^\circ<\alpha<90^\circ$ the situation is the opposite.

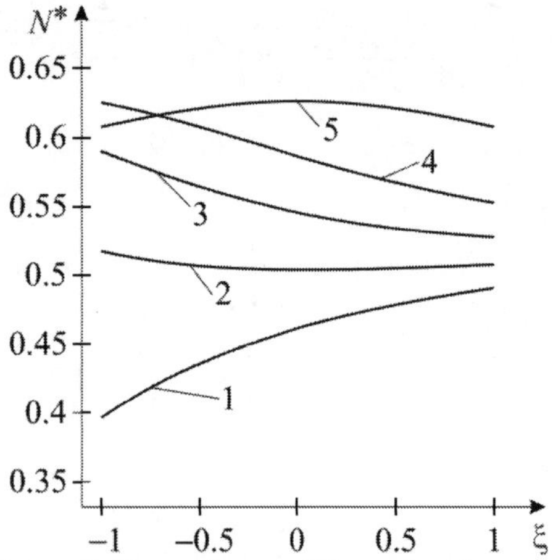

Figure 2. Dependence of the reduced contact force at various slopes of crack α

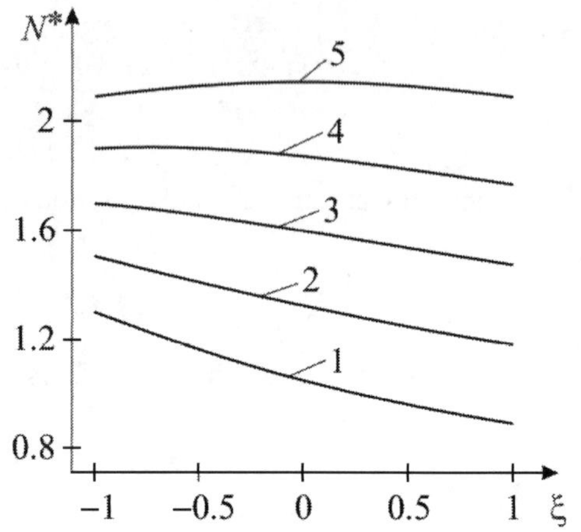

Figure 3. Dependence of the reduced contact force at various slopes of crack α

Figure 3 corresponds to the $\varepsilon = d/R = 1$, $\lambda = l/R = 0.5$, $\rho = M_x^\infty / M_y^\infty = 3$. It demonstrates that under such a load N^* is greater at the tip a than at the tip b, and increases with increasing angle α.

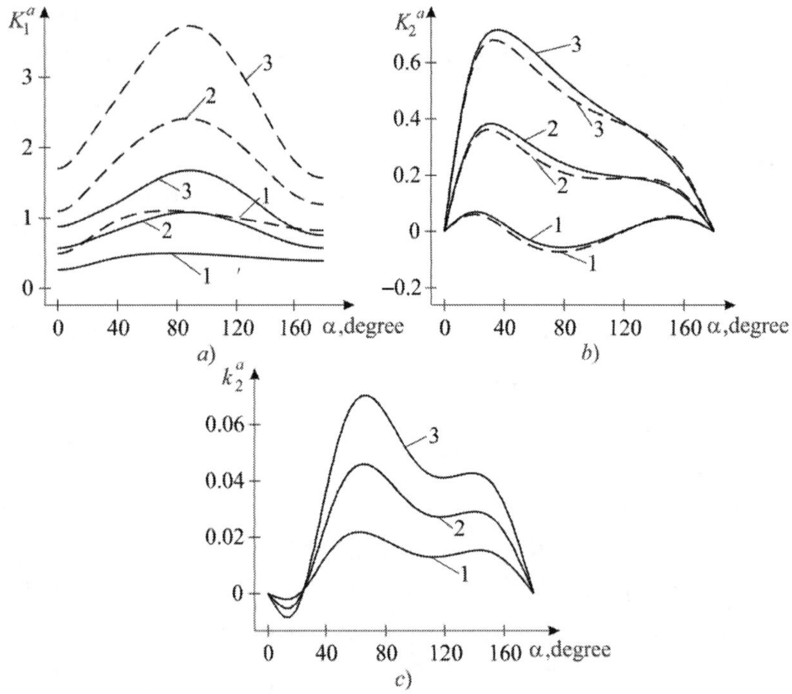

Figure 4. Dependences of the moment intensity factor and force intensity factor on slope of crack α at various values $\rho = M_x^\infty / M_y^\infty$

Figure 4 demonstrates the graphical dependences of moment intensity factors $K_M^* = K_M / \left(M_y^\infty \sqrt{l}\right)$ and force intensity factor $k_N^* = hK_N / \left(M_y^\infty \sqrt{l}\right)$ at point a on the slope of crack α at $\varepsilon = d/R = 1$, $\lambda = l/R = 0.8$. Plot lines labeled by 1, 2 and 3 are obtained at $\rho = M_x^\infty / M_y^\infty = 1$, $\rho = 2$ and $\rho = 3$, respectively. Dashed lines correspond to the case when crack closure is neglected and solid ones – to the case of strip contact between crack faces. We can come to the conclusion that the moment intensity factor K_1 for the strip contact case is smaller than that without taking into account the crack closure. As for the moment intensity factor K_2, it is greater for the strip contact case at $\rho=1$, $0<\alpha<125^\circ$, at $\rho=2$, $0<\alpha<127^\circ$ and at $\rho=3$, $0<\alpha<128^\circ$ and is greater for the no crack closure case at $\rho=1$, $125^\circ<\alpha<180^\circ$, at $\rho=2$, $127^\circ<\alpha<180^\circ$ and at $\rho=3$, $128^\circ<\alpha<180^\circ$.

References

[1] Berezhnytsky L T, Delyavskyy M V and Panasyuk V V 1979 *Bending of thin plates with crack-like defects* (Kyiv: Naukova dumka) p 400
[2] Muskhelishvili N I 1966 *Some basic problems of the mathematical theory of elasticity* (Moskow: Nauka) p 708
[3] Panasyuk V V, Savruk M P and Datsyshyn A P 1976 *Stress propagation near the cracks in plates and shells* (Kyiv: Naukova dumka) p 444
[4] Prusov I A 1975 *The method of conjugation in the theory of plates* (Minsk: Belarus. Univ.) p 256
[5] Savruk M P and Zelenyak V M 2009 *Two-dimensional problems of the theory of elasticity for piecewise homogeneous bodies with cracks* (Lviv: RASTR) p 212
[6] Kalthof J F, Shatsky I P and Burgel 1999 A *Experimental confirmation of the contact of the crack edges during plate bending* (*Mechanics of fracture of materials and strength of structures* issue 2 vol 1) (Lviv: Kamenyar) pp 72–74
[7] Shatsky I P 1988 *Bending of a plate weakened by a crack with contacting faces* (*Rep. of AS of USSR, Series Phys. Math. and Tech. Sci.* vol 7) pp 49–51
[8] Young M. and Sun C 1992 *Influence of crack closure on the stress intensity factor in bending plates* (*International Journal of Fracture* vol 55) pp 81–93
[9] Opanasovych V K and Slobodyan M S 2006 *Biaxial bending of a plate with a circular hole and two radial cracks, the shores of which are in contact* (*Mathematical methods and physical and mechanical fields* vol 49(3)) pp 106–119
[10] Dempsey J P, Shektman I I and Slepyan L I 1998 *Closure of a through crack in a plate under bending* (*International Journal or Solids and Structures* vol 35) pp 4077–4089
[11] Slepyan L I, Dempsey J P and Shekhtman I I 1995 *Asymptotic solutions for crack closure in an elastic plate under combined extension and bending* (*Journal of the Mechanics and Physics of Solids* vol 43) pp 1727–1749
[12] Opanasovych V K 2007 *Bending of a plate with a through rectilinear crack taking into account the width of the contact area of its surfaces* (*Scientific Notes of Lutsk Technical University* vol 20(2)) pp 123–127
[13] Opanasovych V K, Slobodian M S and Zvizlo I S 2015 *Biaxial bending of a plate with a circular hole and an edge radial crack taking into account the width of the contact area of its shores* (*Visnyk of Zaporizhzhya National University. Physical and mathematical sciences* vol 1) pp 127–135
[14] Opanasovych V K, Slobodian M S and Bedriy V Y 2015 *Biaxial bending of an isotropic plate with a rectangular rigid washer and a rectilinear through crack, taking into account the width of the contact area of its shores* (*Bulletin of the Taras Shevchenko National University of Kyiv. Physical and mathematical series* vol 3) pp 21–24

IOP Conf. Series: Materials Science and Engineering 1277 (2023) 012007 doi:10.1088/1757-899X/1277/1/012007

Assessment of elliptical defect dimensions influence on the metal structure contact strength properties

Z A Stotsko [1], O A Kuzin [2], V I Kopylov [2], M O Kuzin [1, 3, 4, 5]

[1] Department of Machine Design and Automotive Engineering, Lviv Polytechnic National University, Ukraine
[2] National Technical University of Ukraine "Igor Sikorsky Kyiv Polytechnic Institute", Ukraine
[3] Lviv Institute of Ukrainian State University of Science and Technology, Ukraine
[4] Lviv Scientific Research Institute of Forensic Expertise, Ukraine
[5] E-mail: kuzin.nick81@gmail.com

Abstract. Based on the approaches of continual mechanics and computer simulation, a computational methodology for establishing the operational parameters of functionally graded parts depending on the dimensional characteristics of elliptical defects is proposed. For the first time, using spatially non-local models of mechanics and the FEniCS finite element package, a calculation scheme was constructed for determining the contact strength of a material depending on the ratio of the sizes of the main radii of an elliptical defect in a structure. On the example of locomotive wheel tyres, it is shown that the most dangerous for operation are elliptical "needle-shaped" hubs with a ratio of the radii of the main axes of more than 40. They can also be a source for the development of "crack-type defects" during the operation of products. The established connection between the change in the properties of railway wheels under the action of loads and the characteristics of stress concentrators indicates that the shape of the concentrator is a more dangerous factor than its volume, and allows you to determine ways to control the operational parameters of parts with the presence of stress concentrators by surface engineering methods.

1. Introduction

Under the heavy-duty operation of critical machine assemblies, equipment and other devices in order to ensure their performance specifications, including the strength of structural elements, various technological practices of parts surface engineering are employed.

Such technologic practices allow modifying operational properties of parts over a wide range and providing the required specifications of critical parts' contacting layers.

In the majority of cases, technological solutions are applied to parts that do not contain geometric stress concentrators as well as surface layers, in which no physical nonlocality (heterogeneity of mechanical properties) exists.

At the same time, one of the characteristic features of transport engineering is the occurrence of geometric stress concentrators close to the contact surfaces, either of constructive, technological or in-service origin. The operational characteristics of parts significantly depend on the parameters of stress concentrators [1], which allow reducing their negative effect.

2. Literature review

Issues of the stress concentrator contribution to the formation of operational properties of parts began to appear in the scholarly literature as early as in the first half of the 20th century. To establish the stress state around concentrators in simplest shaped bodies, the analytical apparatus of mathematical modelling was applied. Stress concentrators and force overloads were properly constrained by their shape and spatial location in order to use analytical mathematics [2]. With the development of finite element method and computer technology, numerical mathematical modelling is more frequently applied to solve such problems [3].

Technological problems of strengthening of parts showing stress concentrators are complicated and require considering a great number of factors. Their solution is based not only on the application of nonlocal models of continuum mechanics, but also on the methods of nonlinear functional analysis, which considerably complicates the formulation of such problems and the search for their solution [1, 4].

Currently, the optimization of technologies for surface strengthening of circular stress concentrators has been examined [5] and approaches to the influence of elliptical defect orientation angle, in particular, on the serviceability of railway rolling stock wheels, have been proposed [1].

However, the issues of dimensional correlations of elliptical defects and their influence on the formation of operational characteristics of functionally graded structures have not been presented in the literature. Such studies are essential for optimizing the stress-strain state of parts in terms of the criteria for providing a pre-set range of operational parameters.

3. Formulation of the problem of determining optimal functionally graded structure parameters under changing dimensions of elliptical stress concentrators

Computer and mathematical modelling is one of the productive ways of predicting the reliability of a structure given the lack of possibility to carry out in-situ experiments. A generalized formulation of the problem of determining operational parameters of parts with nonlocal mechanical properties and arbitrary orientation of elliptical stress concentrators is presented in [1]. Relying on the results of the considered work, let us propose the following two-dimensional problem of determining optimal parameters of functionally graded structures under variable sizes of elliptic stress concentrators.

Let the studied body (a part) occupy an spatial region X, and let us denote the elements of its surface as ∂X. The problem will be considered in the two-dimensional formulation. Let us label the body regions with elliptic stress concentrators as $Y_1, \dots, Y_n$. Each region Y_i, $i=1..n$ is characterized by parameters R_1^i, R_2^i, α^i, where R_1^i, R_2^i are radii of the principal axes, α^i is the rotation angle of elliptic stress concentrators with respect to the axis perpendicular to the studied object plane. On the body surface ∂X let us specify a set of sub-regions $\partial X^* = \bigcup_{i=1}^{m} \partial X^i$, in which the force load is applied.

A mathematical model of a spatially nonlocal continuum body is provided in [6]. As a result of force loads, a stress state is formed in a body, which is characterized by tensor $\hat{\sigma}(x)$, $x \in X$.

The body (part) contact strength analysis is determined by the following relation:

$$k(x) = 1 - \frac{\sigma^*(x)}{\sigma_m(x)}, \tag{1}$$

where $k(x)$ is a safety factor, $\sigma^*(x)$ is a scalar equivalent of the stress tensor representation, $\sigma_m(x)$ is a material scalar strength equivalent.

The relation (1) suggests that at $0<k(x) \leq 1$ no softening is present, at $k(x) \leq 0$ softening occurs.

In the structural analysis, let us assume the total relative level of the material with a contact strength value below zero as a functional:

$$L = \frac{\underset{x \in X}{dim}(k(x) \leq 0)}{\underset{x \in X}{dim}(x \in X)}. \tag{2}$$

In general, the functional depends on geometric parameters of Y_i.

In this paper let us assume that

$$\beta^i = \frac{R_1^i}{R_2^i}, \tag{3}$$

and then calculate following parameters β^i, which minimize the functional (2):

$$\beta^1, \dots, \beta^n : L(\beta^1, \dots, \beta^n) \to min. \tag{4}$$

As a computer tool of establishing optimal parameters β^i we use FEniCS finite element analysis software and its implementation in the Python language [7], to display the results we use MatPlotLib software [8].

4. Using computational mechanics to establish the influence of elliptical defect dimensional parameters

As an example of establishing elliptical defect dimensional parameters let us consider their role in structure softening under the conditions of contact interaction of rolling surfaces (rolling stock wheel - rail).

For the purpose of modelling, let us consider a 2-dimensional contact problem for a body with gradient mechanical properties, containing a single subsurface elliptic stress concentrator (**Figure 1**) [1].

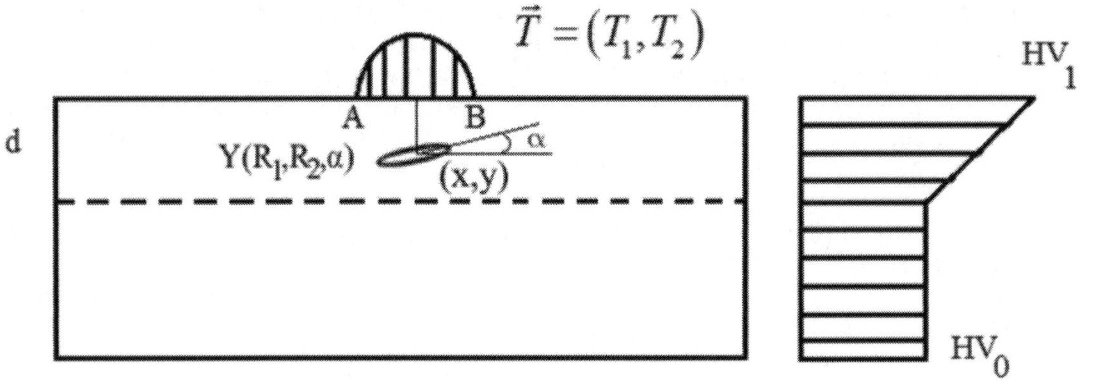

a b

Figure 1. Schematic representation of the 2-dimensional contact state analysis – a) study object: *AB* – contact site (area of contact interaction), (T_1,T_2) – load vector, *d* – strengthening zone thickness, R_1, R_2 – main radii of elliptic stress concentrator, α – rotation angle of the normal-wise concentrator to the load plane, (x, y) – coordinates of elliptic stress concentrator centre; b) epure of microhardness distribution with the depth: HV_1 – surface microhardness value, HV_0 – microhardness value in the material depth

Following the results of [1], let us assign angle α to be 60° as the most unsafe angle of rotation for this type of stress concentrator.

This problem formulation is a classical one for analysing the body behaviour under single frictional contact conditions [9].

For the purpose of softening level immediate calculation, as an example, let us assume that HV_1 = 3,900 *MPa*, HV_0 = 2,600 *MPa*, *d* = 2 mm, the output power load value is $12.5 \cdot 10^4$ *N*, friction coefficient is 0.25, α=60°, the ellipse centre is at a depth of 5 mm from the centre of a contact interaction region. The dimensional parameters of an elliptical concentrator are assumed to be R_1=4.5 mm, R_2=1.5 mm. Initial data for calculations are standard for the locomotive wheel–rail contact pairs [10]. To calculate mechanical parameters using microhardness values we consider the results of work [11].

As a result, we obtain a stress field, the general view of which is shown in **Figure 2**.

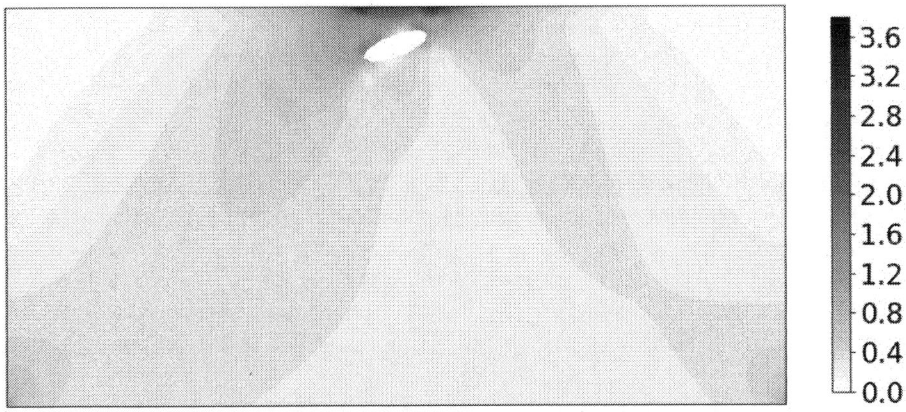

Figure 2. Mises stress distribution in the studied structure given the presence of a stress concentrator, 10^7 *Pa*

The stress state analysis (**Figure 2**) indicates that the highest stress level originates directly in the region of contact interaction and near the stress concentrator, which significantly changes the stress field pattern as compared to the region with no such concentrator present.

Figure 3 presents findings from the study of changes in functional L depending on the parameter $\beta = \frac{R_1}{R_2}$, where R_1, R_2 are ellipse main radii, where $\beta\epsilon[1;150]$. At $\beta=1$ the stress concentrator is circular, at $\beta=150$ it becomes elliptical, looking like a crack needle-shaped defect.

Consequently, the following variation of the contact strength value depending on the parameter β is received (**Figure 3**).

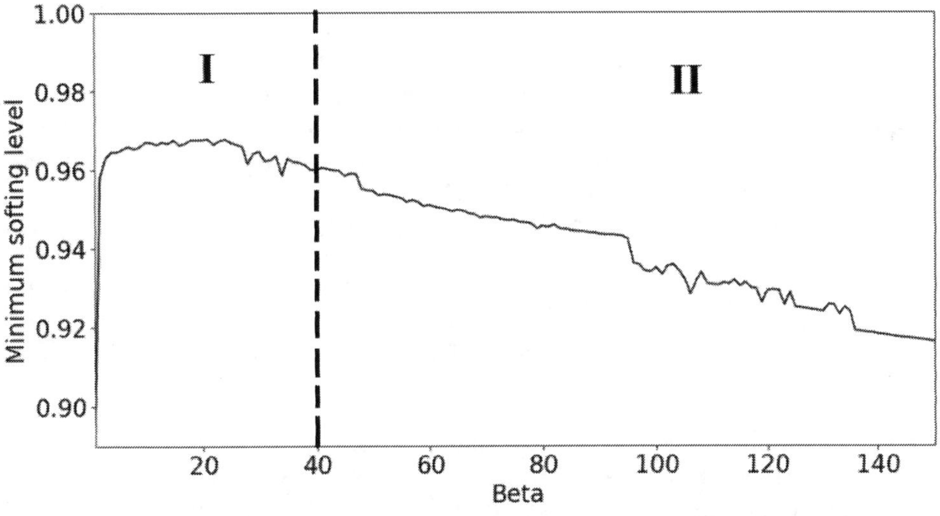

Figure 3. Dependence of the contact strength level on the parameter β

Let us analyse the dependence of the contact strength level of a structural element with a stress concentrator on the parameter β. At β=1 the contact strength level is minimal, then with the growth of β it initially increases and then decreases again. The first segment (region I, **Figure 3**) of the contact strength level change is explained by the fact that the volume taken up by the stress concentrator decreases with the growth of this parameter and therefore the contact strength level of the material increases. The second segment (region II, **Figure 3**) of the contact strength level decrease at increasing of β is explained by the fact that when the second main radius of the elliptic stress concentrator R_2 decreases, the influence of ellipse shape on the contact strength begins to increase, since it becomes needle-shaped, thereby influencing the level of stress state in the neighbourhood of its vertices, and to some extent compensating even the volume parameters of the defect in question.

For this reason, it should be noted that one of the most damaging stress concentrators in terms of operation are needle-shaped or crack-like elliptical concentrators, since the stress state with the worst possible operational parameters is formed exactly in the area around their vertices.

5. Conclusions

Based on the approaches of modern computational and applied mechanics the problem of assessing the contact strength parameters of parts housing elliptical stress concentrators has been considered.

Based on the finite element analysis, it has been shown that the stress concentrator shape influences the contact strength of parts far more significantly than the volume that this concentrator covers.

The results of the study of softening parameters depending on the characteristics of stress concentrators have shown that the most damaging conditions for operation are elliptic needle-shaped concentrators with the principal axis radius ratios of more than 40, and these are likely to be the source for further development of crack-type defects.

6. References

[1] Stotsko Z, Kuzin O, Kuzin M and Mechnik V 2022 Estimation of strength properties of functionally graded structures with elliptical stress *Archives of Materials Science and Engineering* **113 (1)** 35-41

[2] Pluvinage G 2003 *Fracture and Fatigue Emanating from Stress Concentrators.* (Springer) p. 240

[3] Murakami Y 2017 *Theory of elasticity and stress concentration.* (Wiley) p. 446

[4] Tikhonov A, Kalner V and Glasko V 1990 *Mathematical modeling of technological processes and the method of inverse problems in mechanical engineering* (Moscow: Mashinostroenie) p. 264

[5] Lyashenko B, Stotsko Z, Kuzin O and Kuzin M 2020 The use of computational contact mechanics approaches to access the performance of parts bearing stress concentrators *Journal of Achievements in Materials and Manufacturing Engineering.* **103 (1)** 25-32.

[6] Kuzin N 2015 A Mathematical Model Describing the Variation in Material Properties *International Applied Mechanics.* **51** 474–479.

[7] Langtangen H and LoggA 2016 *Solving PDEs in Python. The FEniCS Tutorial I* (Springer Open) p. 160

[8] Johansson R 2019 *Numerical Python: Scientific Computing and Data Science Applications with Numpy, SciPy and Matplotlib* (Apress) p. 709

[9] Popov V 2017 *Contact Mechanics and Friction. Physical Principles and Application* (Springer), p. 322

[10] 2001 *Guidelines to best practices for heavy haul railway operations: wheel and rail interface issues* (International Heavy Haul Association). p. 408

[11] Bulychev S. 1999 The ratio between the restored and unrecovered hardness when testing with nanomicroindentation *Journal of Technical Physics.* **69** 42-48. (in Russian)

Study of characteristics of PWM signals in electromagnetic vibration drive control systems

R V Chubyk[1], I D Zelinskyi[2], P M Lub[3]

[1] Department of Information Security, Lviv Polytechnic National University, 12 S. Bandera Street, Lviv, 79013, Ukraine
[2] Department of Robotics and Integrated Mechanical Engineering Technologies, Lviv Polytechnic National University, 12 S. Bandera Street, Lviv, 79013, Ukraine
[3] Department of information technologies, Lviv National Environmental University, 1 V. Velykiho Street, Dublyany, 80381, Ukraine

Corresponding author's e-mail address: ihor.d.zelinskyi@lpnu.ua

Abstract. The paper studies three forms of the pulse-width modulated (PWM) signal which is used to power the electromagnetic vibration drives of resonance-type vibrating machines. On the basis of a comparative analysis of the characteristics of PWM signals using the root mean square deviation for the time domain, the nonlinear harmonics (distortions) coefficient for the spectral domain, and the structure of the power part of the switches that form these signals, conclusions have been drawn on the feasibility of using different types of PWM in the control systems of electromagnetic vibratory drives.
Keywords: electromagnetic vibration drive, PWM, controlled vibration machines, adaptive vibration machines, resonant vibration machines.

1. Introduction

The use of control systems for the dynamic parameters of the vibratory drive [1, 2] based on pulse-width modulated voltage in vibrating technological machines (VTM) allows to expand their technological capabilities, and in some cases to qualitatively implement the given technological processes. However, at the current stage of the development of vibration technology, adaptive vibration machines [3, 4, 5] whose control systems have feedback on the amplitude and frequency of oscillations of the working body of the vibration machine are clearly distinguished. Control (adaptation to the resonant mode of operation) occurs due to the tracking of the phase shift between the amplitude of oscillations of the working body and the amplitude of the cyclic disturbing force and the constant correction of the frequency of the cyclic driving force of the electromagnetic vibration drive by changing the frequency of the PWM signal in order to keep this phase shift equal $\pi/2$. In [3], it is shown that the introduction of sparsity in the PWM signal allows to control the amplitude of oscillations of the working body of the vibrating machine at the resonant frequency of the adaptive vibrating machine, thereby providing optimal technological parameters with minimal energy consumption in the resonant mode of operation. The development of modern power electronics allows creating PWM signals of various types when programming controllers at the stage of creating the electromagnetic vibration drive control system itself. The question arises of choosing criteria for evaluating the parameters of PWM signals of various types that feed the electromagnetic vibration drive during the operation of the VTM. In the case of a rectangular form of the PWM signal as in work

[6], you can use the analytical description of this signal (which due to the approximation, has smoothed sharp corners and is as close as possible to the real rectangular signal supplied to the electromagnetic vibration drive) and draw up the differential equation of motion of the working body of the vibrating machine and establish analytical connections between the current, the form of voltage, displacements of the working body and other parameters of the vibratory drive. However, even in the case of a simple rectangular form of the PWM signal on an electromagnetic vibration drive, the analytical establishment of such a relationship requires a lot of resources for the numerical solution of the differential equation, and the expressions themselves are cumbersome and not very informative. Based on this, when designing the control system at the stage of selecting the type of PWM signal for the electromagnetic vibration drive, it will be advisable to use a different approach for the selection of criteria. In particular, it should be noted that the mechanical oscillating system of the VTM will not work out the high harmonics that will be part of any PWM signal. That is, a vibrating machine with an electromagnetic vibrating drive will essentially be a low-pass filter for this high-frequency signal. Therefore, when choosing the optimal form of a PWM signal for analysis, the lower part of the spectrum of these signals should be used, and it will be advisable to carry out an assessment both in the time domain and in the spectral domain, using for this the root mean square deviation and the nonlinear distortions (harmonics) coefficient in comparison with the test classical sinusoidal supply voltage.

2. Material and method

When studying the root mean square deviation and coefficient of nonlinear distortions (harmonics) in PWM signals, we will take a sinusoidal signal with a frequency of 50 Hz and an operating voltage of 220 V as a test signal. Accordingly, to ensure equivalence, our simulated PWM signals will have the same operating voltage and frequency. In order to be able to use the discrete Fourier transform function, which is in the MathCAD environment, let us divide the PWM signals and the test signal into the number of readings, which is equal to $N = 2^{11} = 2048$. Based on the above, the test signal can be written as

$$u_i = 220 \cdot \sqrt{2} \cdot \sin\left(\frac{2 \cdot \pi}{N-1} \cdot i\right),\ i = 0 \ldots N-1. \tag{1}$$

One of the most common PWM signals used in the control systems of electromagnetic vibratory drives is a signal in which during one period of oscillation, which is coded by pulse-width modulation, it is divided into a large number of small intervals Δt . And already, accordingly, on each interval Δt , two different-polarity pulses $\Delta t1$ and $\Delta t2$ amplitudes $\pm U\max$ are formed, respectively, but of different duration in time, and the sum of their time limits is limited by the duration of the interval Δt . It is possible to determine the duration of a positive pulse $\Delta t1$ at each interval Δt using the expression

$$\Delta t1_k = INT\left(\frac{\Delta t}{2} \cdot \left(\frac{u_{k \cdot \Delta t}}{220 \cdot \sqrt{2}} + 1\right) - 0.5\right),\ k = 0 \ldots K-1. \tag{2}$$

In this case, K is the number of intervals Δt that form the period in our case of sinusoids by the pulse-width modeling method. For example, if the duration of the interval in counts is $\Delta t = 32$, and the number of counts with which we form our PWM period of the sine wave is $K = 64$ then $\Delta t \cdot K = 32 \cdot 64 = 2048 = N$. To calculate the duration of the pulse of negative polarity during the interval Δt , the expression should be used

$$\Delta t2_k = \Delta t - \Delta t1_k . \tag{3}$$

When designing electromagnetic vibration drive control systems, it should be taken into account that the power part of the commutator, which forms a PWM signal described by expressions (2) and (3) and shown in figure 1 a) consists of four power transistors (keys), and the turn-on time of these transistors will be determined for the positive pair by expression (2) for the negative pair by expression (3). At the time of operation of such a commutator, switching of the electromagnetic vibration drive takes place between a pair of transistors that switches to a positive potential $+U\max$ for time $\Delta t1$ and a pair of transistors that switches to a negative potential $-U\max$ for time $\Delta t2$ during each half-cycle of the coded sinusoid.

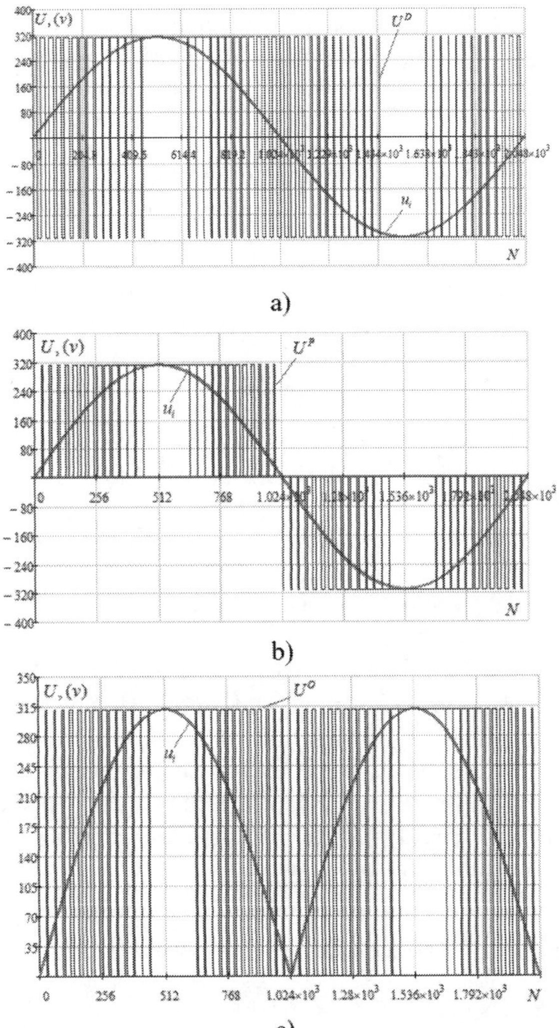

c)

Figure 1. Pulse-width modulated voltage for powering the electromagnetic vibration drive.

The voltage at the output of such a switch U^D which consists of four transistors and powers an electromagnetic vibration drive and is shown in figure 1 a) can be conditionally represented by the following expression

$$U^D_{k\cdot\Delta t} = \begin{vmatrix} 220\cdot\sqrt{2} & if \quad i < \Delta t1_k \\ -220\cdot\sqrt{2} & otherwise \end{vmatrix}, \; k = 0\ldots K-1, \; i = 0\ldots\Delta t-1. \tag{4}$$

In order not to cause a short circuit in such a power part of the control of the electromagnetic vibration drive, it is necessary to introduce a pause (delay) between the switching of the transistors that switch the transition of the positive potential $+U\max$ and the negative potential $-U\max$ when passing through the zero potential at the program level.

However, there are variants of PWM signals that are used to power an electromagnetic vibration drive, the power part of which does not require the introduction of a program delay in order to eliminate a short circuit during each period of oscillation. In particular, the variant when the PWM signal is formed by coding the duration of pulses in each half cycle period separately (figure 1 b)). That is, when generating a PWM signal U^P in accordance with expression (5), two pulses $\Delta t1$ and $\Delta t2$ are also formed at each interval Δt with an amplitude of $\pm U\max$, respectively, but in the case of encoding a positive half-period of a sine curve, we get pulse amplitudes $\Delta t1$ equal to , and pulse amplitudes $\Delta t2$ equal to $U=0$.

$$U^P_{k\cdot\Delta t}=\left|\begin{array}{l}220\cdot\sqrt{2}\cdot\frac{\Delta t1_k}{|\Delta t1_k|} \quad if \quad i<|\Delta t1_k| \\ 0 \quad otherwise\end{array}\right., k=0\ldots K-1,\ i=0\ldots\Delta t-1. \tag{5}$$

The duration of the $\Delta t1$ interval in this case is determined from the expression on each Δt interval

$$\Delta t1_k=\left(\frac{u_{k\cdot\Delta t}}{220\cdot\sqrt{2}}\right)\cdot\Delta t,\ k=0\ldots K-1. \tag{6}$$

When encoding a negative half-period of a sine curve per revolution, the amplitudes of pulses $\Delta t1$ are equal to $U=0$, and the amplitudes of pulses $\Delta t2$ are equal to $-U\max$. Thanks to this algorithm of work in a four-transistor switch, two transistors operate during pulse-width coding of a positive half-cycle and, accordingly, two transistors when forming a negative half-cycle. At the moment of transition of the coded sine wave through zero, the duration of the pulses is small and the switching time of the transistors is small, therefore, the switching of the positive and negative pair of transistors are separated in time.

Another form of the PWM voltage signal that is used in control systems of an electromagnetic vibration drive is a unipolar pulse-width modulated sinusoidal voltage U^O over two half-cycles, shown in figure 1 c). The peculiarity of this voltage is that only one transistor is needed for its formation in the power part of the switch, unlike four in the previous two cases. During the formation of the output voltage in each half-period of the pulse-width-coded sine wave, the output transistor of the commutator also forms two pulses $\Delta t1$ and $\Delta t2$ at each interval Δt, and the amplitude of the pulses $\Delta t1$ is equal to and the amplitude of the pulses $\Delta t2$ is equal to U^O. The duration of the interval $\Delta t1$ in this case on each interval Δt is determined by expression (6). The law of formation of U^O unipolar PWM during two half periods can be represented by the following expression

$$U^O_{k\cdot\Delta t}=\left|\begin{array}{l}220\cdot\sqrt{2} \quad if \quad i<\Delta t1_k \\ 0 \quad otherwise\end{array}\right., k=0\ldots K-1,\ i=0\ldots\Delta t-1. \tag{7}$$

Considering the fact that all three PWM signals (figure 1) which feed the electromagnetic vibratory drive is a set of rectangular pulses of different durations, therefore, the spectrum of these signals will contain high-frequency harmonics that are multiples of the frequency of the test coded by the method of pulse-width modulation sine wave u_i. Undoubtedly, due to its inertia, the oscillating mechanical system of the vibrating machine will not be able to react to the high harmonics of this spectrum, which are present in these three signals. Therefore, the main contribution to the formation of the traction properties of the electromagnetic vibration drive at the frequency of the test coded sinusoid u_i using the method of pulse width modulation will be the creation of low harmonics from the data spectra of the three signals U^D, U^P and U^O. In essence, a vibrating technological machine with an electromagnetic vibratory drive can be considered as a low-pass filter in relation to the PWM voltage that feeds its vibratory drive. Therefore, it will be appropriate to decompose each of the three PWM signals U^D, U^P, U^O and the test signal u into the sum of multiple harmonics $\sum_{i=0}^{N-1} H_i^D = (a_i \cdot \cos(i \cdot \omega \cdot t) + b_i \cdot \sin(i \cdot \omega \cdot t))$, $\sum_{i=0}^{N-1} H_i^P$, $\sum_{i=0}^{N-1} H_i^O$ and $\sum_{i=0}^{N-1} u_i^h$ in this case, $i \cdot \omega$ is a multiple harmonic to the main angular frequency ω coded by the method of pulse width modulation of the sine wave in each of the four signals U^D, U^P, U^O and u.

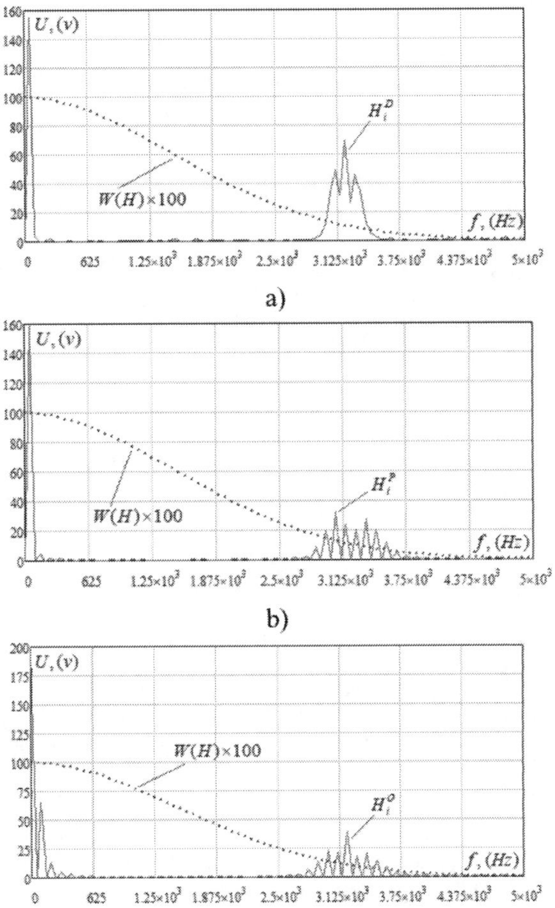

c)

Figure 2. Spectral characteristics of the PWM voltage feeding the electromagnetic vibrator.

Applying the spectral filter $W(H)$ (or spectral window) (8) and expression (9), it is advisable to remove the upper spectrum and significantly reduce the amplitudes in the middle frequency range and definitely skip all frequencies from the lower data range of the three PWM signals.

$$W(H)=\exp\left(\frac{-[(i+1)\cdot(15/N)]^2}{0.1}\right),\ i=0\ldots N/2. \tag{8}$$

$$h_i^D=H_i^D\cdot W(H_i),\ h_i^P=H_i^P\cdot W(H_i),\ h_i^O=H_i^O\cdot W(H_i). \tag{9}$$

After that, from the received spectral characteristics h_i^D, h_i^P and h_i^O (which are created on the basis of the low-frequency components of their original signals U^D, U^P and U^O) thanks to the inverse Fourier transform, reproduce the sinusoidal voltage coded by the method of pulse-width modulation and compare it with the test coded by the method of pulse-width modulation sinusoidal voltage u_i in temporal and spectral domains.

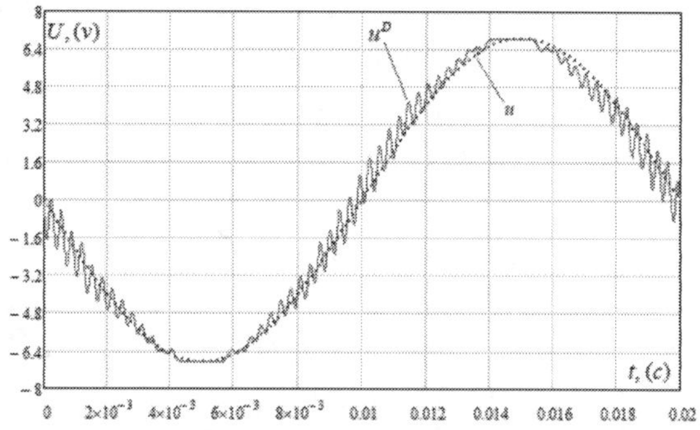

a)

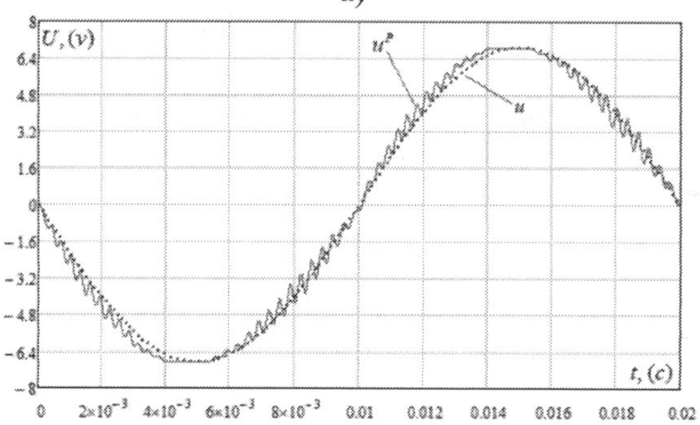

b)

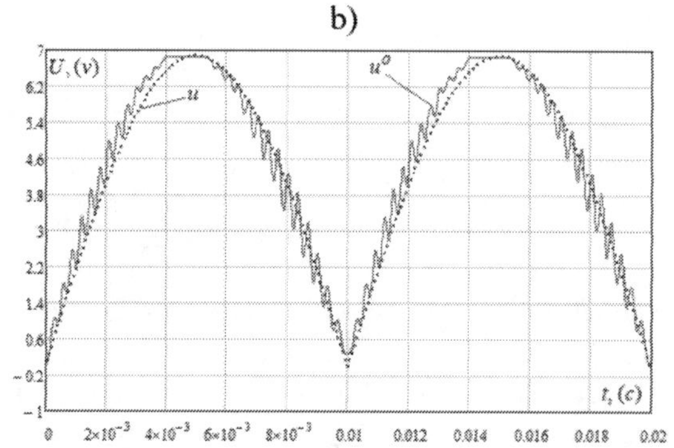

c)

Figure 3. Reproduced PWM signal based on 100 harmonics multiples of the resonant frequency of the vibrating machine.

3. Results

Taking into account the fact that in the Fourier expansion we get the number of harmonics equal to $N/2=1024$, then, taking into account the multiplicity of all harmonics of the frequency of the first harmonic for the test, which is 50 Hz for all three PWM signals, we get 51.2 kHz. Therefore, it makes no sense to us to display the entire spectrum, and considering that most vibromachines basically have their own resonance frequency close to 50 Hz, it will be appropriate to consider the first 100 harmonics in our pulse-width modulated 50 Hz signals. As a result, the frequency of the highest harmonic that we will consider and take into account when forming spectral characteristics h_i^D, h_i^P and h_i^O is $f=50\cdot 100=5000$ Hz. According to this condition, the spectral filter $W(H)$ in expression (8) is adjusted. The initial spectral characteristics H_i^D, H_i^P and H_i^O PWM signals U^D, U^P and U^O together with the spectral window (8) that cuts the signal spectrum after 5 kHz is shown in figure 2. The spectral filter (8) passes 50 percent of the initial harmonic of the original PWM signal at a frequency close to 2 kHz (figure 2), which is 40 times higher than the frequency of natural oscillations of the oscillating mechanical system of the vibrating machine in the case of feeding it with a PWM voltage from a frequency coded sine wave of 50 Hz.

As a result of the analysis of figure 2 a), b), c) it can be seen that the spectral composition H_i^D, H_i^P and H_i^O of three PWM signals U^D, U^P and U^O differs both in the range up to 625 Hz and around 3 kHz. Taking this into account after processing (9) with the spectral filter $W(H)$ (8) in the filtered spectral characteristics h_i^D, h_i^P and h_i^O we will get different (practically unchanged) spectrum variations in the range up to 625 Hz and significantly reduced amplitudes of harmonics around 3 kHz. By applying the inverse discrete Fourier transform to calculate the value of each i reading of analog signals with a finite spectrum to the filtered spectral characteristics h_i^D, h_i^P, h_i^O and to the spectral characteristic of the test signal u_i^h, we will calculate the value of each i reading of the three PWM signals $u_i^D=\frac{1}{N}\cdot\sum_{i=0}^{N-1} h_i^D\cdot e^{j\cdot\frac{2\cdot\pi\cdot i\cdot n}{N}}$ where $n=0\ldots N-1$, u_i^P, u_i^O and the test sinusoidal voltage u_i.

Given that the frequency of the test sinusoidal voltage u is 50 Hz, the price in the time of one count will be $1/f\cdot N=9.76\cdot 10^{-6}$ c, thanks to which it is possible to display (figure 3) the reconstructed test sinusoidal voltage u using the inverse Fourier transform and the reconstructed three PWM signals u^D, u^P, u^O based on the first 100 harmonics from their spectrum in the time domain. Analyzing figure 3 shows that the reconstructed three PWM signals u^D, u^P, u^O have waveform deviations (non-linear distortions) compared to the test signal.

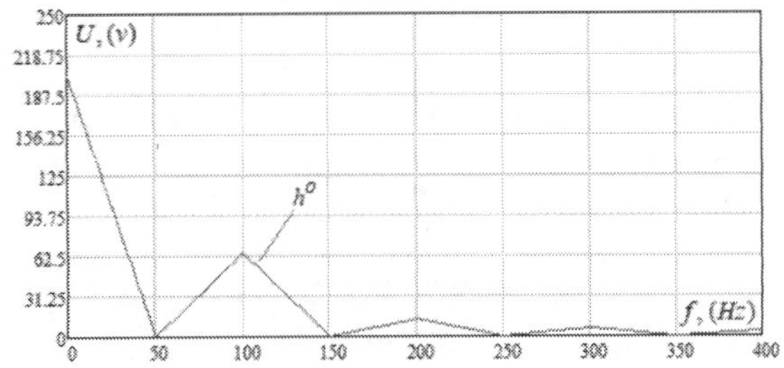

Figure 4. Spectral characteristic of a unipolar pulse-width modulated sinusoidal voltage over two half-cycles for powering an electromagnetic vibration drive.

For numerical evaluation and comparison between them in the time domain, it is advisable to use the root mean square deviation σ of each signal relative to the test signal as a criterion. That is, the root mean square deviation of each of the three PWM signals will be expressed as

$$\sigma^D = \sqrt{\frac{1}{N} \cdot \sum_{i=0}^{N-1} \left(u_i - u_i^D\right)^2}, \ \sigma^P, \ \sigma^O.$$

The calculations of mean square deviation for all three forms of PWM signals show that $\sigma^D = 0.4789$, $\sigma^P = 0.3437$ and $\sigma^O = 0.3457$.

In order to evaluate the three PWM signals u^D, u^P, u^O and the test sinusoidal voltage u reconstructed using the inverse Fourier transform in the spectral domain, it is advisable to apply the coefficient of nonlinear distortions, which in this case will be equal to the ratio of the root mean square sum of harmonics (h_i^D, h_i^P, h_i^O and u_i^h) in the spectrum of each signal except the first, which carries the basic frequency of the PWM sine wave coded by each signal and corresponds to the resonant frequency of the vibration machine to the root mean square sum of the harmonics of the entire spectrum that form these PWM signals. In particular, in our case, the coefficient of nonlinear distortions for the reconstructed three PWM signals and the test sinusoidal voltage will be expressed as

$$Kn^D = \frac{\sqrt{\sum_{i=2}^{100} \left|h_i^D\right|^2}}{\sqrt{\sum_{i=1}^{100} \left|h_i^D\right|^2}}, \ Kn^P, \ Kn^O, \ Kn^u.$$

The calculation of the nonlinear distortion coefficient shows that for our signals reconstructed using the inverse Fourier transform, it is $Kn^u = 0.001$ the test sinusoidal voltage u, and, accordingly, for three PWM signals $Kn^D = 0.086$, $Kn^P = 0.054$, $Kn^O = 1.0$. Additionally, the level of non-linear distortions can be expressed through the harmonic ratio, taking into account that it is the ratio of the root mean square sum of higher harmonics h_i^D, h_i^P, h_i^O and u_i^h except for the first (h_1^D, h_1^P, h_1^O and u_1^h) to the amplitude value of the first harmonic. In our case, the vibration machine works on the first harmonic (50 Hz) and it is the coding using PWM of this frequency that takes place. The coefficient of harmonics and the coefficient of nonlinear distortions are related by the following relation $Kh^D = Kn^D \big/ \sqrt{1 - \left[Kn^D\right]^2} = 0.086$, $Kh^P = 0.054$, $Kh^O = 441.38$. However, the unipolar sinusoidal voltage (7) for powering the electromagnetic vibration drive during two half-periods of PWM has a high coefficient of harmonics in the spectrum (figure 4) a zero harmonic h_0^O is present in this PWM signal, which is responsible for the constant component in this signal because only one potential is switched by the power switch (transistor) during two half-cycles. However, one test sinusoid with a given frequency of 50 Hz is used as the basis of formation in all three PWM signals, but only in the case of signal formation, with a unipolar sinusoidal voltage over two half-periods of PWM, which is described by expression (7), we get the operating frequency for the vibrating machine (figure 4) 100 Hz. In the two cases when a PWM sinusoidal voltage with alternating sign during each half cycle is formed, which is described by expression (4) and a PWM sinusoidal voltage that does not change polarity during a half cycle, which is described by expression (5), the frequency of the first harmonic will correspond to the operating frequency of the vibrating machine, that is, the operating frequency of the resonant vibrating machine will match the PWM frequency.

4. Conclusions

Conducted studies in the time and spectral domain of three pulse-width modulated signals for powering the electromagnetic vibration drive of resonant vibration technological machines show that the most promising for the design of vibration drive control systems is the PWM sinusoidal voltage, which does not change polarity during each half-cycle of the sinusoidal voltage coded by the PWM method, the frequency of which corresponds to the frequency of oscillations of the oscillating mechanical system of the resonant vibrating machine.

It should be noted that the unipolar sinusoidal voltage during two half-periods of PWM is due to the simplicity of the schematic solution of the power part (one switching transistor) during its formation and the creation of control systems for electromagnetic vibration drives, controlled vibration machines and adaptive vibration machines. However, the constant component in the spectrum of this signal will be the cause of the magnetization of the electromagnetic vibration drive and as a result of the reduction of the gap between the anchor and the yoke, but the second multiple harmonic will be working for resonant vibration machines. Therefore, such a signal form due to the simplicity of the circuitry of its power part requires further research and development of the engineering calculation method of the parameters of the electromagnetic vibration drive.

References

[1] Zelinskyi I D 2003 The control system of the electromagnetic vibration exciter *Automation of production processes in machine building and instrument building Ukrainian interdepartmental scientific and technical collection* (Lviv: LPNU) Issue 37 p 3 - 7. [In Ukrainian]

[2] Despotovic Z and Ribic A 2012 The increasing energy efficiency of the vibratory conveying drives with electromagnetic excitation *International Journal of Electrical Power & Energy Systems* 6(1) p 38-42

[3] Chubyk R 2009 Optimal ruling system of electromagnetic vibrodrive of adaptive vibrational technological machines *Vibroengineering 2009 Proc. of the 8th Int. Conf.* (Klaipeda, Lithuania, September 16-18) pp 10-14

[4] Cherno A A 2017 Control of electromagnetic vibratory drive using a phase difference between current harmonics *Journal of automation and information sciences* Vol 49 Issue 7 pp. 58 – 76.

[5] Sinik V, Despotovic Z and Palinkas I 2016 Optimization of the operation and frequency control of electromagnetic vibratory feeders *Elektronika ir Elektrotechnika* №1, vol 22. pp. 24 – 30.

[6] Chubyk R 2011 Investigation of current parameter in electro-magnettic drive of resonance adaptive vibromachines while using the duopolar latitude-impulsive voltage for its power supply *Journal Of Vibroengineering* Vol 13 Issue 2 pp. 295 - 301.

Research of the influence of car suspension characteristics on the dynamics of its sprung part

[1]M.H. Hrubel, [2] R.A. Kovalchuk, [3] O.V. Lanets

[1]Department of Cars and Vehicles Fleet, Hetman Petro Sahaidachnyi National Army Academy, 79026, Lviv, Heroes of Maidan street 32.
[2]Department of Engineering Mechanics (Weapons and Equipment of Military Engineering Forces), Hetman Petro Sahaidachnyi National Army Academy, 79026, Lviv, Heroes of Maidan street 32.
[3]Department of Technical Mechanics and Dynamics of Machines, Lviv Polytechnic National University,79013, Lviv, Ukraine, Bandera street 12.

Abstract. A method of determining the parameters of the car suspension has been developed, which allows taking into account the influence of the main characteristics of the suspension. In particular, the influence of static deformation, nonlinearity parameter, and such external factors as the movement of a car over rough terrain, stability of movement are taken into account. Mathematical models of vertical oscillations – dynamics of the sprung part of the vehicle with non-linear force characteristics – have been improved for the specified method. With this aim, a mathematical apparatus was developed for the analytical research of the influence of vertical oscillations on traction-speed characteristics, performance, smoothness, and stability of movement.

Analysis of the problem and formulation of research problems. The analysis of various literary sources showed that for now there is no optimal method of researching the dynamics of sprung and unsprung masses from the conditions of formation of perspective parameters of car suspensions [1]. Under such conditions, the formation of perspective parameters of car suspensions cannot be carried out without conducting research aimed at developing new [2–4] or improving existing mathematical models of their dynamics; creating methods or techniques of theoretical research of their developed models.

The research on the main parameters that describe the force characteristics of elastic elements and shock absorbers of the suspension system of cars and their influence on operational properties was carried out. As for the latter, here, first of all, it is necessary to keep in mind smoothness of movement, stability, controllability, performance [5–8]. Regardless of the types of suspensions (active or passive), the main physical parameters that determine the dynamics of the vehicle during off-road driving, and therefore almost all operational characteristics, are the restoring force of elastic elements and the resistance force of damping devices. The experience of operating vehicles with active suspensions in which the elastic element is hydraulic, pneumatic, electromagnetic, etc. shows that they have a number of disadvantages. This is primarily their significant inertia, the complexity of the mathematical model of the dynamics of sprung and unsprung parts, and therefore the problems, which are related to the creation of a software product that would determine the reaction of controlled (changeable) suspension parameters to one or another type of road irregularities (suspension stiffness, damping law, body height, etc.). All of this is the basis for the development of fundamental principles for the modernization of passive suspensions: the development of a methodology, and therefore, on its base, the provision of practical recommendations for the selection of its main parameters, which would maximally ensure operational characteristics depending on the type of the vehicle. Thus, the problem of researching the influence of the main parameters that represent the elastic force of shock absorbers and the resistance force of damping devices is solved under the condition of nonlinear laws of change depending on the deformation of the elastic element or the rate of deformation of the shock absorber. Why the non-linear characteristics of the specified elements of the suspension system? The essence of the problem is as follows: the experience of operating certain types of cars that use elastic shock absorbers with a linear law of restorative force change significantly reduces the average speed of movement along a road with bumps, and the resource of separate units exploitation decreases, and the weariness of the driver and crew increases significantly. At the same time, almost all theoretical research on the dynamics of car samples was conducted using linear force characteristics of elastic shock absorbers and damping devices. This is primarily due to the lack of a perfect mathematical apparatus for the construction and analytical research of nonlinear mathematical models of the dynamics of car samples adequate to the dynamic process. That is why, in order to describe the nonlinear elastic force of shock absorbers, and therefore to develop refined mathematical models of the dynamics of car samples, the article proposes a universal ratio that describes with sufficient accuracy a wide range of elastic forces of shock absorbers under progressive or regressive laws of their change. For vehicles with the specified characteristics of the suspension system, a wide range of research was conducted on the dynamics of the sprung part and its influence on the main operational characteristics [9–11].

Main part. The article considers four main parameters of the suspension: the value of the elastic force F, the nonlinear static deformation $\Delta_{ст}$, he amplitude of oscillations a, the nonlinearity parameter v, and the constant of proportionality c. Analytical methods cover the dependence of operational characteristics on the specified parameters, at the same time, they do not allow the research of the entire range of changes of the specified parameters. At the same time, even for the movement of cars over rough terrain, numerical methods can't determine the most dangerous speed characteristics of the movement.

Research of the dynamics of the sprung part of vehicle is based on:

the choice of a physical model of the vehicle, which takes into account the influence of the most important factors on the vertical and transverse-angular oscillations of the sprung part;

building an adequate mathematical model (differential equations) that takes into account the nonlinear characteristics of shock absorbers and damping devices;

the application of such well-known and tested analytical methods of solutions development to the above-mentioned differential equations, which would describe the laws of change of the determining parameters of the sprung mass oscillations.

As for the physical model of the car, for the research of vertical and transverse-angular oscillations, it [5] is presented in Fig.1 and represents unsprung part -1 and sprung part - 2 They interact with each other with shock absorbers - 3 and damping devices - 4.

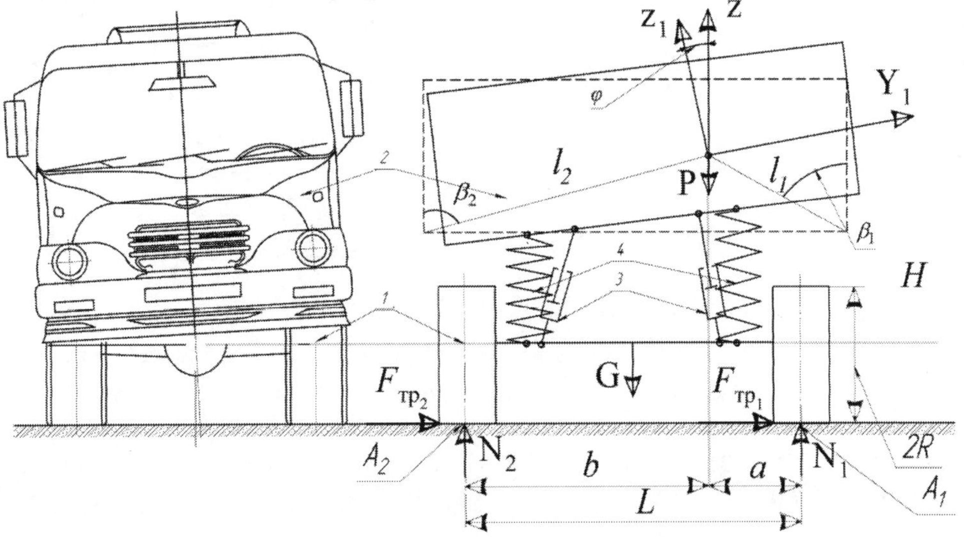

Fig. 1 Physical model for research the oscillations of a car

The main limitations regarding the physical model and movement of the researched object under study are given in [12]. The given limitations make it possible to state that the relative movement of the sprung mass of the car is plane-parallel, and therefore its position is definitely determined by the position of the center of mass O and the angle of rotation of the specified part about O. The position of the center of mass at an arbitrary moment in time (taking into account the above restrictions) is most conveniently determined relative to the static position by the coordinate – $z(t)$. Thus, to build a mathematical model of the relative motion of a sprung mass, it is possible to use the second-order Lagrange equation. According to the selected generalized coordinates, the kinetic energy of the sprung part in its relative motion takes the form:

$$T = M\frac{\dot{z}^2(t)}{2} + I_C\frac{\dot{\varphi}^2(t)}{2}, \tag{1}$$

where M – sprung mass, and I_C – its moment of inertia relative to the transverse axis, which passes through the center of mass and is perpendicular to the velocity vector of the transfer motion.

As for the generalized forces that correspond to the selected generalized coordinates, to find them, first of all, let's express the vectors of relative movements ($\vec{\Delta}_1$, $\vec{\Delta}_2$) of the joint points of the elastic shock absorbers to the sprung mass from its initial position to an arbitrary one through the adopted generalized coordinates and geometric parameters of the vehicle. The specified magnitudes take next values

$$\begin{aligned} \vec{\Delta}_1 &= \vec{j}l_1\varphi\cos\left(\frac{\varphi}{2}+\beta_1\right)+\vec{k}\left[z+l_1\varphi\sin\left(\frac{\varphi}{2}+\beta_1\right)\right], \\ \vec{\Delta}_2 &= \vec{j}l_2\varphi\cos\left(\beta_2-\frac{\varphi}{2}\right)+\vec{k}\left[z-l_2\varphi\sin\left(\beta_2-\frac{\varphi}{2}\right)\right]. \end{aligned} \tag{2}$$

At the same time, the specified ratios also determine the relative velocities of the points that coincide with the ends of the specified vectors ($\dot{\vec{\Delta}}_1$, $\dot{\vec{\Delta}}_2$) –

$$\begin{aligned} \dot{\vec{\Delta}}_1 &= \vec{j}l_1\dot{\varphi}\cos\left(\frac{\varphi}{2}+\beta_1\right)-\vec{j}l_1\varphi\frac{\dot{\varphi}}{2}\sin\left(\frac{\varphi}{2}+\beta_1\right)+\vec{k}\left[\dot{z}+l_1\dot{\varphi}\sin\left(\frac{\varphi}{2}+\beta_1\right)+l_1\dot{\varphi}\frac{\varphi}{2}\cos\left(\frac{\varphi}{2}+\beta_1\right)\right], \\ \dot{\vec{\Delta}}_2 &= \vec{j}l_2\dot{\varphi}\cos\left(\beta_2-\frac{\varphi}{2}\right)+\vec{j}l_2\varphi\frac{\dot{\varphi}}{2}\sin\left(\beta_2-\frac{\varphi}{2}\right)+\vec{k}\left[\dot{z}-l_2\dot{\varphi}\sin\left(\beta_2-\frac{\varphi}{2}\right)+l_2\dot{\varphi}\frac{\varphi}{2}\cos\left(\beta_2-\frac{\varphi}{2}\right)\right]; \end{aligned} \tag{3}$$

and deformations of elastic elements δ_1, δ_2 will be

$$\begin{aligned} \delta_1 &= \sqrt{(z-\Delta_{cm})^2+l_1^2\varphi^2+2(z-\Delta_{cm})l_1\varphi\sin\left(\frac{\varphi}{2}+\beta_1\right)}, \\ \delta_2 &= \sqrt{(z-\Delta_{cm})^2+l_2^2\varphi^2-2(z-\Delta_{cm})l_2\varphi\sin\left(\beta_2-\frac{\varphi}{2}\right)} \end{aligned} \tag{4}$$

where $l_1, l_2, \beta_1, \beta_2$ are geometric parameters that characterize the placement of elastic elements relative to the center of mass, $\vec{k}, \vec{j}$ are unit vectors are directed, respectively, along the axes OZ and OY of the fixed coordinate system (see Fig. 1). In case of symmetrical placement of the center of gravity of the sprung part relative to the elastic elements $l_1 = l_2$, a $\beta_1 = \beta_2$.

Ratios (2) - (4), as well as assumptions about the force characteristics of the suspension system, make it possible to present the active components ($\vec{F}_1$ and $\vec{F}_2$) of the forces acting on the sprung part in the form

$$\vec{F}_1 = \delta_1^{v} c_1 \left(\vec{j} l_1 \varphi \sin\left(\frac{\varphi}{2} + \beta_1\right) + \vec{k}\left(z - \Delta_{cm} + l_1 \varphi \sin\left(\frac{\varphi}{2} + \beta_1\right)\right)\right),$$
$$\vec{F}_2 = \delta_2^{v_2} c_2 \left(\vec{j} l_2 \varphi \cos\left(\beta_2 - \frac{\varphi}{2}\right) + \vec{k}\left[z - \Delta_{cm} - l_2 \varphi \sin\left(\beta_2 - \frac{\varphi}{2}\right)\right]\right) \quad (5)$$

The sum of the projections of the specified forces and the gravity force of the sprung part on the vertical axis will be the active component of the generalized force, which corresponds to the generalized coordinate z, i.e.

$$Q_z^a = \delta_1^{v} c_1 \left(z - \Delta_{cm} + l_1 \varphi \sin\left(\frac{\varphi}{2} + \beta_1\right)\right) - \delta_2^{v} c_2 \left(z - \Delta_{cm} - l_2 \varphi \sin\left(\beta_2 - \frac{\varphi}{2}\right)\right) - P. \quad (6)$$

The active component of the generalized force, which corresponds to the generalized coordinate φ i.e. Q_φ^a will be determined on the basis of the following considerations [13]: on the one hand, the work of active forces acting on the sprung part on the possible displacement of the specified generalized coordinate $\delta\varphi$ is determined by the ratio $\delta A_\varphi^a = Q_\varphi^a \delta\varphi$, and on the other hand - $\delta A_\varphi^a = M_C^a \delta\varphi$, де M_C^a - the sum of moments of active forces acting on the sprung part relative to the center of mass. The last magnitude, according to Varignon's theorem [13], is defined by dependence $M_C^a = F_{1z}^a y_1 - F_{1y}^a z_1 + F_{2z}^a y_2 - F_{2y}^a z_2$, where F_{1y}^a, F_{1z}^a та F_{2y}^a, F_{2z}^a – projections on the OY and OZ axis of the active forces acting on the right and left parts of the sprung mass of the car; y_1, z_1 and y_2, z_2 – the coordinates of the points of application of these forces, i.e. $y_1 = l_1 \sin\beta_1$, $z_1 = -l_1 \cos\beta_1$; $y_2 = -l_2 \sin\beta_2$, $z_2 = -l_2 \cos\beta_2$. As follows $Q_\phi^a = F_{1z}^a y_1 - F_{1y}^a z_1 + F_{2z}^a y_2 - F_{2y}^a z_2$. If the corresponding values will be substituted in the last dependence, we get

$$Q_\phi^a = c_1 \left((z - \Delta_{cm})^2 + l_1^2 \varphi^2 + 2(z - \Delta_{cm}) l_1 \varphi \sin\left(\frac{\varphi}{2} + \beta_1\right)\right)^{\frac{v_1}{2}} \left\{ l_1 \sin\beta_1 \left(z - \Delta_{cm} + l_1 \varphi \sin\left(\frac{\varphi}{2} + \beta_1\right)\right) + l_1 \cos\beta_1 \left(z - \Delta_{cm} + l_1 \varphi \sin\left(\frac{\varphi}{2} + \beta_1\right)\right)\right\} +$$
$$+ c_2 \left((z - \Delta_{cm})^2 + l_2^2 \varphi^2 - 2(z - \Delta_{cm}) l_2 \varphi \sin\left(\beta_2 - \frac{\varphi}{2}\right)\right)^{\frac{v_2}{2}} \left\{ -l_2 \sin\beta_2 \left(z - \Delta_{cm} - l_2 \varphi \sin\left(\beta_2 - \frac{\varphi}{2}\right)\right) + l_2 \cos\beta_2 \left(z - \Delta_{cm} - l_2 \varphi \sin\left(\beta_2 - \frac{\varphi}{2}\right)\right)\right\}. \quad (7)$$

Proceeding in a similar way as for the active components of the restoring force, we will obtain the passive components of the generalized forces

$$Q_{iz}^{\Pi} = -\gamma_i \left| \bar{f}_i(\varphi, \dot{z}, \dot{\varphi}) \right| \times \left[2\dot{z} + \dot{\varphi}\left(l_1 \sin\left(\frac{\varphi}{2} + \beta_1\right) - l_2 \sin\left(\beta_2 - \frac{\varphi}{2}\right)\right) + \varphi \frac{\dot{\varphi}}{2}\left(l_1 \cos\left(\frac{\varphi}{2} + \beta_1\right) + l_2 \cos\left(\beta_2 - \frac{\varphi}{2}\right)\right)\right],$$
$$Q_\phi^{\Pi} = -\gamma_i \left| \bar{f}_i(\phi, \dot{z}, \dot{\varphi}) \right| \{ l_1 \left[\dot{z} + \dot{\varphi} l_1 \left(\sin\left(\frac{\varphi}{2} + \beta_1\right) + \frac{\varphi}{2}\cos\left(\frac{\varphi}{2} + \beta_1\right)\right)\right]\left(\sin\beta + \varphi \cos\left(\frac{\varphi}{2} + \beta_1\right)\right) -$$
$$- l_1^2 \left[\dot{\varphi}\left(\cos\left(\frac{\varphi}{2} + \beta_1\right) - \frac{\varphi}{2}\sin\left(\frac{\varphi}{2} + \beta_1\right)\right)\right]\left(\cos\beta - \varphi \sin\left(\frac{\varphi}{2} + \beta_1\right)\right) +$$
$$+ l_2 \left[\dot{z} - \dot{\varphi} l_2 \left(\sin\left(\beta_2 - \frac{\varphi}{2}\right) - \frac{\varphi}{2}\cos\left(\beta_2 - \frac{\varphi}{2}\right)\right)\right]\left(\sin\beta_2 - \varphi \cos\left(\beta_2 - \frac{\varphi}{2}\right)\right) -$$
$$- l_2^2 \left[\dot{\varphi}\left(\cos\left(\beta_2 - \frac{\varphi}{2}\right) + \frac{\varphi}{2} ssn\left(\beta_2 - \frac{\varphi}{2}\right)\right)\right]\left(\cos\beta_2 + \varphi \sin\left(\beta_2 - \frac{\varphi}{2}\right)\right)\}. \quad (8)$$

In the above relations $\bar{f}_i(\varphi,\dot{z},\dot{\varphi})$ are known functions, which are determined according to the assumptions concerning the resistance force of damping devices. Thus, the oscillating process of a sprung mass is described by a system of differential equations

$$M\frac{d^2z}{dt^2}=Q_z^a+Q_z^{\Pi},\qquad I_C\frac{d^2\varphi}{dt^2}=Q_\varphi^a+Q_\varphi^{\Pi} \tag{9}$$

Of considerable practical and theoretical interest are the oscillations of the sprung part caused by the impact of the right and left wheels on individual bumps. In this case, in accordance with the accepted limitations and the calculation model of the car, the sprung part will perform vertical oscillations. To obtain a mathematical model of their oscillations, first of all, we will simulate the inequalities of the path

$$h(x)=\begin{cases}h_j\sin^{2s_j}\dfrac{2\pi x}{d_j}, & npu\quad x_j\le x\le x_j+d_j\\ 0, \quad npu\quad x_j+d_j\le t\le x_{j+1}, & j=1,2,3,...,\end{cases} \tag{10}$$

(h_j – height, d_j – length of jth inequalities, s_j – parameter that describes its shape). If we take into account that the deformation of the tires is much smaller than the deformation of the shock absorbers and can be neglected, then the differential equation of the disturbed motion of the sprung part will take the form

$$\ddot{z}^*+\frac{c}{M}\left(z^*+h(x)\right)^{\nu+1}=g-\frac{\gamma_1+\gamma_2}{M}\left|\dot{z}^*\right|^s\dot{z}^* \tag{11}$$

In the case when the car moves along a path with a speed of a constant magnitude, and the height of the bump is limited, then the path bumps can be represented as a function of time in the form

$$\bar{h}(t)=\begin{cases}h_j\sin^{2s_j}\dfrac{2\pi Vt}{d_j}, & npu\quad t_j\le t\le t_j+\dfrac{d_j}{V}\\ 0,\quad npu\quad t_j+\dfrac{d_j}{V}\le t\le t_{j+1}+\dfrac{d_{j+1}}{V}, & j=1,2,3,...,\end{cases}$$

and therefore the differential equation (11) for the first approximation is transformed to the form

$$\ddot{z}^*+\frac{c}{M}z^{*\nu+1}=-\frac{c}{M}(\nu+1)\left(z^*\right)^{\nu}\bar{h}(t)-\frac{\gamma_1+\gamma_2}{M}\left|\dot{z}^*\right|^s\dot{z}^*, \tag{12}$$

In the same way as for damped oscillations, the general ideas of excitation methods can be used to research the non-autonomous equation (12), and therefore the amplitude-frequency characteristic of the excited motion can be presented in the form

$$\frac{da}{dt}=\frac{sa(1,\nu+1,\psi)}{\omega(a)M}\times$$

$$\times\begin{cases}c(\nu+1)a^{\nu}ca^{\nu}(\nu+1,1,\psi)\left(\sin^{2s_j-1}\dfrac{2\pi V}{d_j}t-\Delta_{cr}\right)-Mg-\alpha\left(\omega(a)sa(1,\nu+1,\psi)\right)^{2s+1} & npu\quad t_j\le t\le t_j+\dfrac{d_j}{V};\\ -Mg-\alpha\left(\omega(a)sa(1,\nu+1,\psi)\right)^{2s+1}\ npu\quad t_j+\dfrac{d_j}{V}\le t\le t_{j+1}+\dfrac{d_{j+1}}{V}, & \end{cases}$$

$$\frac{d\psi}{dt}=\omega(a)+\frac{(\nu+2)ca(\nu+1,1,\psi)}{2a\omega(a)M}\times \tag{13}$$

It is known that the dynamic process of the sprung part is determined not only by the characteristics of the suspension system and unevenness of the road, but also by the conditions of entry into the unevenness. For the system of differential equations (13), they degenerate into initial conditions. For the general case, they take the form

$$a(t)\big|_{t=0} = a_0, \quad \psi(t)\big|_{t=0} = \psi_0, \tag{14}$$

where respectively a_0 and ψ_0 amplitude and initial phase of vibrations of the sprung part (amplitude and phase of entry into the first unevenness). As for the conditions for exiting inequalities, based on the physical meaning of the process of the sprung part oscillations, they must satisfy the conjugation conditions, i.e.

$$a(t)\big|_{t_j-0} = a(t)\big|_{t_j+0}, \quad a(t)\big|_{t_j+\frac{d_j}{V}-0} = a(t)\big|_{t_j+\frac{d_j}{V}+0}, \tag{15}$$

where respectively $a(t)\big|_{t_j-0}$ – amplitude of oscillations of the sprung part of the vehicle during exit from the jth unevenness, $a(t)\big|_{t_j+0}$ – he amplitude of oscillations when entering the area between j^{th} and $j+1$ unevenness.

Below in Fig. 2, a – 2, l the results of the numerical integration of the system of differential equations obtained above are presented.

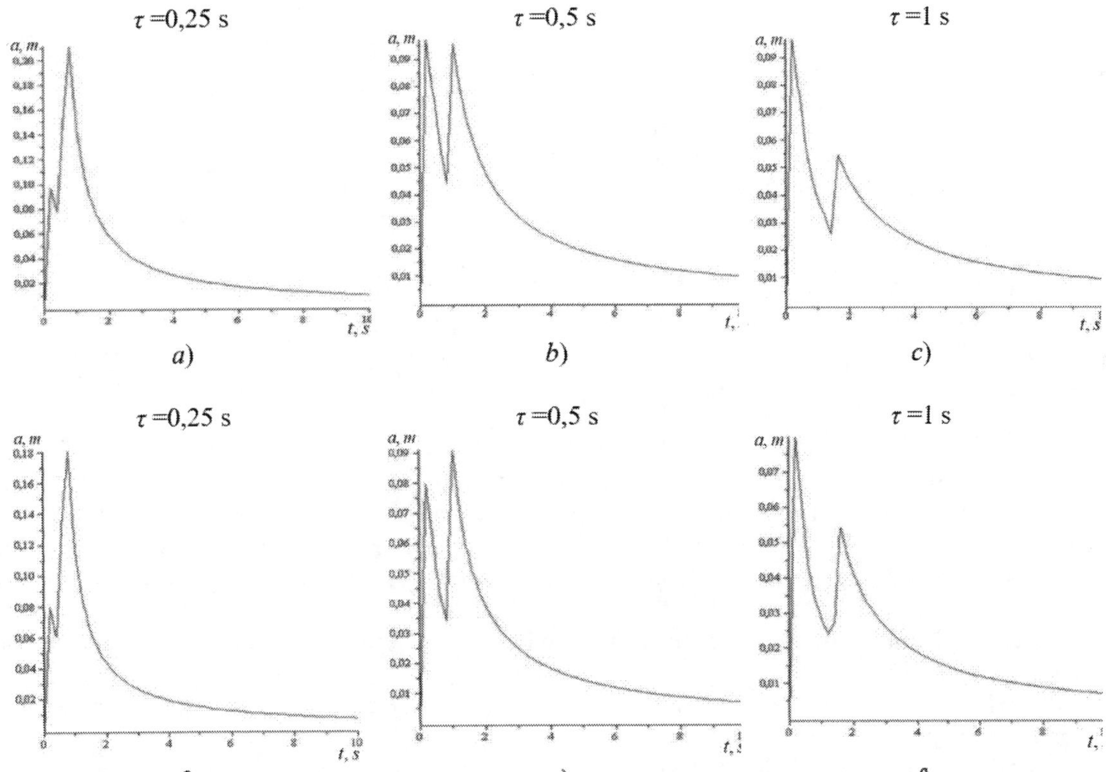

IOP Conf. Series: Materials Science and Engineering 1277 (2023) 012009 doi:10.1088/1757-899X/1277/1/012009

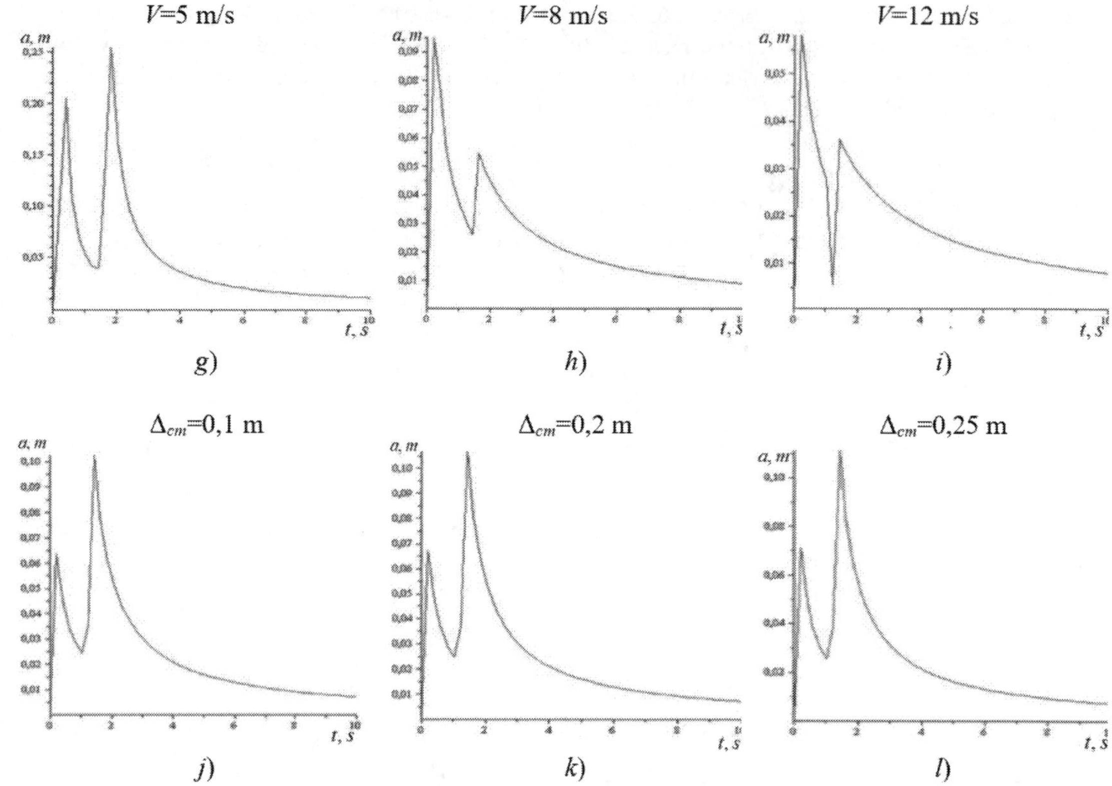

Fig. 2 Changes in the amplitude of oscillations during the movement of the car along a path with two bumps under the following parameters of the researched system:

Fig.2 *a)–c)* P=5000 N; h=0,2 m; l=3 m; V=8 m/s; τ – travel time to the next hill; for the progressive characteristic of the elastic force of the suspension;

Fig.2 *d)–f)* P=5000 N; h=0,2 m; l=3 m; V=8 m/s; τ – travel time to the next hill; for the regressive characteristic of the elastic force of the suspension;

Fig.2 *g)–i)* P=5000 N ; h=0,2 m; l=3 m; $\nu = 2/5$; $d_j = 4$ m; $\Delta_{cm} = 0,2$ m; for the progressive characteristic of the elastic force of the suspension;

Fig.2 *j)–l)* P=5000 N; h=0,2 m; l=3 m; $\nu = -2/7$; V=10 m/s $d_j = 4$ m; for the regressive characteristic of the elastic force of the suspension;

The amplitude of oscillations of the sprung part of a car when exiting the unevenness of the same profiles under the same static deformations of the sprung part for the suspension system with the progressive law of changing the elastic force is greater than for the case of the regressive law of changing the elastic force of shock absorbers.

Conclusions. As a result of mathematical modelling of oscillatory phenomena in a vehicle during movement along a road with two bumps, it was established that:

at small distances between adjacent irregularities, the amplitude of the exit from the next unevenness is greater than from the previous one;

for suspension systems with progressive or regressive laws of changing the elastic force of shock absorbers, static deformation has a minor effect on the amplitude of movement of a car, in particular, an increase in static deformation from 0,2 m to 0,25 m causes an increase in the amplitude of the exit from the second unevenness up to 4% – for progressive law and up to 2% – for the regressive law of changes in the elastic force of shock absorbers;

The developed methodology for determining car suspension parameters can be used to improve the dynamic characteristics of vehicles for design work.

References:

[1]. Boldovskyi V. N. Issledovanie protsessov vzaimodeistviia khodovykh sistem avtomobylei s deformiruemoi opornoi poverkhnostiu. Vostochno-Evropeiskii zhurnal peredovykh tekhnolohii. 2012. №5/4(59). S. 50–52.

[2]. Hrubel M. G., Nanivskyi R. A., Sokil M. B. *Influence of characteristics of wheeled vehicle suspensions of its road-holding along curved stretches of track.* Science & military. 2014. Vol. 9. No 1. P. 15–19.

[3]. Hrubel M. H., Sokil M. B. Kolyvannia pidresorenoi chastyny transportnykh zasobiv iz nekonservatyvnoiu kharakterystykoiu amortyzatoriv. Viiskovo-tekhnichnyi zbirnyk. Lviv, 2015. №13. S. 30–33.

[4]. Hrubel M. H., Sokil M. B. Dynamika pidresorenoi chastyny transportnykh zasobiv iz nekonservatyvnoiu sylovoiu kharakterystykoiu pidvisky. Visnyk mashynobuduvannia ta transportu. Vinnytsia, 2016. №1. S. 16–23.

[5]. Artiushchenko A.D. Doslidzhennia vplyvu kharakterystyk pidvisky avtomobilia maloho klasu na plavnist khodu ta yii modernizatsiia / Artiushchenko A.D., Suiarkov O.H. // visnyk NTU "KhPI".- 2013. –№ 32 (1004). – S. 21–27.

[6]. *Ustoichivost koliosnykh mashyn protiv zanosa v protsesse tormozheniia i puti yeio povysheniia / M.A. Podryhalo, V.P. Volkov, A.A. Boboshko, V.A. Pavlenko, M.V. Baitsur, A.Y. Nazarov, V.O. Alekseev / pod red. M.A. Podryhalo.- Kharkov: KhNADU,* 2006. – 337p.

[7]. Podryhalo M.A. Otsinka dynamichnoi stiikosti avtomobilia / Podryhalo M.A., Korbko M.I., Klets D.M. //"Avtomobile- ta traktorobuduvannia ". Visnyk NTU "KhPI" 58, 2008. – S.134-137

[8]. Litvinov A.S. Upravliayemost i ustoichivost avtomobilia – M.: Mashynostroenie, – 1971. – 416 p.

[9]. Rotenberg R.V. Podveska avtomobil. / Rotenberg R.V. – M.: Mashynostroienie. – 1972. – 392 p.

[10]. Velychko L.D. Rozrobka metodu rozrakhunku neliniinykh pozdovzhno-kutovykh kolyvan husenychnykh transportnykh zasobiv / L.D. Velychko, B.I. Sokil, Yu.A. Chahan // Optymizatsiia vyrobnychykh protsesiv i tekhnichnyi kontrol u mashynobuduvanni ta pryladobuduvanni. – Lviv: Visnyk NU «LP». – 2011. – Vyp. № 702. – S. 49–53.

[11]. Kuzo I.V. Vplyv parametriv pidvisky na neliniini kolyvannia transportnykh zasobiv / I.V. Kuzo, B.I. Sokil, V.M. Paliukh // Visnyk NU "LP" "Dynamika, mitsnist ta proektuvannia mashyn i pryladiv". – 2007. №588 – S. 49–52.

[12]. Hrubel M.H. Teoretychni zasady formuvannia parametriv pidvisky kolisnoi viiskovoi avtomobilnoi tekhniky za umov rukhu bezdorizhzhiam / M.H. Hrubel//Suchasni tekhnolohii v mashynobuduvanni ta transporti. Naukovyi zhurnal. – Lutsk: Lutskyi NTU, 2019. – №2(13). – s. 38-52.

[13]. Hrubel M.G., Nanivskyi R.A., Sokil M.B. *Oscillation of a sprung part of vehicles with non-conservative specifications of shock absorvers.* Zbirnyk naukovykh prats Kharkivskoho natsionalnoho universytetu Povitrianykh Syl. Kharkiv, 2018. № 2(56) S. 132-136.

Research of the characteristics of an air spring under cyclic load

A Y Kuzyshyn [1, 2, 4], Y H Sobolevska [1], S A Kostritsa [3], O M Voznyak [1, 2]

[1] Lviv Institute of Ukrainian State University of Science and Technology, Ukraine
[2] Lviv Scientific Research Institute of Forensic Expertise, Ukraine
[3] Ukrainian State University of Science and Technology, Ukraine
[4] E-mail: kuzyshyn1993@gmail.com

Abstract. The dynamic stiffness, energy dissipation per cycle and the damping coefficient of the air spring of the second stage of spring suspension of high-speed rolling stock are investigated on the basis of a simplified mathematical model. The studies were carried out in the frequency range from 0 to 15 Hz. To determine the parameters of the air spring, its power characteristic was built, the area that characterizes the energy dissipation of the system per cycle. Based on the constructed characteristics, the damping coefficient and air spring stiffness during compression and expansion were determined. The influence of the frequency of the disturbing force and the diameter of the connecting element on the investigated parameters was studied. The obtained ratios of quantities would be made it possible to optimize the parameters of spring suspension of high-speed rolling stock from the point of view of ensuring admissible dynamic performance and traffic safety indicators.

1. Introduction

To ensure smooth running and limit the level of amplitudes of oscillations of high-speed electric trains, it is necessary to have a fairly "soft" spring suspension. For this purpose, air springs are used, which are the main structural element of the spring suspension system on the operation of which the overall dynamic behavior of high-speed rolling stock during its movement on the rail track depends (**Figure 1**). The design of such springs allows providing the standard value of static deflection up to 200 mm for passenger wagons and multiple unit trains, intended for a speed of 200 km/h. In addition, this suspension is characterized by both elastic and dissipative properties, i.e. there is no need to install a special dampener.

At the stage of designing high-speed rolling stock, the question arises regarding the modeling of an air spring and the determination of its main parameters. The dynamic behavior of an air spring can be assessed by the following main parameters: dynamic rigidity, energy dissipation per cycle, damping coefficient, etc. Determining these parameters is important because it allows, at the design stage, for assessing the dynamic indicators of rolling stock and establishing the optimal operational conditions of moving along the rail track [1, 2]. Currently, the verification of the respective indicators of rolling stock is performed mainly experimentally after selecting a definite model of an air spring. However, taking into account a lot of different factors makes such an experiment multifactorial and complex.

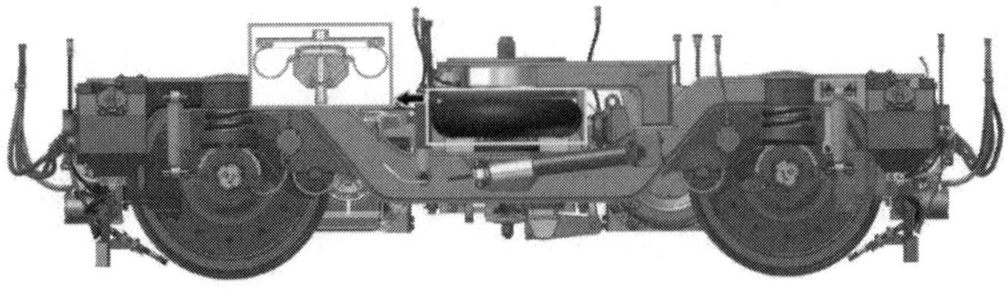

Figure 1. General appearance of an air spring

Thus, the analysis of the relationship between the parameters of an air spring and the factors that characterize its design and the conditions in which it operates is a relevant task and needs further solving.

2. Analysis of literary references and problem statement

Depending on the chosen mathematical model of an air spring, the methods of calculating dynamic rigidity, energy dissipation per cycle and damping coefficient of an air spring are chosen.

In work [3], the rigidity of an air spring was determined as the sum of adiabatic rigidity and rigidity of the air spring due to changes in its effective area:

$$\mathrm{k_s} = \frac{\gamma \cdot \mathrm{A_н} \cdot \mathrm{p}}{\mathrm{H}}, \tag{1}$$

$$\mathrm{k}_l = l \cdot \mathrm{p_g}, \tag{2}$$

where $\mathbf{k_s}$ – adiabatic rigidity of the air spring; $\mathbf{k}_l$ – rigidity of the air spring due to changes in its effective area; $\mathbf{A_н}$ – effective area of the air spring at the design height; $\mathbf{p_g}$ – air pressure in the air spring; $\boldsymbol{\gamma}$ – ratio of specific heats of air; $\mathbf{p}$ – absolute air pressure in the air spring; $\boldsymbol{l}$ – constant change of the effective area of the air spring with change of its height; $\mathbf{H}$ – design height of the air spring.

In works [4, 5], the vertical dynamic rigidity, which consists of three components, and viscous damping, which involves energy loss due to air flow between the air spring and the additional reservoir, are determined by the formulas:

$$K_1 = n \cdot A_e^2 \cdot \frac{p_0}{V_r};\ K_2 = n \cdot A_e^2 \cdot \frac{p_0}{V_b};\ K_3 = (p_0 - p_a) \cdot \frac{dA_e}{dz}, \tag{3}$$

$$C = R_f \cdot A_e^2 \cdot \rho_0 \cdot \mathrm{g} = \frac{0{,}126}{d_s^3} \cdot A_e^2 \cdot \rho_0 \cdot \mathrm{g}, \tag{4}$$

where $\boldsymbol{K_1}$ – rigidity of the air spring, appearing due to air compression in case of a blocked equalizing pipeline, N/m; $\boldsymbol{K_2}$ – rigidity appearing due to air compression in the additional reservoir, N/m; $\boldsymbol{K_3}$ – rigidity of the surface of the air spring which changes due to changes in its effective area during compression or expansion, N/m; $\boldsymbol{R_f}$ – coefficient of hydraulic resistance; $\boldsymbol{A_e}$ – effective area of the air spring, m^2; $\boldsymbol{\rho_0}$ – air density, kg/m^3; $\mathbf{g}$ – free fall acceleration, m/s^2; $\boldsymbol{d_s}$ – diameter of the connecting pipeline, m.

In paper [6] a method of determining the equivalent rigidity of an air spring was proposed, which additionally takes into account the rigidity of the emergency spring:

$$k_e = k \cdot \frac{\beta_2 \cdot \left(\beta_1 + \frac{\alpha}{1+\alpha}\right)}{\beta_1 + \beta_2 + \frac{\alpha}{1+\alpha}}, \tag{5}$$

$$k = \frac{A_{\text{еф.}}^2 \cdot p \cdot n}{V};\ \alpha = \frac{V}{V_{\text{дод.}}};\ \beta_1 = \frac{k_1}{k};\ \beta_1 = \frac{k_2}{k},$$

where $\boldsymbol{k}$ – rigidity of the air spring; $\boldsymbol{k_1}$ – rigidity defined by the change in effective area of the air spring; $\boldsymbol{k_2}$ – rigidity of the emergency spring; $\boldsymbol{\alpha}$ – ratio of volumes of the air spring and the additional reservoir.

In work [7], which is a continuation of works [8-10], on the basis of a viscoelastic mechanical model of an air spring, formulas for calculating dynamic rigidity and damping coefficient were obtained:

$$K_{ez} = \left(\frac{1}{\frac{p_0 \cdot A_e^2 \cdot n}{V_{b0} + V_{r0}} + p_g \cdot \frac{dA_e}{dz}} + \frac{1}{K_{\text{доп.}}} \right)^{-1}, \quad (6)$$

$$K_{vz} = \left(\frac{1}{\frac{p_0 \cdot A_e^2 \cdot n}{V_{b0}} + p_g \cdot \frac{dA_e}{dz}} + \frac{1}{K_{\text{доп.}}} \right)^{-1} - K_{ez}, \quad (7)$$

$$C_{z,\beta} = \frac{1}{2}\rho \cdot k_t \cdot A_s \cdot \left(\frac{A_e}{A_s} \cdot \frac{V_{r0}}{V_{b0} + V_{r0}} \right)^{1+\beta}, \beta = 2 \quad (8)$$

where $\boldsymbol{l_s}$ – length of the connecting element; $\boldsymbol{A_s}$ – cross section of the connecting element; $\boldsymbol{\rho}$ – air density at pressure p_0; $\boldsymbol{A_e}$ – effective area of the air spring; $\boldsymbol{V_{r0}}$ – volume of the additional reservoir; $\boldsymbol{V_{b0}}$ – volume of the air spring; $\boldsymbol{p_0}$ – initial absolute pressure of the air spring; $\boldsymbol{p_g}$ – manometric pressure; $\boldsymbol{k_t}$ – loss coefficient in the connecting element; $\boldsymbol{M}$ – mass of air.

In works [11, 12], the laws of thermodynamics were used under the assumption that the process is adiabatic, and the heat exchange between an air spring and the environment was neglected. The dynamic rigidity of the air spring was found by the formula:

$$K_d = (P_s - P_0)\frac{\partial A_e}{\partial z} + \frac{k \cdot P_s \cdot A_e}{m_s} \cdot \frac{G}{\dot{z}} - \frac{k \cdot P_s \cdot A_e}{V_s} \cdot \frac{\partial V_s}{\partial z}, \quad (9)$$

where $\boldsymbol{z}$ – vertical deformation of the air spring, m; $\boldsymbol{A_e}$ – effective area of the air spring, m^2; $\boldsymbol{k}$ – adiabatic index.

In paper [13], the formula of dissipation coefficient was obtained using a dimensionless model of an air spring with an additional reservoir connected by a hydraulic support:

$$\xi = R_F \frac{V_{20}}{nRT_0} \cdot \frac{nP_{10}\nu A - \alpha V_{10}(P_{10} - P_A)}{nP_{10}\nu A - \alpha(V_{10} + V_{20})(P_{10} - P_A)} \sqrt{\frac{\frac{nP_{10}\nu A}{V_{10}} - \alpha(P_{10} - P_A)}{m}}, \quad (10)$$

where $\boldsymbol{R_F}$ – orifice resistance, $\frac{Pa}{kg/s}$; $\boldsymbol{P}$ – absolute pressure, Pa; $\boldsymbol{\alpha}$ – derivative of the effective area, m; $\boldsymbol{A}$ – effective area of the air spring, m^2; $\boldsymbol{C_2}$ – pneumatic capacity of the additional reservoir, kg/Pa; $\boldsymbol{m}$ – sprung mass, kg; subscripts: **1** – spring; **2** – reservoir; **0** – static conditions.

In work [14], the dynamic vertical rigidity of the spring was determined taking into account the change in its effective area and volume due to the deformation of the spring:

$$K_z = n(P + P_a)\frac{S_e^2}{V} + 2\pi P R_e A_z^R. \quad (11)$$

Additional consideration of the rigidity of the air spring shell, performed in work [15], allowed obtaining the following formula for determining the vertical rigidity of the spring at a fixed height:

$$K = k(P_{z0} + p_a)\frac{A_{z0}}{V_{z0}}\frac{dV_z}{dh}\bigg|_{h=h_0} - P_{z0}\frac{dA_z}{dh}\bigg|_{h=h_0} + K_B, \tag{12}$$

where $\frac{dV_z}{dh}$ – rate of change of the effective volume; $\frac{dA_z}{dh}$ – rate of change of the effective area; $\boldsymbol{K_B}$ – rigidity of the air spring shell; $\boldsymbol{k}$ – polytrope indicator.

Thus, the analysis of the carried out works showed a significant number of approaches to determining the characteristics of rigidity and dissipation of air springs. However, the issue of discrepancy between the values of spring rigidity during its compression and expansion was not analyzed. Also, the considered models did not take into account the phenomenon of heat exchange between the spring and the environment.

3. The purpose and tasks of the study

The purpose of this work is to study the dynamic rigidity of an air spring at different modes of its operation and energy dissipation depending on the design and parameters of the connecting element and the frequency spectrum of system oscillations.

Tasks of the study:

– analysis of the developed models and approaches to determining the characteristics of an air spring;

– obtaining the dependences of the dynamic rigidity of an air spring on the deformation at different modes of its operation (compression and expansion);

– research of the relationships between dynamic rigidity, energy dissipation per cycle, damping coefficient and perturbation force frequency;

– assessment of the influence of the diameter of the connecting element on the characteristics of the spring.

4. Determination of the parameters of an air spring

Determination of the main parameters of an air spring, namely dynamic rigidity, energy dissipation per cycle and damping coefficient, previously involves obtaining the dependence of force on deformation (**Figure 2**).

From the obtained dependence «force-deformation», rigidity and damping coefficient are determined by the formulas:

$$S = \frac{F_0}{x_0}, \tag{13}$$

$$D = \frac{E}{F_0 \cdot x_0}, \tag{14}$$

where $\boldsymbol{F_0}$ – amplitude of force; $\boldsymbol{x_0}$ – amplitude of deformation; $\boldsymbol{E}$ – energy dissipation per cycle.

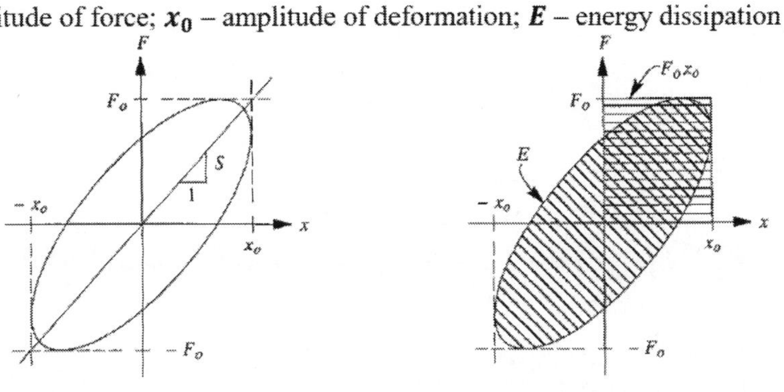

Figure 2. Dependence of force on deformation of an air spring for determining rigidity and damping

When constructing the «force-deformation» dependence, the ideal behavior of gas was considered, where the force was determined by the internal pressure of the air spring:

$$\dot{P}_1(t) = -\dot{h}(t)\frac{P(t)}{h(t)} + \dot{m}(t)\frac{R \cdot T(t)}{h(t) \cdot A_{a.s.}} + \dot{T}(t)\frac{m(t) \cdot R}{h(t) \cdot A_{a.s.}}, \tag{15}$$

where $\boldsymbol{P}, \boldsymbol{V}, \boldsymbol{T}$ – pressure, volume and temperature of the working medium of the air spring, respectively; $\boldsymbol{m}$ – mass of air; $\boldsymbol{R}$ – universal gas constant; $\boldsymbol{h(t)}$ – height of the air spring in the modeling process; $\boldsymbol{A_{a.s.}}$ – cross-sectional area of the air spring.

The dissipation of energy per cycle during the operation of the air spring is characterized by the area of the figure limited by the graph of the dependence «force-deformation» (**Figure 2**).

5. Analysis of the obtained results

Study of the parameters of an air spring were carried out with the following initial data: perturbing force frequency f = 0...15 Hz; diameter of the connecting element d = 0...40 mm and length l = 0,5 m, respectively; initial pressure of the air spring p_0 = 0,7 MPa; amplitude value of the sinusoidal irregularity of the rail track H = 10 mm; universal gas constant R = 287 J/(kg·K); specific heat for the process of constant volume c_v = 718 J/(kg·K) and pressure c_p = 1005 J/(kg·K), respectively. The solving of differential equations of the mathematical model of an air spring was performed in the system of computer algebra «Mathcad». Based on the produced results, the "force-deformation" characteristics of the air spring were built and the dependences of dynamic rigidity during compression and expansion, energy dissipation per cycle, damping coefficient on the perturbing force frequency and the diameter of the connecting element were obtained (**Figure 3-8**).

From the obtained dependence «force-deformation», rigidity and damping coefficient are determined by the formulas:

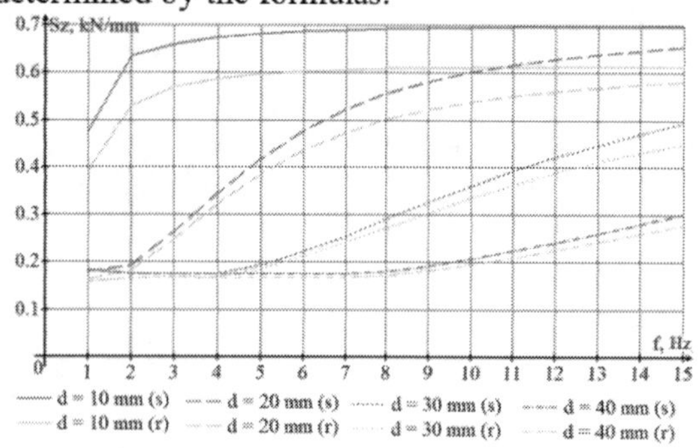

Figure 3. Dependence of rigidity of the air spring on frequency of perturbing force

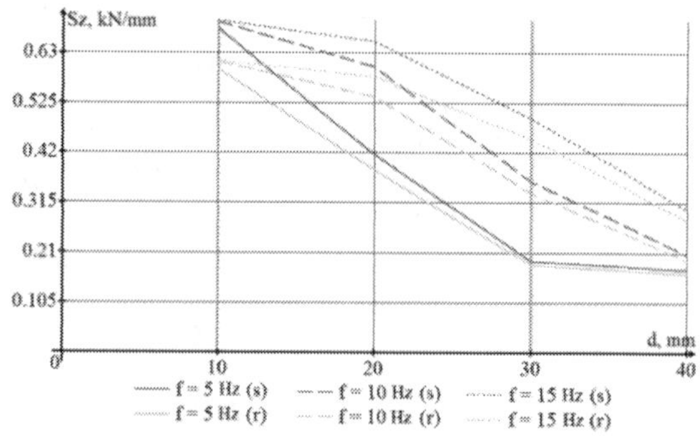

Figure 4. Dependence of rigidity of the air spring on diameter of the connecting element

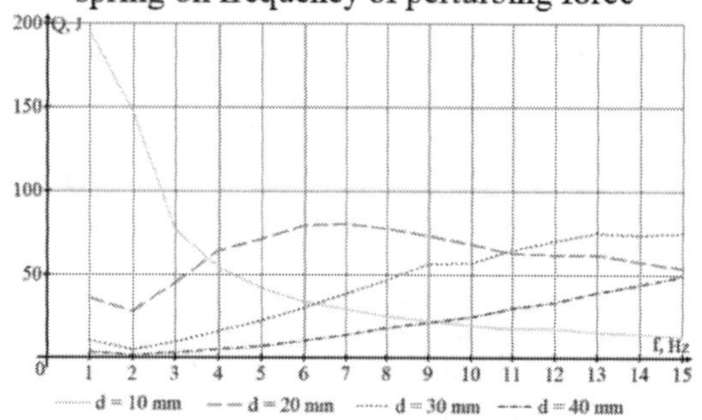

Figure 5. Dependence of energy dissipation of the air spring on frequency of perturbing force

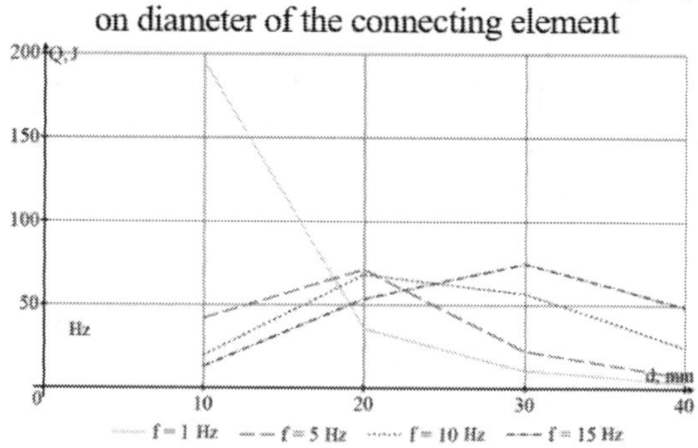

Figure 6. Dependence of energy dissipation of the air spring on the diameter of the connecting element

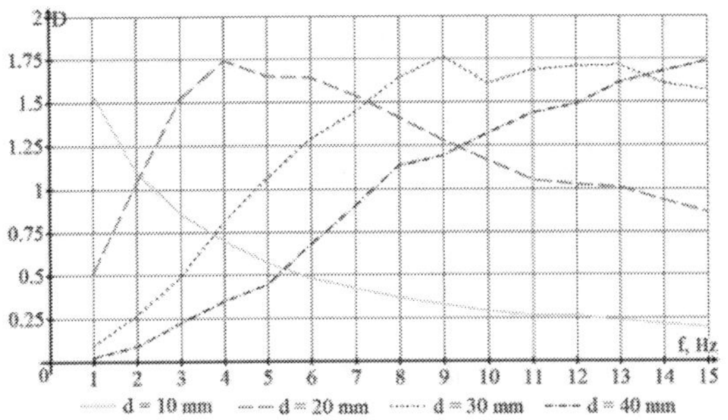

Figure 7. Dependence of damping coefficient of the air spring on frequency of perturbing force

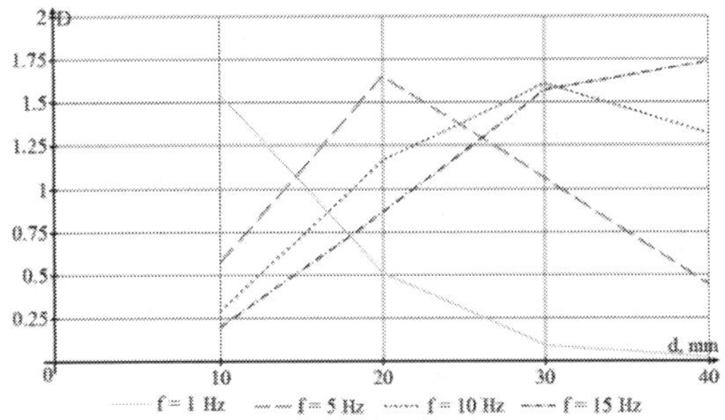

Figure 8. Dependence of damping coefficient of the air spring on the diameter of the connecting element

From the analysis of (**Figure 3**), it is established that the dynamic rigidity of an air spring during its compression is bigger than during expansion in the considered range of frequencies. As the diameter of the connecting element increases, the frequency to which the spring rigidity corresponds to static one (at d = 20 mm - to the frequency of 2 Hz, at d = 30 mm - to the frequency of 4 Hz, at d = 40 mm - to the frequency of 8 Hz) rises. Also, the rigidity of the air spring decreases. With the constant diameter of the connecting element, the increase in frequency f leads to a greater difference between the rigidity of the air spring when it is compressed and expanded.

The graphs in (**Figure 4**) confirm the previously made conclusion that as the diameter of the connecting element increases, the rigidity of the air spring decreases. As the diameter increases, the difference between rigidity at compression and expansion decreases.

Analysis of the correlations in (**Figure 5-6**) shows that with increasing frequency of the perturbing force, the energy dissipation of the air spring increases to a certain maximum value and decreases with further rise of f. The obtained maximum energy dissipation depends on the diameter of the connecting element. The smaller the diameter, the lower the frequencies at which the maximum energy dissipation is observed. If the diameter of the connecting element is small enough (up to 10 mm), the air spring will effectively dissipate energy only at low oscillation frequencies (up to 5 Hz). At higher frequencies f > 5 Hz, the flow of air into the additional reservoir is virtually stopped and energy dissipation is significantly reduced Q < 40 J. At the frequency of oscillations f = 4 Hz, the most effective is an air spring with a diameter of the connecting element d = 10÷20 mm, which corresponds to the results of work [16], and a spring with d = 40 mm virtually does not provide the damping of oscillations. In the frequency range of the perturbing force from 5 to 15 Hz, the change in diameter in the range from 20 to 30 mm has almost no effect on the maximum dissipation value.

The dependences of damping coefficient on f and d (**Figure 7-8**) are similar in nature to the dependences obtained for energy dissipation. They also have a marked maximum of damping coefficient, shifted in frequency, depending on the diameter of the connecting element. But in contrast to energy dissipation, the value of the maximum damping coefficient is almost the same for all calculated cases.

6. Conclusions

Approaches to determining the characteristics of an air spring were considered in the work. On the basis of the obtained dependences of the force on the deformation, the dynamic rigidity of the spring during compression and expansion, the dissipation of energy by the spring and the damping coefficient were determined. The dependences of the listed parameters on the frequency of the perturbing force and the diameter of the connecting element of the spring with an additional reservoir were investigated.

It was shown that the dynamic rigidity of the spring during compression and expansion is different, and in the considered frequency range the rigidity at compression is greater. A decrease in the diameter or a growth in the frequency of the perturbing force leads to an increase in the discrepancy between the rigidity at compression and the one at expansion. As the diameter of the connecting element increases, the frequency to which the rigidity of the spring corresponds to static one rises. The dependences of the dissipation energy and the damping coefficient on the frequency of the perturbing force have maxima that occur at lower frequencies for smaller diameters of the connecting element.

7. References

[1] Kuzyshyn A, Batig A, Kostritsa S, Sobolevska J, Dovhaniuk S and Dzhus V 2020 Study of the Dynamic Behavior of Rolling Stock using a Computer Experiment. *IOP Conference Series: Materials Science and Engineering. 15th Intern. Sci. and Techn. Conf. «Problems of the railway transport mechanics» (PRTM 2020)* **985** doi:10.1088/1757-899X/985/1/012002.

[2] Kuzyshyn A, Batig A, Kostritsa S, Sobolevska J, Kovalchuk V, Dovhanyuk S and Voznyak O 2018 Research of safety indicators of diesel train movement with two-stage spring suspension *BulTrans 2018: 10th Intern. Sci. Conf. on Aeronautics, Automotive and Railway Engineering and Technologies* **234** https://doi.org/10.1051/matecconf/201823405003

[3] Dixon GK and Pearson J 1967 Automatically controlled air spring suspension system for vibration testing *National Aeronautics and Space Administration (NASA)*

[4] Oda N and Nishimura S 1970 Vibration of air suspension bogies and their design *JSME Int J.* **13(55)** 43–50

[5] Building templates in ADAMS/Rail *Component Descriptions* 51–73.

[6] Spiroiu MA 2018 Railway vehicle pneumatic rubber suspension modelling and analysis *J Mater Plastic* **55(1)** 24–27

[7] Haupt P and Sedlan K 2001 Viscoplasticity of elastomeric materials: experimental facts and constitutive modeling *Archive of Applied Mechanics* **71** 89–109

[8] Berg M 2000 Three-dimensional airspring model with friction and orifice damping *Vehicle System Dynamics* **33(1)** 528–539

[9] Berg M 1999 A three-dimensional airspring model with friction and orifice damping *16th IAVSD Symposium, The Dynamics of Vehicles on Roads and on Tracks*

[10] Berg M 1999 An air spring model for dynamic analysis of rail vehicles *TRITA-FKT Report, Division of Railway Technology, Dep. of Vehicle Engineering, Royal Institute of Technology*

[11] Zhang Z and Shi B 1982 Technique of Refrigeration and Low Temprature *China Machine Press*

[12] Zhu S, Wang J and Zhang Y 2008 Research on theoretical calculation model for dynamic stiffness of air spring with auxiliary chamber *IEEE Vehicle Power and Propulsion Conference (VPPC)* 2–7

[13] Quaglia G and Sorli M 2003 Air suspension dimensionless analysis and design procedure *Vehicle Syst Dyn.* **35(6)** 443–475

[14] Xu W, He L and Shuai CG 2005 Stiffness calculation and dynamic simulation of air spring *Proceedings of the ASME International Design Engineering Technical Conferences and Computers and Information in Engineering Conference – DETC2005* **1** 1395–1399

[15] Guo KH, He JH and Chen JJ 2017 Theoretical modelling and experimental analysis of the vertical stiffness of a convoluted air spring including the effect of the stiffness of the bellows *Proc Inst Mech Eng Part D: J Auto Eng.* **232(4)** 547–561

[16] Lobachev N A 1983 Pneumatic system of supports and connection of the locomotive body with trolleys *[Manuscript] dissertation of candidate of technical sciences* 164

Dynamics of mobile robot equipped with inertial vibration exciter and unidirectionally rotating wheels

V M Korendiy [1, 2], **O Y Kachur** [1] **and V M Gurskyi** [1]

[1] Department of Robotics and Integrated Mechanical Engineering Technologies, Lviv Polytechnic National University, 12 S. Bandera Street, Lviv, 79013, Ukraine
[2] E-mail: vitalii.m.korendii@lpnu.ua

Abstract. Vibration-driven locomotion principles are currently of significant interest among designers and researchers dealing with mobile robotics. Among a great variety of robots chasses, the wheeled ones are most commonly used. The major purpose of this study consists in defining the dynamic characteristics of the wheeled vibration-driven robot equipped with the centrifugal (inertial) vibration exciter (unbalanced rotor) and overrunning clutches ensuring the robot's wheels rotation in one direction. The research methodology is divided into three basic stages: developing the robot's dynamic diagram and deriving the motion equations followed by numerical modelling in the Mathematica software; designing the 3D-model and simulating the robot motion in the SolidWorks software; creating the experimental prototype and conducting the full-scale tests. The obtained results show the time dependencies of the robot's body acceleration, speed, and displacement at certain operational conditions. The main scientific novelty of the paper resides in substantiating the relationships between the robot's design parameters and its dynamic characteristics under different operational conditions. The performed investigations can be useful for researchers and designers dealing with the vibration-driven robots, capsule-type locomotion systems, pipelines inspecting vehicles, etc.

1. Introduction

Mobile robotics is a relatively young branch of scientific research. In most cases, mobile robots and other vehicles use wheeled or tracked (caterpillar) chasses with an active transmission connecting the motor (or engine) with the corresponding actuators (wheels, tracks, etc.). In distinction to the classical vehicles' designs, mobile robots are sometimes equipped with the vibration-driven locomotion systems. One of the most widespread locomotion principles is based on the anisotropic friction conditions occurring when two bodies are in contact and move relative to one another. This phenomenon is comprehensively studied in numerous publications, particularly in [1, 2]. The mentioned papers are dedicated to the theoretical and experimental research on the dynamic behavior of the mobile robot under the friction-induced stick-slip locomotion conditions.

Another interesting vibration-driven locomotion principle currently being widely studied is implemented in so-called brush-robots or bristle-robots. They use elastic elements, similar to brushes (bristles), for interacting with the supporting surface and generating the stick-slip locomotion. The novel design of the Brushbot is proposed in [3], where the authors carried out the mathematical modeling and experimental investigations on its dynamics. A thorough analysis of the possibilities of providing the bidirectional motion conditions of the bristle-robot is performed in [4]. Another prospective design of the hybrid soft actuator converting the centrifugal forces generated by the unbalanced vibration exciter into the propulsion force of the elastic spikes (bristles) is studied in [5]. The paper [6] is devoted to studying the motion conditions of the bristle-robot consisting of the piezoelectric film placed between the metalized electrodes and of two pairs of the inclined flexible bristles. In [7], the authors derived the mathematical model describing the dynamics of the brush-robot locomotion and considered its basic motion modes: purely stick, stick-slip, purely slip, and jumping (hopping). The paper [8] presents comprehensive research on the multistable dynamic behavior of two rigid bodies pinned together with the help of the linear torsion spring under the action of nonholonomic constraints and with the introduction of damping forces and periodic excitation.

In distinction to the brush-robots and bristle-robots, the stick-slip locomotion principles can be implemented in the vibration-driven capsule-type systems in three major cases: under the anisotropic friction conditions in the contact of the sliding body and the unmovable surface; when the vibro-impact operational conditions are used; when applying the controllable excitation force between the internal disturbing member and robot's body. The third case is comprehensively studied in the papers [9, 10], where the authors studied the influence of the excitation conditions (the frequency and duration of the pulse-width signal) on the average translational speed of the vibration-driven capsule-type robot. The bifurcation analysis of the stick-clip locomotion of a similar vibration-driven system is studied in [11], taking into account different dry friction conditions. In [12, 13], the authors considered the possibilities of optimizing the control algorithms of the vibration-driven robots in order to maximize their translational speed and minimize the power consumption. The papers [14, 15] are devoted to improving the locomotion systems of the vibration-driven capsule-type robots and to investigating the friction influence on the robot's oscillatory system dynamic response. In [16], the capsule-type robot's motion conditions were studied for different inclination angles of the track.

The anisotropic friction phenomenon is another factor ensuring the stick-slip locomotion of the vibration-driven robots. The papers [17, 18] are devoted to investigating the dynamic behavior of the vibration-driven locomotion systems under the influence of the piecewise smooth nonholonomic constraints. One of the most common fields of implementing the vibration-driven robots is inspecting and cleaning the pipelines. In [19], the authors proposed the enhanced design of the vibration-driven walking (stepping) robot actuated by two unbalanced rotors being synchronized with the help of the non-circular gears. The initial idea of this paper is based on the investigations presented in [20, 21, 22], in which an interesting design of the wheeled robot driven by the centrifugal vibration exciter was proposed. This design idea acquired further development in the papers [23, 24], where the improved crank-type exciter was implemented in the robot's drive in order to provide the efficient vibro-impact locomotion conditions. The possibilities of equipping the in-pipe vibration-driven robot with the centrifugal (inertial) exciter in the form of the unbalanced rotor are thoroughly studied in [25]. The major purpose of this study consists in defining the dynamic characteristics of the wheeled vibration-driven robot equipped with the centrifugal (inertial) vibration exciter (unbalanced rotor) and overrunning clutches ensuring the robot's wheels rotation in one direction.

2. Research Methodology

2.1. General design of the wheeled robot

The proposed design of the wheeled vibration-driven robot is shown in figure 1. The robot consists of four rubber wheels 1 which are in contact with the horizontal supporting surface. The front and rear wheels 1 are installed on the corresponding axes 2 with the help of the overrunning (free-wheel) clutches 3 using the bushings (bosses) 4. The clutches 3 allow the unidirectional (single-way) rotation of the wheels 1. Additionally, the ball bearings 5 are used to join the wheels 1 with the axes 2 and to hold the extra loading. In order to eliminate the wheels sliding along the axes 2, the clamps 6 are used. The axes 2 are fixed to the angles 8 of the robot's body with the help of the clamps 7. The DC electric motor 9 with the unbalanced rotor 10 is installed in the housing 11, which is fixed to the angles 8. The centrifugal forces generated due to the unbalanced masses rotation act through the motor 9 upon the housing 11 and cause the pushing effect on the robot's body. When the line of action of the centrifugal forces coincides with the direction, along which the clutches 3 allow the robot to move, the wheels 1 start rotating. On the other hand, when the centrifugal forces are characterized by the opposite direction, the clutches 3 block (restrain) the wheels 1, and the robot remains at rest.

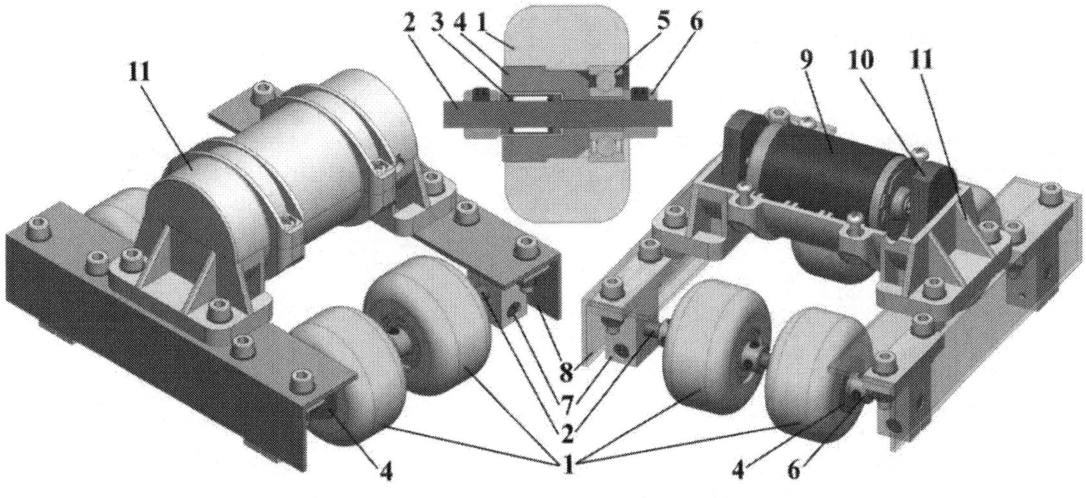

Figure 1. General design of the wheeled vibration-driven robot.

2.2. *Dynamic diagram of the robot's mechanical system and differential equations of its locomotion*

The simplified dynamic diagram of the robot's mechanical system is presented in figure 2. The robot of the mass m_1 moves along a straight horizontal surface due to the action of centrifugal forces generated by the mass m_2 rotating around the hinge A at the constant angular speed ω. The inertial coordinate system xOy is related with the unmovable surface. The generalized coordinate x_1 is adopted for describing the robot horizontal displacement. The wheels are connected to the robot's body with the help of the overrunning (free-wheel) clutches being engaged in one direction and free in the other. Therefore, the wheels are allowed to rotate in a clockwise direction, and the robot can move to the right. The leftward motion of the robot is not possible because the wheels become blocked (restrained) by the overrunning clutches. Let us assume that the contact between the wheels and the surface is ideal, i.e., there is no rolling friction for the clockwise rotation of the wheels, while the counterclockwise rotation is restricted. Herewith, the sliding friction coefficient is equal to f_{fr}. Any energy losses and viscous friction effects are also assumed to be negligibly small.

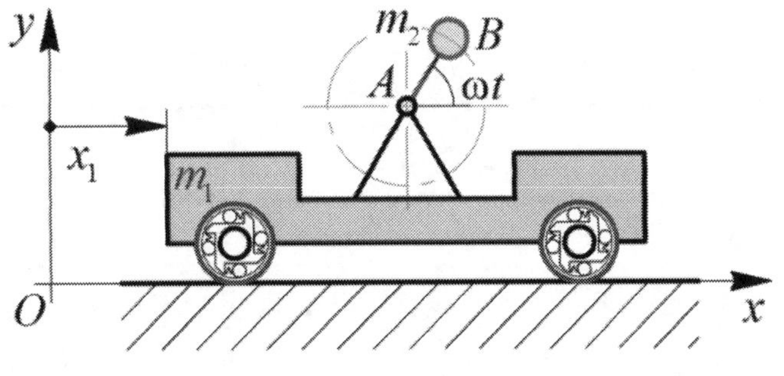

Figure 2. Dynamic diagram of the robot's mechanical system.

The differential equation describing the robot locomotion conditions can be derived using the D'Alembert's principle:

$$(m_1 + m_2) \cdot \ddot{x}_1 = m_2 \cdot r \cdot \omega^2 \cdot \cos(\omega \cdot t) - F_{fr}, \tag{1}$$

where r is the length of the eccentric rod AB (see figure 2); F_{fr} is the sliding friction force that can be determined by the following formula:

$$F_{fr} = \begin{cases} 0, & \text{if } \dot{x}_1 \geq 0; \\ m_2 \cdot r \cdot \omega^2 \cdot \cos(\omega \cdot t) - f_{fr} \cdot N, & \text{if } \dot{x}_1 < 0 \text{ and } m_2 \cdot r \cdot \omega^2 \cdot \cos(\omega \cdot t) \geq f_{fr} \cdot N; ; \\ m_2 \cdot r \cdot \omega^2 \cdot \cos(\omega \cdot t), & \text{if } \dot{x}_1 < 0 \text{ and } m_2 \cdot r \cdot \omega^2 \cdot \cos(\omega \cdot t) < f_{fr} \cdot N. \end{cases} \quad (2)$$

Considering the case of non-detachable (non-jumping) motion conditions, the total normal force N exerted by the robot's wheels upon the supporting surface is considered to be larger that zero at any time moment. In such a case, the magnitude of the force N can be calculated as follows:

$$N = (m_1 + m_2) \cdot g - m_2 \cdot r \cdot \omega^2 \cdot \sin(\omega \cdot t). \quad (3)$$

Substituting the formulas (3) and (2) into the equation (1), the simplified mathematical model describing the robot locomotion can be derived.

3. Results and Discussion

3.1. Numerical modelling of the robot locomotion in the Mathematica software

Let us carry out the numerical modelling of the robot locomotion in the Mathematica software using the integrated Runge-Kutta methods. The following input parameters have been adopted: $m_1 = 1.05$ kg, $m_2 = 0.1$ kg, $r = 0.028$ m, $\omega = 89$ rad/s (850 rpm), $f_{fr} = 0.7$. The obtained results are shown in figure 3 in the form of time response curves, and time dependencies of the robot speed and acceleration.

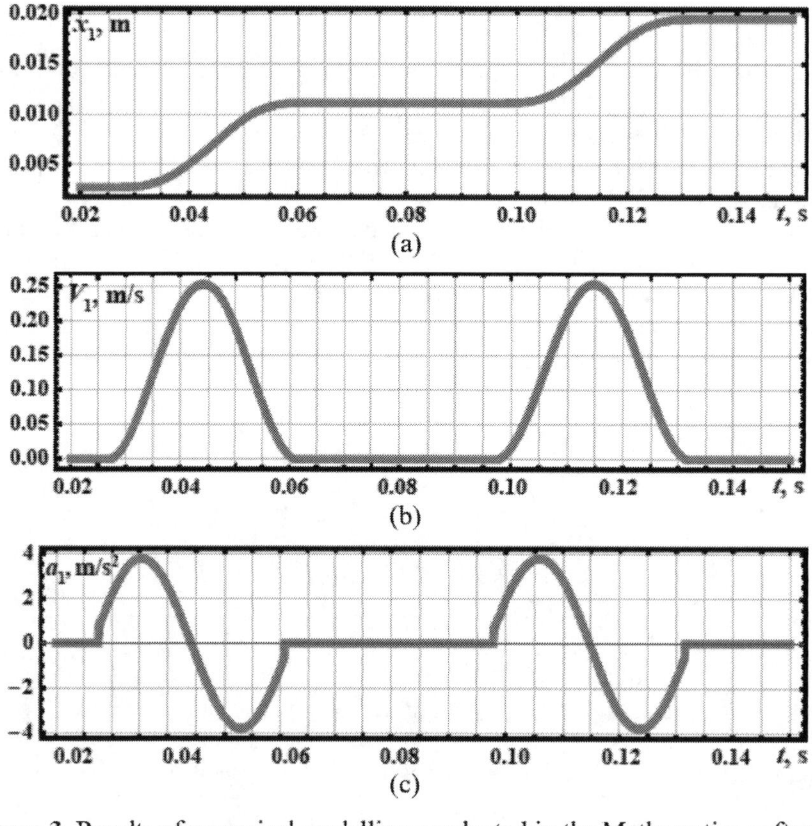

Figure 3. Results of numerical modelling conducted in the Mathematica software: a – displacement; b – velocity; c – acceleration.

The numerical modelling showed that the robot passed the distance of about 0.02 m during two cycles of its stepwise locomotion, i.e., during 0.15 s. The speed varies in the range of 0…0.26 m/s, whilst its average value equals 0.13 m/s. The acceleration changes from –4 m/s^2 to 4 m/s^2. It is observed that the robot speed does not take negative values. This means that the centrifugal force generated by the inertial exciter is smaller than the sliding friction force acting upon the robot's wheels. It is necessary to mention that the derived mathematical model does not take into account the rolling friction. This fact will cause some differences between the modelling and experimental results.

3.2. Simulation of the robot motion in the SolidWorks software
In order to verify the correctness of the numerical modelling results obtained in the Mathematica software, let us perform the simulation of the robot locomotion in the SolidWorks Motion software. The corresponding simulation model is presented in figure 4. The robot's geometrical and inertial characteristics, as well as the operational parameters, are similar to those adopted earlier. The robot can move to the right, whilst its leftward motion is functionally restricted. The friction forces are applied between each wheel and the supporting surface only in the case when the wheels attempt to move backward. The unbalanced rotor rotates at the forced frequency of 850 rpm in the clockwise direction. During the time interval of 0.3 s, the robot's body has passed the distance of about 40 mm. This allows for concluding that its average speed exceeds 130 mm/s. The maximal and minimal values of the robot translational speed are equal to 90 mm/s and 175 mm/s, respectively. The acceleration of the robot's body changes in the range of –3643…3643 mm/s^2. The simulation results satisfactorily agree with the ones obtained during the numerical modeling in the Mathematica software. The only difference is the velocity range. The simulation model took into account the inertia characteristics of the robot's wheels during their rolling, whilst the mathematical model considers only the sliding locomotion conditions.

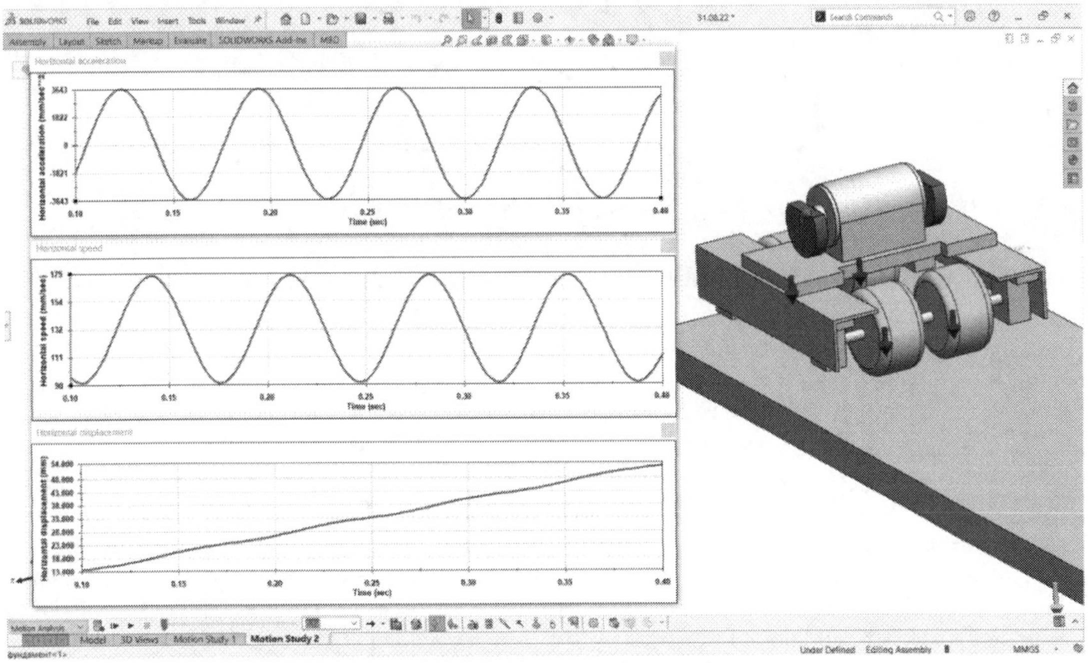

Figure 4. Simulation results obtained in the SolidWorks Motion software.

3.3. Experimental investigations of the robot motion
The experimental prototype of the wheeled vibration-driven robot has been implemented in practice at the Vibroengineering Laboratory of Lviv Polytechnic National University. It totally corresponds to the 3D-model designed in the SolidWorks software and described above (see figure 1). The general photo of all the equipment used for conducting the experimental studies on the robot locomotion conditions is presented in figure 5. In order to register the data obtained from the acceleration sensor 3 (WitMotion

WT901CL), the corresponding WitMotion software has been installed on the laptop 1. The accelerometer uses the USB Type-C connection interface and allows for registering the robot's acceleration at the retrieval rate of 200 measurements per second and the baud rate of 115200. The robot 2 moves along the flat horizontal surface 5 due to the application of the periodic disturbing force generated by the inertial exciter (unbalanced rotor). The forced frequency (the angular speed of the rotor) is changed by means of regulating the voltage supplied to the DC motor that actuates the unbalanced rotor. In this case, the controllable voltage adapter 4 (ZY-009) has been used to change the voltage from 0 V to 24 V and to regulate the robot translational speed. The corresponding relationships allowing for determining the forced frequency (the unbalanced rotor angular speed) at the given voltage value have been established by performing numerous starts of the DC electric motor.

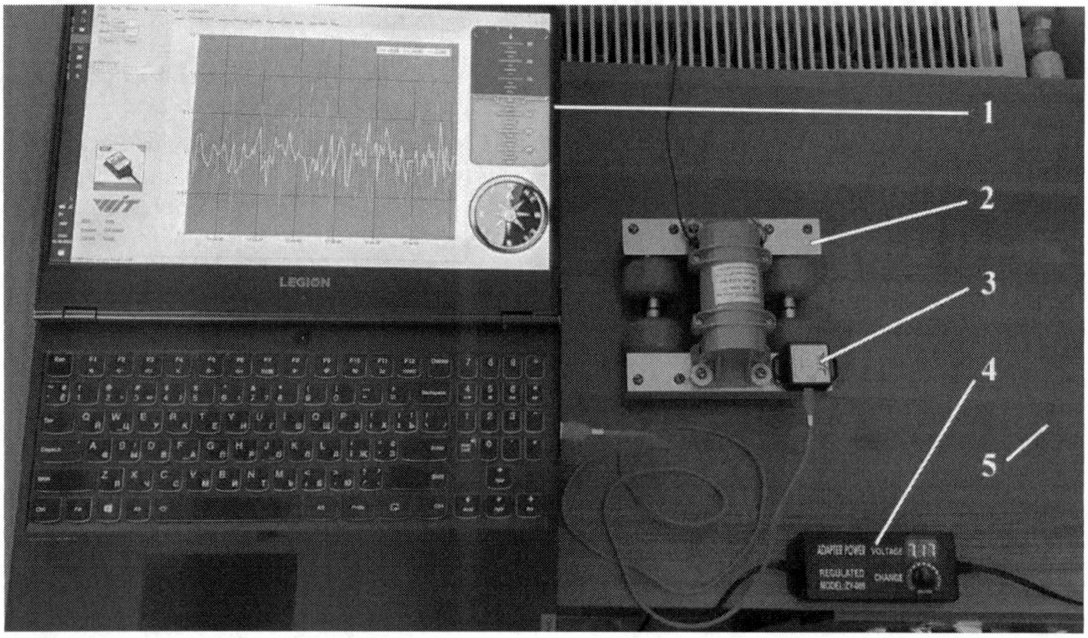

Figure 5. Experimental equipment used for investigating the robot motion conditions.

In order to perform experimental studies, following forced frequencies have been accepted: 48 rad/s (460 rpm), 73 rad/s (700 rpm), 103 rad/s (980 rpm), 167 rad/s (1600 rpm). The experimental results showing the time dependences of the robot's body accelerations at different forced frequencies are presented in figure 6 (a). The curves 1, 2, 3, 4 present the data obtained from the accelerometer and registered in the WitMotion software. The curves 5, 6, 7, 8 correspond to the first four ones and show the experimental data that have been numerically interpolated in the Mathematica software. The interpolated curves will be used for further numerical integration and determining the time dependencies of the robot speeds and displacements at different excitation parameters. Obviously, the larger values of accelerations and decelerations of about 16…20 m/s^2 are obtained at the largest forced frequency of 167 rad/s (1600 rpm) (curves 4 and 8 in figure 6 (a)). The smallest values that do not exceed 1…1.5 m/s^2 correspond to the smallest forced frequency of 48 rad/s (460 rpm) (curves 1 and 5 in figure 6 (a)).

The numerical interpolation and integration of the experimental data, i.e., the robot's body acceleration, were performed in the Mathematica software. This allowed for obtaining the time dependencies of the robot speed and displacement presented in figures 6 (b) and (c). Similarly to the acceleration, the largest speeds of about 0.2…0.25 m/s are observed at the forced frequency of 167 rad/s (1600 rpm) (curve 4 in figure 6 (b)). The smallest speeds do not exceed 0.02 m/s at $\omega = 48$ rad/s (460 rpm) (curve 1 in figure 6 (b)). Considering the time interval of 20…20.3 s, when the robot reached the steady-state locomotion conditions, it passed the distance of about 0.038 m at the largest forced frequency, and the distance of 0.003 m at the smallest one. Therefore, based on the experimental investigations, the robot's average speed varies in the range of 0.01…0.127 m/s.

Comparing the robot average speed of about 130 mm/s obtained by numerical modelling and computer simulation with the experimental results, where the average speed did not exceed 127 mm/s, the following conclusion must be drawn. The major influence on the obtained error is imposed by the fact that the rolling friction has been neglected both in the mathematical and simulation models. Therefore, the prospects of further investigations on the subject of this paper are to be focused on studying the effect of different friction types on the robot average speed and operational efficiency.

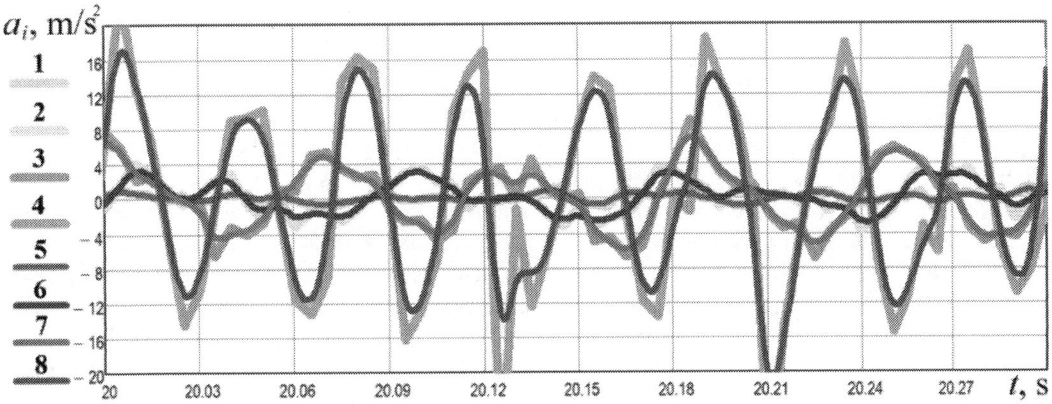

(a)

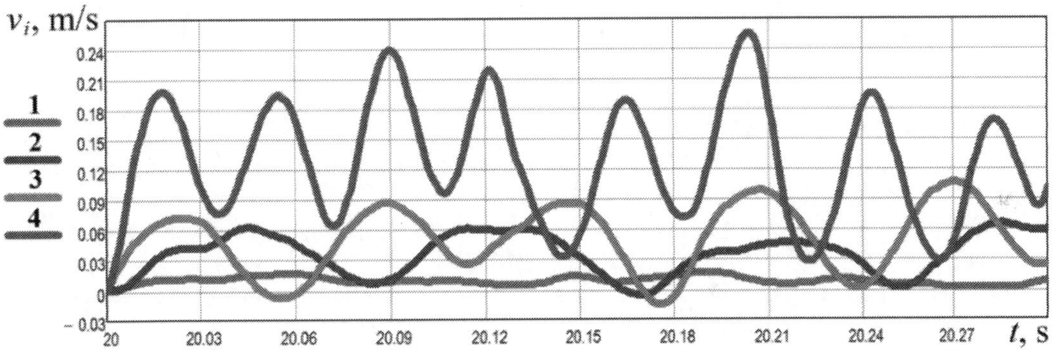

(b)

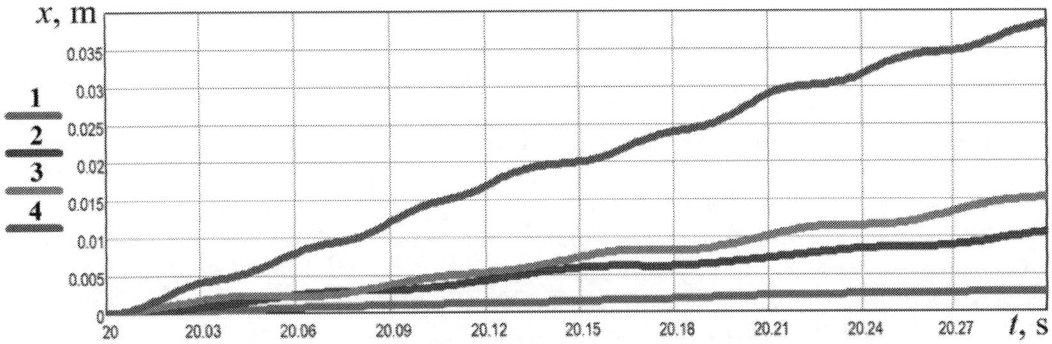

(c)

Figure 6. Results of experimental investigations: a – accelerations; b – velocities; c – displacements; 1, 5 – at ω = 48 rad/s (460 rpm); 2, 6 – at ω = 73 rad/s (700 rpm); 3, 7 – at ω = 103 rad/s (980 rpm); 4, 8 – at ω = 167 rad/s (1600 rpm).

4. Conclusions

The presented research is devoted to investigating the dynamic behavior and kinematic characteristics of the vibration-driven robot equipped with the inertial (centrifugal) exciter and unidirectionally rotating wheels. While conducting the studies, the general design (3D-model) of the wheeled robot, as well as its simulation model, have been developed in the SolidWorks software. The mathematical model that describes the robot locomotion has been derived on the basis of the D'Alembert's principle and numerically solved in the Mathematica software. The results of numerical modeling and computer simulation showed that the robot average locomotion speed is about 130 mm/s at the forced frequency of 89 rad/s (850 rpm). The robot's body acceleration changes in the approximate range of –4…4 m/s^2. The experimental prototype of the robot has been implemented in practice at the Vibroengineering Laboratory of Lviv Polytechnic National University. The obtained experimental data proved the correctness of the modelling and simulation results and conclusions about the robot stepwise locomotion conditions. The robot's average speed varied in the range of 0.01…0.127 m/s at different forced frequencies used during experimental tests. The main scientific novelty of this study resides in substantiating the relationships between the robot's design parameters and its dynamic characteristics under different operational conditions. The performed investigations can be useful for researchers and designers dealing with the vibration-driven robots, capsule-type locomotion systems, pipelines inspecting vehicles, etc. The prospects of further research on this paper's subject consist in substantiating the robot's inertial and excitation parameters in order to maximize its translational speed and minimize the power being consumed by its drive. In addition, it is necessary to improve the developed mathematical and simulation models by considering the friction forces acting upon the robot's wheels.

References

[1] Du Z, Fang H, Zhan X and Xu J 2018 Experiments on vibration-driven stick-slip locomotion: A sliding bifurcation perspective *Mech. Syst. Signal Process.* **105** 261–75
[2] Xu J and Fang H 2019 Improving performance: recent progress on vibration-driven locomotion systems *Nonlinear Dyn.* **98** 2651–69
[3] Notomista G, Mayya S, Mazumdar A, Hutchinson S and Egerstedt M 2019 A Study of a Class of Vibration-Driven Robots: Modeling, Analysis, Control and Design of the Brushbot *2019 IEEE/RSJ Int. Conf. Intell. Robot. Syst. (IROS), Macau, China* 5101–6
[4] Kim D, Hao Z, Mohazab A R and Ansari A 2020 On the forward and backward motion of milli-bristle-bots *Int. J. Non. Linear. Mech.* **127** 103551
[5] Choi D W, Lee C W, Lee D Y, Lee D W and Yoon H U 2020 A hybrid soft actuator inspired by grass-spike: Design approach, dynamic model, and applications *Appl. Sci.* **10** 1–15
[6] Robles-Cuenca D, Ruíz-Díez V, Sanchez-Rojas J and Hernando-Garcia J 2020 Sub-gram in-plane vibration-driven robot with inclined legs *Proceedings* **64** 8487
[7] Tallapragada P and Gandra C 2021 A mobile Mathieu oscillator model for vibrational locomotion of a Bristlebot *J. Mech. Robot.* **13** 054501
[8] Rodwell C and Tallapragada P 2022 Induced and tunable multistability due to nonholonomic constraints *Nonlinear Dyn.* **108** 2115–26
[9] Bolotnik N N, Nunuparov A M and Chashchukhin V G 2016 Capsule-type vibration-driven robot with an electromagnetic actuator and an opposing spring: Dynamics and control of motion *J. Comput. Syst. Sci. Int.* **55** 986–1000
[10] Nunuparov A, Becker F, Bolotnik N, Zeidis I and Zimmermann K 2019 Dynamics and motion control of a capsule robot with an opposing spring *Arch. Appl. Mech.* **89** 2193–208
[11] Li P and Jiang Z 2018 Bifurcation Analysis of Stick-Slip Motion of the Vibration-Driven System with Dry Friction *Math. Probl. Eng.* **2018** Article ID 2305187
[12] Diao B, Zhang X, Fang H and Xu J 2021 Bi-objective optimization for improving the locomotion performance of the vibration-driven robot *Arch. Appl. Mech.* **91** 2073–88
[13] Diao B, Zhang X, Fang H and Xu J 2022 Optimal control of the multi-module vibration-driven locomotion robot *J. Sound Vib.* **527** 116867

[14] Nguyen V Du and La N T 2020 An improvement of vibration-driven locomotion module for capsule robots *Mech. Based Des. Struct. Mach.* **49** 1–15
[15] Nguyen K-T, La N-T, Ho K-T, Ngo Q-H, Chu N-H and Nguyen V-D 2021 The effect of friction on the vibro-impact locomotion system: modeling and dynamic response *Meccanica* **56** 2121–37
[16] Duong T-H, Van C N, Ho K-T, La N-T, Ngo Q-H, Nguyen K-T, Hoang T-D, Chu N-H and Nguyen V-D 2022 Dynamic response of vibro-impact capsule moving on the inclined track and stochastic slope *Meccanica* 1–19
[17] Fedonyuk V and Tallapragada P 2017 Stick–Slip Motion of the Chaplygin Sleigh With a Piecewise-Smooth Nonholonomic Constraint *J. Comput. Nonlinear Dyn.* **12** 031021
[18] Fedonyuk V and Tallapragada P 2020 Locomotion of a Compliant Mechanism With Nonholonomic Constraints *J. Mech. Robot.* **12** 051006
[19] Liu D, Lu J, Cao Y and Jin X 2022 Dynamic characteristics of two-mass impact pipeline robot driven by non-circular gears *Adv. Mech. Eng.* **14** 1–13
[20] Loukanov I A and Stoyanov S P 2015 Experimental determination of dynamic characteristics of a vibration-driven robot *IOSR J. Mech. Civ. Eng.* **12** 62–73
[21] Loukanov I A, Vitliemov V G and Ivanov I V 2016 Dynamics of a mobile mechanical system with vibration propulsion (VibroBot) *Int. J. Res. Eng. Sci.* **4** 44–51
[22] Loukanov I A, Vitliemov V G and Ivanov I V 2016 Dynamics of a vibration-driven one-way moving wheeled robot *IOSR J. Mech. Civ. Eng.* **13** 14–22
[23] Korendiy V, Kachur O, Gursky V, Kotsiumbas O, Dmyterko P, Nikipchuk S and Danylo Y 2022 Motion simulation and impact gap verification of a wheeled vibration-driven robot for pipelines inspection *Vibroengineering Procedia* **41** 1–6
[24] Korendiy V, Kachur O, Gursky V, Gurey V, Pelio R and Kotsiumbas O 2022 Experimental investigation of kinematic characteristics of a wheeled vibration-driven robot *Vibroengineering Procedia* **43** 14–20
[25] Korendiy V, Kotsiumbas O, Borovets V, Gurey V and Predko R 2022 Mathematical modeling and computer simulation of the wheeled vibration-driven in-pipe robot motion *Vibroengineering Procedia* **44** 1–7

Modeling the sequence of part machining to minimize the time spent on the machining center, considering the working tool changing

Yevhen Matviishyn

Department of Regional and Local Development, Lviv Polytechnic National University, Lviv-Brukhovychi, 79-491, Ukraine

Author's e-mail: evhmat@yahoo.com.

Abstract. An algorithm for choosing a rational sequence for machining a part on a machine with a tool magazine is proposed using the drilling operations example. The algorithm is based on the analysis of all possible options and the calculation of the time spent on each of them. In this respect, the working tool (drill) change is considered, if it is necessary to switch to another machining type (drilling a different diameter hole). Three types of time costs are included in the calculation of the total machining time for one part: the duration of each operation, the duration of the transition to the next operation, and the duration of the working tool change. The result of the algorithm application is the sequence of machining the part with the least total time spent on transitions to the machining areas and the working tool changing.

1. Introduction

Modern machine-building enterprises are equipped with high-performance machines with computer numerical control. The so-called machining centers have become widespread. They are machines with many functions that can perform various programs. The use of such a technique allows not only to machine parts with high quality but also to set the desired sequence of operations that need to be performed. To reduce the time spent on parts machining, the user has the opportunity to plan a rational sequence of operations in advance. Appropriate planning requires preliminary calculations, which include options analysis. There are best and worst options in terms of time consumption. There is a need to apply an algorithm that facilitates the selection of the best option.

2. Methodology

Methods for optimizing the sequence of production operations are based on well-known theoretical principles, which are considered by the scientific and applied direction "Operations Research", as well as the so-called mathematical programming. There are several "typical" problems with available solution algorithms. Practical problems associated with machining operations in mechanical engineering can be reduced to the form of some of these problems. Particularly, the simulation of the sequence of the operations for the machining of a particular part can be performed, considering it as a Traveling Salesman Problem. In turn, this problem can be solved by several methods. Available approximate methods: "sweep algorithm" [1], "ant colony method" [2] or their combination [3], and others.

The Sweep Algorithm has long been used mainly for rational transport route planning. It divides the single terminal vehicle dispatch problem into two subproblems; one of assigning locations to routes and the other of minimizing the length of each route using a traveling salesman algorithm [4]. In general, this algorithm can be used for the task of planning the sequence of a part machining on a machine. However, its accuracy (closeness to the optimum) and efficiency (time spent on calculations) depend on the number of points on the route. The sweep algorithm generally produces results that are significantly better than those produced by Gaskell's savings approach [5] and are generally slightly better than Christofides and Eilon's results; however, the sweep algorithm is not as computationally efficient as Gaskell's and is slightly less so than Christofides and Eilon's [6]. This was stated by the algorithm developers themselves [7]. The advantages and disadvantages of several heuristic methods used to build a route between given points are detailed in the Rand's article [8]. These methods continue to be developed. A team of scientists proposed a Parallel Ring Method for solving a large-scale Traveling Salesman Problem (TSP), where a clustering algorithm produces a set of small TSP problems that can be executed in parallel to generate partial solutions [9]. In general, heuristic methods make it possible to obtain a solution, but not always an optimal one. Modern computing resources make it possible to apply exact methods for solving such problems. Scientists pay attention to the improvement of the corresponding algorithms. In particular, methods based on dynamic programming are common [10]. So, in the textbook [11] analytic methods for solving planning problems in continuous time are represented.

A special case of the practical task of planning the sequence of the operations for machining parts is the problem of finding a rational operations sequence using interchangeable tools on a machine of the "machining center" type. Such machines are equipped with a so-called tool magazine, which allows changing quickly (within a few seconds) the working tool for the next operation. Here, to determine a rational sequence of operations, it is necessary to consider not only the duration of the transition from one area part to another but also the duration of the tool change, if it is necessary for the next operation. There are special algorithms for planning the sequence of certain operation types. For example, Abbas and Megahed developed an algorithm for planning the drilling of holes spaced on the part evenly. This algorithm may be used for creating user-defined cycle and/or subroutines for drilling holes lying in a normal matrix, a staggered matrix, a single row, and/or column [12]. There are more often practice cases when the holes are spaced unevenly. An example of the practical problem described is the drilling of holes of different diameters in a part; this requires changing the drill to operate with a hole of a different diameter. For such a case Ivanov and Vavilkina offer to divide the holes into groups with the same diameters. Further, for each group, the planning of the sequence of the operations is performed as per the algorithm for solving the TSP. First, a rational sequence of drilling the holes of one diameter is planned, then another diameter, and so on [13]. But it may turn out to be more expedient, instead of drilling a distant hole of the same diameter, to change the working tool (drill) and machine a hole of a different diameter, located next to the previous one. This can be rational if the time required to change the tool (drill) is less than the time to move to another drilling location. On a large production scale, there can be significant savings in part machining time by optimizing the workflow to accommodate tool changes.

The article's purpose is to present an algorithm for planning a rational sequence of part machining operations, considering the possibility of multiple changes to the working tool.

3. Results and discussion

The initial data description for modeling the sequence of the operations using replaceable tools covers information about the duration of the working tool transitions from one area of the part machining to another, considering the necessary time to change the tool, as well as the time for each area machining. For clarity, an example of operations for drilling the holes of different diameters located in different areas on the part surface is considered (figure 1).

The numbering of areas (holes) on the part can be optional. It is needed to describe the initial data and the interpretation of the results. The transition duration from one area of part machining to another includes the time to change the position and the duration of the working tool changing (if the diameters of the previous and subsequent holes are different). Example data under consideration are given in table 1.

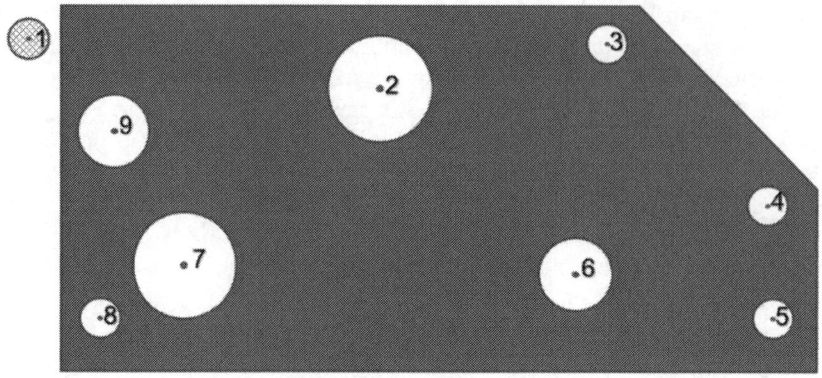

Figure 1. An example of the machining areas location (future holes) on the part surface (number 1 – the area of the working tool "parking" (base point); numbers 2-9 – areas where the holes will be with diameters of three sizes).

Table 1. Initial data – the transition duration between the areas of part machining (including changing the tool if necessary) and machining the holes of various diameters, in seconds.

From \ To	1 (base point)	2 (16 mm)	3 (8 mm)	4 (8 mm)	5 (8 mm)	6 (12 mm)	7 (16 mm)	8 (8 mm)	9 (12 mm)	Machining
1 (base point)	x	3,5	2,0	2,5	3,0	4,0	3,5	1,5	1,0	-
2 (16 mm)	3,5	x	3,5	4,0	4,5	3,5	1,5	4,5	4,0	15
3 (8 mm)	2,0	3,5	x	1,0	1,5	3,5	4,5	2,5	4,5	5
4 (8 mm)	2,5	4,0	1,0	x	0,5	2,5	4,5	3,0	5,0	5
5 (8 mm)	3,0	4,5	1,5	0,5	x	2,5	4,5	3,0	5,5	5
6 (12 mm)	4,0	3,5	3,5	2,5	2,5	x	3,5	4,5	1,5	10
7 (16 mm)	3,5	1,5	4,5	4,5	4,5	3,5	x	1,0	1,5	15
8 (8 mm)	1,5	4,5	2,5	3,0	3,0	4,5	1,0	x	2,5	5
9 (12 mm)	1,0	4,0	4,5	5,0	5,5	1,5	1,5	2,5	x	10

In the above example, the table is symmetrical concerning the diagonal ("x" cells), since the transition duration between the machining areas does not depend on the direction of the transition (for example, the transition duration from hole 3 to hole 5 is the same as from hole 5 to hole 3, it amounts 1.5 seconds). However, the transition duration between the two holes may differ. The proposed algorithm provides the possibility of calculations for such asymmetric data.

Simulation of the task modeling of planning a rational operations sequence is proposed to be performed as an algorithm-based for consideration of all possible options for the passage of a working tool. The first area is the base point (the tool "parking" area). In the given example, this is the item under the first number. The algorithm assumes that this is both the beginning of a transition for machining parts and the completion of the tool "route", which is necessary for the smooth installation of the next part to be machined. This approach (planning to start from the base point and return to it at the end of part machining) is an advantage of the proposed algorithm. After all, this considers the time spent on starting and completing machining.

The algorithm is developed in the VBA environment (Visual Basic for Applications) as a macro that is run in an Excel file containing initial data. In the same electronic table, the algorithm results are displayed – the optimal option of the part machining sequence from the point of view of the shortest total time spent on the start of work, transitions between areas of machining, including tool changes if necessary, and completion of machining. The optimal alternative is displayed as a list of numbers starting and ending with the number 1, and the other numbers correspond to the number of the part machining area (drilling a hole). The algorithm is presented as a program code part of a VBA macro (table 2).

Table 2. Algorithm for finding the optimal sequence of part machining.

Code	Explanation
Sub Alg()	
Dim Z, Zmin, A1(), Am(), L() As Currency	
Dim X(), Xopt(), NumberInList() As Integer	
Dim i, j, p, k, kk As Long	
maxN = 50	The largest number of machining areas.
n = 0	
For i = 1 To maxN	Definition of the item number in the general list of the database.
If Cells(5 + i, 26).Value <> 0 Then	
n = n + 1	
NumberInList(n) = i	
End If	
Next i	
For i = 1 To n	Downloading data on the transition duration.
For j = 1 To n	
L(i, j) = Cells(5 + NumberInList(i), 27 + NumberInList(j)).Value	
Next j	
Next i	
p = 1 : k = n	Counter of possible combinations.
For i = 2 To k	
Xopt(i) = 0	
A1(i) = 1 : X(i) = 1 : Am(i) = n ' Am(i)	
p = p * (Am(i))	
Next i	
Cells(2, 4) = p	
Zmin = 999999 : kk = 0	
3:	Block of options feasibility, calculation, and memorization of the current optimum.
For i = 2 To n	
For j = 1 To i - 1	
If X(i) = X(j) Then GoTo 2	Condition: if the combination includes different numbers.
End If	
Next j	
Next i	
Z = 0	
For i = 1 To n - 1	
Z = Z + L(X(i), X(i + 1))	
Next i	
Z = Z + L(X(n), X(1))	The time to return to the starting point is added.
If Z <= Zmin Then	

```
Zmin = Z
kk = kk + 1 : Cells(17 + kk, 1) = kk          Results with current optima are displayed.
 For j = 1 To n
 Cells(17 + kk, j + 2) = X(j)
 Xopt(j) = X(j)                               Remember the optimal route.
 Next j
 Cells(17 + kk, 2) = Zmin
 Cells(17 + kk, 2).Select                     The active cell with the current result.
End If
2:
q = 0
For j = 1 To n
 If X(j) = Am(j) Then
 q = q + 1
 End If
 If q = n Then                                All options are considered.
 GoTo Endd
 End If
Next j
End If
End If
X(1) = X(1) + 1
If (Not X(1) > Am(1)) Then
GoTo 3
End If
X(1) = A1(1) : i = 1
4:
If i = n Then
GoTo 3
End If
i = i + 1 : X(i) = X(i) + 1
If Not X(i) > Am(i) Then
GoTo 3
End If
X(i) = A1(i)
GoTo 4
Endd:
End Sub
```

The algorithm result is displayed on the screen (figure 2) as a table with a list of areas where the working tool passes, starting from 1 and ending with 1 (the area where the working tool is “parked”). For each area, the duration of this hole machining is indicated, and between the points – the transition duration. For convenience, the physical time of operations passage is also given.

	A	B	C–D	E	F	G	H
34			in seconds				Time
35			Transitions	Machining	Total		
36	1	base point					9:20:00 AM
37			1,5				
38	8	hole 8 (8 mm)		5,0	6,5		9:20:06 AM
39			1,0				
40	7	hole 7 (16 mm)		15,0	22,5		9:20:22 AM
41			1,5				
42	2	hole 2 (16 mm)		15,0	39,0		9:20:39 AM
43			3,5				
44	3	hole 3 (8 mm)		5,0	47,5		9:20:48 AM
45			1,0				
46	4	hole 4 (8 mm)		5,0	53,5		9:20:54 AM
47			0,5				
48	5	hole 5 (8 mm)		5,0	59,0		9:20:59 AM
49			2,5				
50	6	hole 6 (12 mm)		10,0	71,5		9:21:12 AM
51			1,5				
52	9	hole 9 (12 mm)		10,0	83,0		9:21:23 AM
53			1,0				
54	1	base point					9:21:24 AM
55							
56		Time spent on transitions	14,0 s				
57		and the tool changing					

Figure 2. Screenshot with the algorithm result.

On the graphic interpretation of the obtained optimal alternative of the part machining sequence (figure 3), the order of the working tool passage is displayed with arrows.

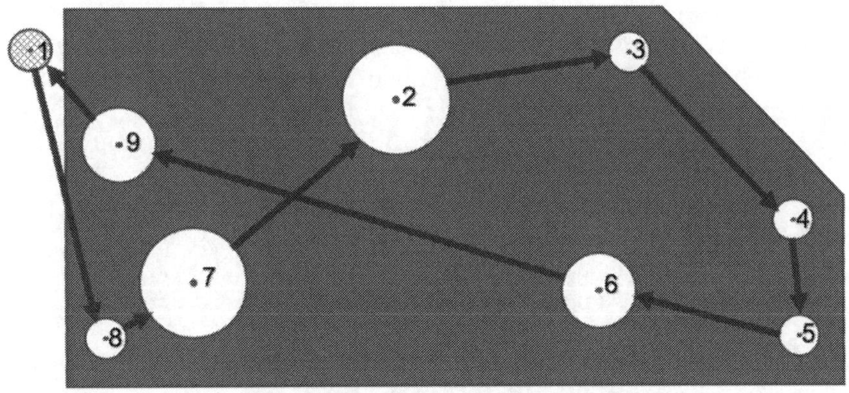

Figure 3. Optimal sequence of part machining

In conditions of symmetry (when the transition duration between two areas (holes) is the same in the forward and reverse directions), the return route will also be optimal. So, for the given example, the second optimal option of the part machining sequence is 1-9-6-5-4-3-2-7-8-1.

An additional advantage of the proposed algorithm is the consideration of the drilling time (the working tool standing in the given area). It allows calculating the total duration of a part machining in a "machining center", which is important for planning the production process of a machine-building enterprise.

The number of possible options for the part machining sequence (drilling holes) depends on the number of machining locations. The power of modern computing resources makes it possible to obtain the optimal alternative relatively quickly. For example, for the given example with 8 machining areas (holes), the calculation takes about 2 seconds. However, for the option with 12 machining areas, the computation time is already about 5 minutes on a computer with a 2.5 GHz processor. For parts with a large number of machining areas, it is advisable to use heuristic methods for modeling the sequence of their machining. In the developed VBA macro in the Excel environment for cases when the number of machining areas exceeds 12, the available Little's Algorithm is used [14]. This allows to quickly get a rational (though not necessarily optimal) option of the part machining sequence.

4. Conclusion

The proposed algorithm provides a simulation of all possible options of the part machining sequence and the selection of the optimal alternative with the shortest machining time. Not only the duration of position changes is considered, but also the time spent on changing the tool, if necessary. The result of the algorithm's work reflects the sequence of numbers of areas where the part is machined, as well as the transition duration between them and the time spent on machining. The algorithm limitation is the maximum number of machining areas, which leads to a significant increase in the time spent on calculations; in this case, a heuristic algorithm for finding a rational sequence for machining the part is applied. Implementation of both algorithms in the form of a VBA macro in the Excel allows the practical application of the developed software.

References

[1] Nurcahyo G W, Alias R A, Shamsuddin S M and Sap M N 2002 Sweep algorithm in vehicle routing problem for public transport *Jurnal Antarabangsa (Teknologi Maklumat)* **2** 51-64

[2] Dorigo M and Birattari M 2010 Ant colony optimization. *Encyclopedia of Machine Learning* (Springer) 36-39

[3] Yousefikhoshbakht M and Sedighpour M 2012 A combination of sweep algorithm and elite ant colony optimization for solving the multiple traveling salesman problem *Proceedings of the Romanian Academy. Series A: Mathematics, Physics, Technical Sciences, Information Science* **13** 295-301

[4] Gillet B E, Miller L E and Johnson J G 1979 Vehicle dispatching – sweep algorithm and extensions. *Disaggregation* (Springer, Dordrecht) 471-483

[5] Gaskell T J 1967 Bases for vehicle fleet scheduling *Operations Research Quaterly* **18** 281-295

[6] Christofides N and Eilon S 1969 An algorithm for the vehicle routing dispatching problem *Operations Research Quaterly* **20** 309-318

[7] Gillet B E and Miller L E 1974 A heuristic algorithm for the vehicle-dispatch problem *Operations Research* **22** 340-349

[8] Rand G K 2009 The life and times of the savings method for vehicle routing problems *ORiON* **25** 125-145

[9] Bazylevych R, Kuz B, Kutelmakh R, Dupas R, Prasad B, Haxhimusa Y and Bazylevych L 2016 A parallel ring method for solving a large-scale traveling salesman problem *International Journal of Information Technology and Computer Science* **5** 1-12

[10] Datta S 2020 Traveling salesman problem – dynamic programming approach *Baeldung* **2020**

[11] Kamien N I and Schwartz N L 1991 Dynamic optimization: the calculus of variations and optimal control in economics and management *Advanced Textbooks in Economics* (Elsevier Science B. V., Amsterdam) **31**

[12] Abbas A and Megahed S A general algorithm for drilling holes lying in a matrix *Robotics and Computer-Integrated Manufacturing* **21** 235-239

[13] Ivanov E I and Vavilkina V V 2015 Optimization processing of the elementary surfaces of the composite body parts on machines such as «machining center» *Reporter of the Priazovskyi State Technical University. Section: Technical sciences* **30** 81-85

[14] Little J D C, Murty K G, Sweeney D W, Karel C 1963 An algorithm for the traveling salesman problem *Operations Research* **11** 972-989

Methodology of technological design in the restoration of parts

Alexander Permyakov[1*], Yakiv Nemyrovskyi[2], Eduard Posviatenko[3] and Ihor Shepelenko[2*]

[1] National Technical University «Kharkiv Polytechnic Institute», 2 Kyrpychova Street, Kharkiv, 61002, Ukraine
[2] Central Ukrainian National Technical University, 7 Universytetskyi Avenue, Kropyvnytskyi, 25006, Ukraine
[3] National Transport University, 1 Suvorova Street, Kiev, 02000, Ukraine

***E-mail:** perm_a@i.ua, kntucpfzk@gmail.com

Abstract. In the work, from the standpoint of a systematic approach, a methodology for technological design of repair and restoration of worn parts using methods to improve the quality of the working layer is proposed. For the stages of the product life cycle, solving the problem of extending the service life after repair and restoration of worn parts, a general scheme and structure of information support for technological design were developed using restoration methods that ensure and improve the quality of the working layer. A variant of the technological process for the restoration of cylinder liners from modified gray cast iron by a combination of surface plastic deformation (SPD) and finishing antifriction non-abrasive treatment (FANT) is proposed.

1. Introduction
The authors of the article, each having their own scientific experience and accumulated knowledge of the subject area, set out to generalize them from the standpoint of a systematic approach and develop a methodology for technological design in repair and restoration of worn parts processes using methods to improve the quality of the working layer.
Since, by definition, methodology is the doctrine of the structure and logical organization of activities, the task was to describe a system of organizing principles and methods, building theoretical and practical activities for the technological design in repair and restoration of worn parts processes using methods to improve the quality of the working layer.

2. Literature Review
The life cycle of engineering products from the inspiring idea to their creation to disposal is well known (Fig. 1), and each stage has been sufficiently studied and described in the scientific and technical literature.

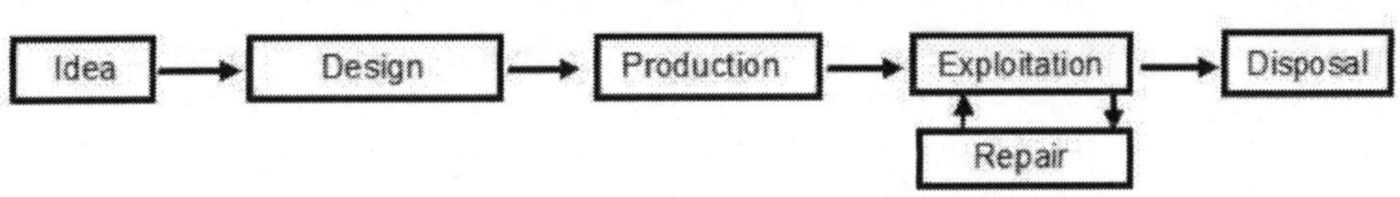

Figure 1. Life cycle of an engineering product.

This general structure can be refined and supplemented with a detailed description of the individual stages and the tasks solved at them, the study of which provides new knowledge and experience that require systematization, generalization and development of recommendations for practical use. Fig. 2 proposes such a variant of the refined structure of the life cycle for the stages that solve the problem of the product service life extending during the repair and restoration of worn parts.

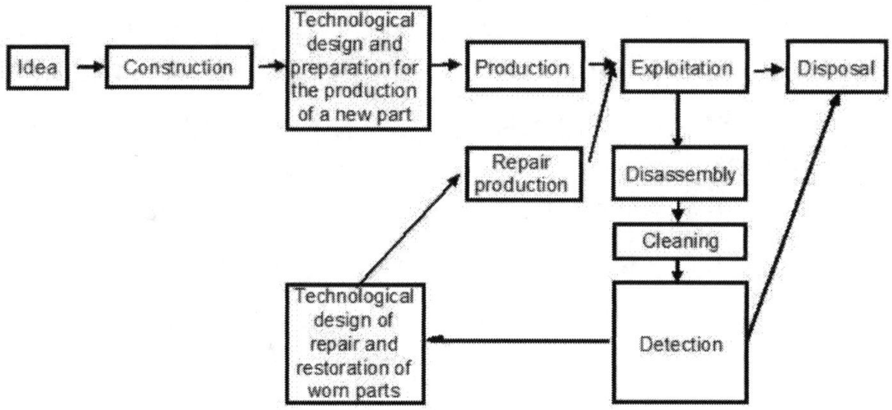

Figure 2. The product life cycle with the specification of extending the service life stages after repair and restoration of worn parts.

Over a long period of mechanical engineering technology development as a science, general principles and recommendations, methodology and methods for designing technological processes for manufacturing and assembling machines have been developed, implemented and described, taking into account the whole variety of production types (single, serial, mass), types of production processes organization (in-line, non-in-line), industry specifics etc.

Development of route, operational, group, typical, etc. technological processes are devoted to scientific works of famous scientists and there is a lot of educational literature [1-2]. In the general case, the tasks that are solved during the production organization, as well as the sequence of their implementation and the main elements of the initial and basic information are presented on the structural-logical scheme of information support for technological design and the production process of machine parts (Fig. 3).

When analyzing the process of technological design and the actual process of manufacturing a new part, one important feature can be noted. In the production process, the workpiece (real object), in accordance with the processing route, sequentially undergoes technological operations in accordance with the principle of successive approximation to the accuracy and quality of the processed surfaces. As a result of removing the total allowance, the final part is formed, which must correspond in shape, size and physical-mechanical parameters of the surfaces to the working drawing, which is established at the stage of technical control. In the process of technological design, the sequence is different. The initial information is the working drawing of the part (virtual object) and the given release program. In the course of technological design based on comprehensive knowledge about processing methods, theories of accuracy, basing, productivity, cutting, etc. processing route is formed. During the development of the operational technological process, basing schemes are determined, operational allowances, cutting conditions are calculated, and technical normalization is carried out. Only after determining the allowances and choosing from the possible options for the method of obtaining the workpiece, its working drawing (virtual object) can be obtained. In the course of technological design, a cutting tool, technological equipment and measurement means are selected and designed, the composition of technological equipment and transport-storage systems is selected, and the necessary technological documentation is drawn up.

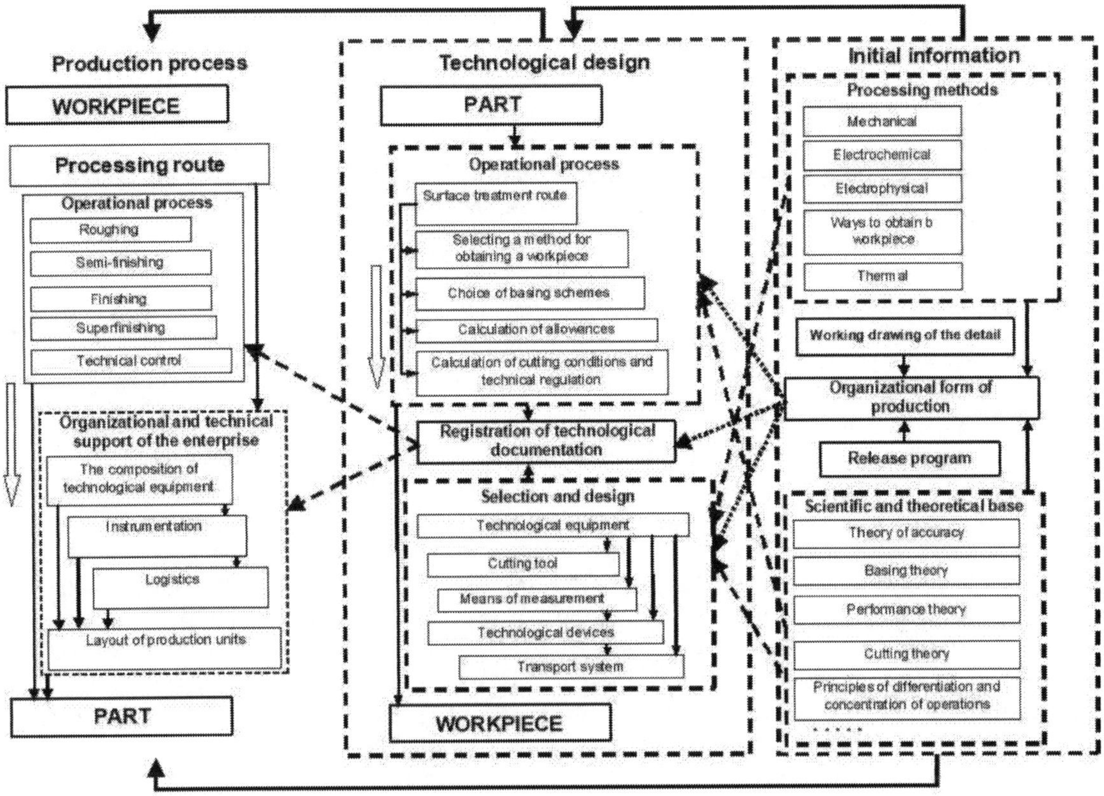

Figure 3. The sequence of production technological preparation.

Operation as a stage in a part life cycle is terminated after reaching the maximum allowable wear, leading to the cessation of the entire mechanism operation. To extend the life cycle of worn parts, technologies for their restoration are used [3]. These technologies are fundamentally different from the technologies for manufacturing a new part, although the end product of production in both cases is the same finished part with its own service functions.

Consider the technological design of worn parts restoration, which begins with the preparatory operations for disassembling a mechanism or assembly after the end of its operation and cleaning the worn parts.

Further, the worn parts are sent to the defect detection operation, at which the wear values of the working surface and the presence of the parts destruction that do not allow their restoration are determined. In the process of defect detection, parts are sorted into three groups [4]:

- suitable for operation;
- requiring repair or restoration;
- unusable, to be replaced.

Details of the first group do not require repair, the third – are rejected. For parts of the second group, it is necessary to assign a rational technological process of repair and restoration.

The works of V.A. Schadrychev [5], V.V. Efremov [6], M.A. Masyno [7], S.O. Luzan [8] and others are devoted to solving the issue of choosing a rational way of restoring parts. In the work [9], three main methods of choosing the technological process of part restoration are highlighted, which vary in different degrees of consideration in technological and economic indicators. The first method is based on calculating the cost of restoring the part and comparing it with the price of a new part. According to the second method, complex values are compared among themselves in the ratio form of cost and price to the resource of the part – new and restored. The third method takes into account the cost and durability of new and restored part.

V.A. Schadrychev proposed partial criteria for evaluating methods of parts restoration: applicability, durability and economy [5]. M.A. Masyno developed a methodology for determining the applicability of methods to restore parts with the determination of the numerical values of the applicability coefficient for specific groups of parts [7]. A.N. Batyschev introduced a combined criterion that reflects the energy intensity, labor intensity, given costs and durability of the part [10].

The authors [11] proposed to evaluate methods of restoration from the point of quality, using structural and technological characteristics of parts. On the basis of these characteristics, possible recovery methods and their specific indicators are determined. The considered approaches to the choice of machine parts restoration method have found their wide application in repair production and can be successfully used in relation to the restoration of mechanical engineering parts.

About two hundred methods of surface engineering are known, which are effectively used in the restoration and strengthening of machine parts. It is advisable to classify these methods as follows: application of coatings, modification of the surface layer, technological and combined (hybrid) methods. At this stage of mechanical engineering development, various methods of applying coatings, which can be wear-resistant, anti-friction, corrosion-resistant, heat-resistant, etc., deserve the greatest attention of researchers [12]. Such coatings can be obtained by gas-thermal and mechano-thermal spraying, vacuum-condensation technologies, surfacing, galvanic and chemical deposition, enameling, plating, cladding, hot coating with metals, application of solid lubricants [13]. Having a number of advantages (thickness of 0.005÷5 mm; high level of physical-mechanical properties; worked out equipment and technological equipment), the coatings are still not sufficiently reliably held on the base, require finishing mechanical processing, and create a large gradient of harmful residual stresses. Modification of the surface layer, which includes surface thermal and chemical-thermal treatment and processes of cold plastic deformation (CPD), is free from the mentioned disadvantages, since structural transformations, diffusion processes and changes in the density of dislocations in the base material serve as a means of influencing the properties of the metal. The disadvantage of modification is the difficulty of controlling the dimensions of the worn parts. Technological methods involve influence on the surface layer of the part in order to change its properties by cutting or related processes. This group includes most methods of obtaining regular macro- and microreliefs. Hybrid methods involve obtaining a new effect from the influence on a part by two or more disparate processes belonging to one or different groups of surface engineering [14].

The most effective processes of surface engineering of machine parts, both in the main and in secondary (repair and restoration) production, are hybrid technologies. Special attention should be paid to the methods of finishing parts by CPD, among which deforming broaching should be singled out. The use of hybrid technologies based on deforming broaching allows combining the advantages of various methods, including surface modification and coating with the achievement of higher operational properties of parts, requiring further research. An example of such a technology creation is the technological process of processing ICE liners developed by the authors [15], which includes the operations of deforming broaching and finishing anti-friction non-abrasive treatment (FANT), which ensures an increase in the quality of the liner working surface with improved physical-mechanical and tribological characteristics.

3. Results

By analogy with the scheme in Fig. 3, a structural-logical scheme of information support for technological design and the process of repair and restoration of worn parts using SPD is proposed (Fig. 4).

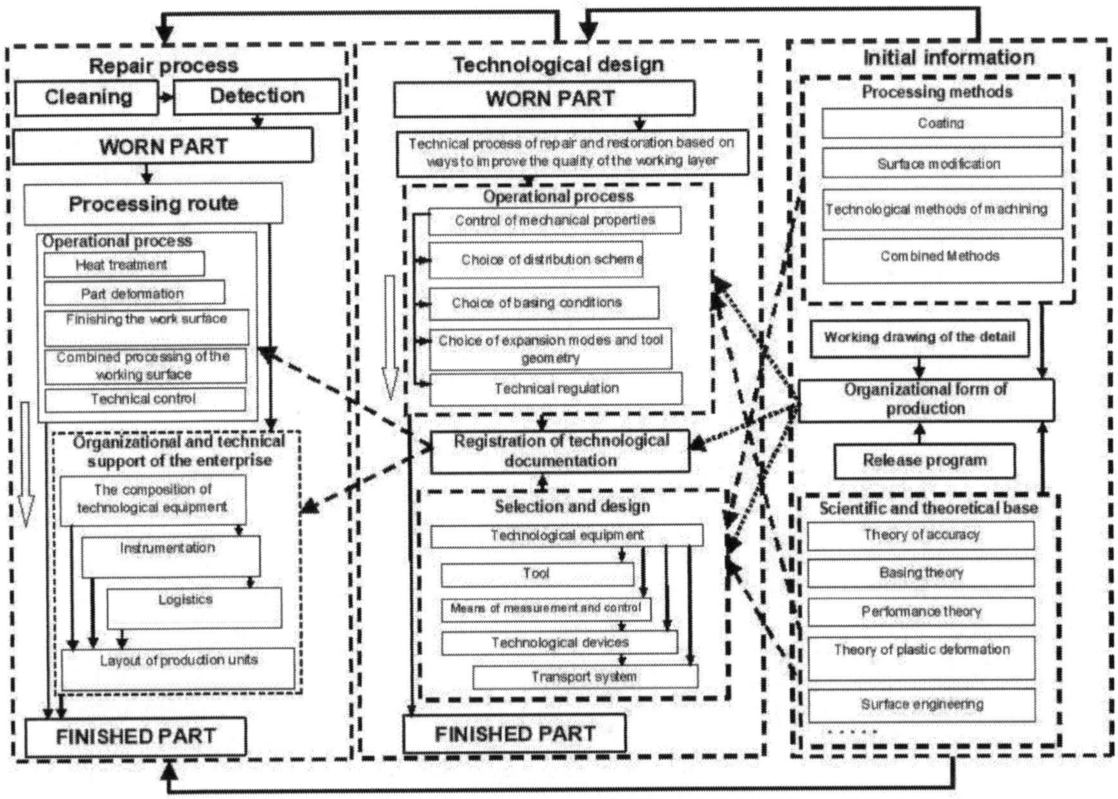

Figure 4. The sequence of technological preparation for the repair production of parts using SPD.

Comparison of two schemes (Fig. 3 and 4) shows the difference and common positions in the organization of the restoration technological process and the process of manufacturing a new part and serves as the basis for constructing the restoration process of worn parts.

At the Institute of Superhard Materials named after V.N. Bakul of the National Academy of Sciences of Ukraine, over the past few decades, deep studies of deforming and combined broaching process have been carried out, which made it possible to obtain important scientific results on the mechanics of forming parts and deformation of the working layer [16]. They formed the basis of combined technologies creation that make it possible to ensure the accuracy of the processed hole by parameter of ellipticity within ± 0.02 mm, and by curvature of the generatrix of 0.15–0.3 mm per 1000 mm of part length. The height parameters of roughness are provided within Ra = 0.05–0.15 µm, and the supporting surface can be 0.85–0.95%.

The conducted studies have shown that depending on the broaching modes, tool geometry, the degree of the surface layer hardening reaches 30–300%, and the hardening depth is from 0.3 to 1 mm. This allows creating new engineering methods of working layer that improve the operational characteristics of the processed part. These methods are summarized in a scheme illustrating the ways to solve this problem, which made it possible to use these advantages when creating combined technologies based on broaching, and made it possible to improve the operational characteristics of processed products (Fig. 5).

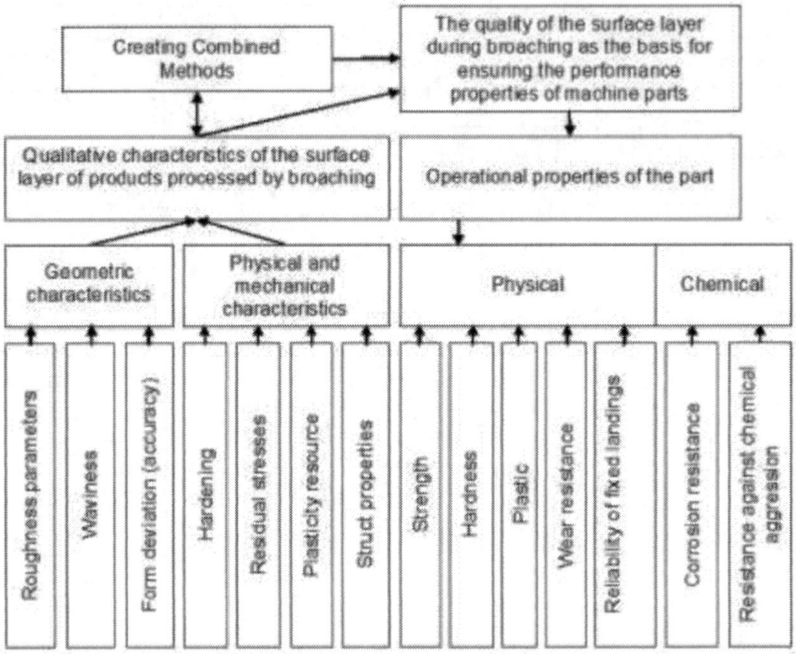

Figure 5. Scheme of creating new technological methods of surface engineering based on broaching.

The use of this scheme makes it possible to develop a technological process for restoring the working surface of ICE worn liners of ZIL brand engines, made of modified graphite-containing cast iron SCH20, to the first repair size (Fig. 6).

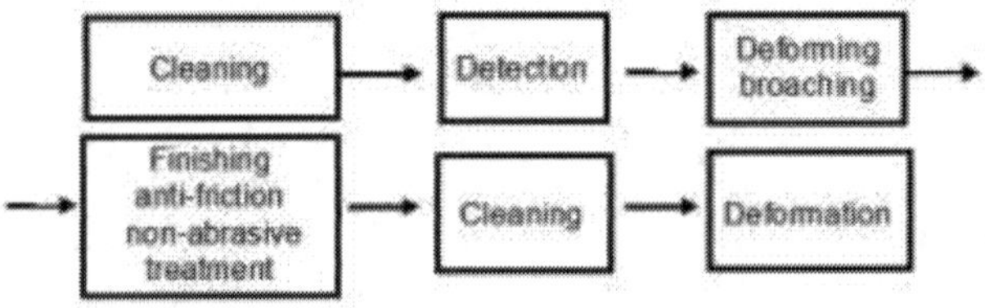

Figure 6. Scheme of the technological process of ICE cylinder liners restoration in the 1st repair size.

The deforming broaching operation allows strengthening the surface layer of the liner material. The depth of strengthening at the same time reaches about 0.3 mm, which guarantees the presence of strengthened material in the friction pair even during long-term operation. Applying an anti-friction coating ensures an increase in the anti-friction properties of the liner working surface.

Thus, the combination of SPD and FANT made it possible to develop a variant of the technological process of processing ICE cylinder liner inner surface made from gray modified cast iron, which is guaranteed to improve the quality of the working layer.

Conclusions

From the standpoint of a systemic approach, the task of summarizing information and developing a methodology for technological design in repair and restoration of worn parts processes using methods to improve the quality of the working layer was solved. The conducted research made it possible to draw the following conclusions:

1. The structure of the product life cycle has been refined and supplemented for the stages that solve the problem of extending the service life after repair and restoration of worn parts.

2. A general scheme and structure of information support for the technological design of worn parts repair using various recovery methods that ensure and improve the quality of the working layer have been developed.
3. Fundamental differences between the schemes of the technological process of manufacturing and restoring the part are established. In the manufacture of the final part, according to its technical requirements, it must correspond to its working drawing. The development of restoration technology is associated with an increase in the quality of the part surface layer and its operational properties.
4. On the basis of the developed scheme, a variant of the technological design of the processing process during the repair and restoration of internal combustion engines worn cylinder liners internal surfaces using SPD by the method of deforming and combined broaching is considered. The use of the combined technology made it possible to improve the quality of the processed part working layer due to the application of an anti-friction coating and its further hardening.

References

[1] Matalin A.A. Engineering technology. Leningrad: Mechanical engineering; 2008. 510 p.
[2] Rudenko P.O. Design of technological processes in mechanical engineering. Kyiv: High School; 2003. 420 p.
[3] Shepelenko Igor, Arifa Warouma, Sherkun Vitalii. Restoration of bronze bushes by the method of surface plastic deformation. International Journal of Engineering & Technology, 2016; 5(1): 29–32. https://www.sciencepubco.com/index.php/ijet/article/view/5651
[4] Sidashenko O.I. Machine repair. Kyiv, Harvest; 2011. 430 p.
[5] Shadrychev V.A. Fundamentals of automotive technology and car repair. Leningrad: Mechanical engineering; 1986. 560 p.
[6] Efremov V.V. Car repair. Moscow: Transport, 1995. 315 p.
[7] Masyno M.A. Increasing the durability of automotive parts during repair. Moscow: Transport, 1992. 148 p.
[8] Luzan S.A. Criteria for choosing a method of restoring machine parts and determining a rational technology route. Bulletin of Kharkiv National Technical University of Agriculture named after Petro P. Vasylenko, 2017; 183: 45–56.
[9] Ivanov V.P. Selection a method for restoring parts. Science and technology, 2016; 1: 9–17.
[10] Batyshchev A.N. Methodology for substantiating a rational way to restore worn parts. Collection of abstracts and reports of the scientific and technical conference "Modern equipment and technological processes for the restoration and hardening of machine parts", 1988; 1: 23–24.
[11] Belyaev A.I., Terentiev A.V., Pushkarev A.E. Evaluation of methods for restoring parts of road-building machines in a multicriteria setting. Bulletin of Civil Engineers, 2018; 1 (66): 123–127.
[12] Chernovol M.I., Shepelenko I.V. Methods of forming antifriction coatings on metal friction surfaces. Kirovogr Natl Tech Univ Sci Work Collect. 2012; 25(1): 3–8.
[13] Shepelenko Ihor, Tsekhanov Yuri, Nemyrovskyi Yakov, Posviatenko Eduard. Improving the Efficiency of Antifriction Coatings by Means of Finishing the Antifriction Non-abrasive Treatment. 2020; 289–298. http://link.springer.com/10.1007/978-3-030-40724-7_30
[14] Posvyatenko E.K. Hybrid processes of machine parts surface engineering. Proceedings of the international scientific and technical conference "Surface Engineering and Product Renovation". 2008; 2: 195–198.
[15] Nemyrovskyi Yakiv, Shepelenko Ihor, Osin Ruslan, Posviatenko Eduard. Improving the processing quality of cylinder liners using combined technology. Cutting and Tools in Technological Systems. 2022; 96: 121–130. http://rits.khpi.edu.ua/article/view/252506
[16] Posvyatenko E.K., Nemyrovskyi Ya.B., Sheykin S.E., Chernyavskyi O.V., Shepelenko I.V. Engineering of the parts processed by broaching. Kropyvnytskyi: Publisher Lysenko V.F., 2021. 466 p.

Prediction of strength of multi-support overground segment of main pipeline

A R Dzyubyk [1, 3], **L W Dzyubyk** [2]

[1] Department of Oil and Gas Engineering and Welding, Lviv Polytechnic National University, 12 S. Bandera Street, Lviv, 79013, Ukraine
[2] Department of Technical Mechanics and Dynamics of Machines, Lviv Polytechnic National University, 12 S. Bandera Street, Lviv, 79013, Ukraine
[3] E-mail: andrii.r.dziubyk@lpnu.ua

Abstract. A problem of strength of a main pipeline in its zone of rectilinear overground passage through an obstacle is considered in the given case. In the given case. It is not the overground multi-support segment alone that is included in the calculation, the adjacent underground parts which are situated at a certain distance from the ground are also taken into account here. The pipeline is described by a beam model with piece-wise constant rigidity, i. e. each segment of the pipeline i of constant rigidity. This allows us to more accurately characterize different thicknesses of pipes of the overground segment of the main pipeline in its construction. There is also assumed that the overground segment of the pipeline is freely positioned and rests on the corresponding supporting elements. The latest can be mutually adjusted for the given altitude. This allows us to evaluate the strength in repairing lifting the pipe over supports. The supporting elements can also experience elastic deformations. Their case is the mechanical compliance of component parts of the supporting assembly and, especially that of the ground under the support. The entrance and the exit of the overground segment into the ground are also considered as supporting elements. Here, they can have vertical displacements and turns. They also take into account the weight of the ground over the pipeline. The calculation of a main oil pipeline with its overground segment resting on two supports has been carried out. There are taken real conditions of operational loading caused to be the following factors: internal pressure, temperature difference, and the weight of the product transported. It has been established that the maximal values of stresses are observed in support sections and between the supports. It is shown that the stresses caused by bending moment can amount to up to half the total value. The influence of altitudes upon the value caused by bending moment stresses in the pipeline is investigated. The change of stresses caused by mechanical compliance of elements of the supports and by adjusting vertical displacements of the supports during repair was studied.

1. Introduction

The erection of overground segments of main pipelines is carried out in the case when it is impossible to implement the underground design of the pipeline, i. e. if there are natural or technological obstacles. This is a more complicated process that calls for the implementation of special designs of construction [1 – 3]. Therefore it is important to have maximally complete information on stresses in the pipeline not only just after its construction, but also in its subsequent operation for a long period [2 – 4]. This is caused by the fact that the pipes of the overground segment which are mounted on supports call for in-detail taking into account the factors which determine their strength in general. There should be included calculation of hot only operating loadings and probable influence of external medium that is characteristic of design. Under complicated circumstances of operation there often emerge unpredicted loads and deviations from the design scheme [2, 4, 5]. In particular, these are shifts of ground in mountainous land, technological influence, etc. As result, they take place redistribution of acting loads and the danger of deformation of the pipeline or destruction of the support element (Figure 1) [2, 4, 7 – 9]. It is important to take into account the mechanical compliance of support elements not only the overground segments, but also the mechanical compliance of the ground in the places of transition of the pipeline into its underground part [5, 10]. This can lead to unpredicted increase in the span and to additional load to material of the pipes. However, in special literature the issue of complex taking into account the mechanical compliance of all the supports of the pipeline is not elucidated completely.

a

b

Figure 1. Damages of overground segment: a) plastic deformation of the pipeline, b) destruction of the mounting group of the pipeline support [2].

Particular attention should be paid to consideration of the issue of adjusting vertical displacements of the supports and that of the pipeline axis in general during repairing-restoring works, etc. [1, 2, 11]. Here often exists the problem of local elevation of overground segment of pipeline by a certain value. With this, the rest of the overground segment is not displaced, and the process of operation is not stopped [11]. These factors lead to unpredicted deviation of the acting loads, bending moments, and shearing forces in the cylindric structure of the overground segment from the rated values. In the works [4, 6], there are presented requirements for the means of mathematics as well as the characteristics of the model that enables us to take into account the aforesaid factors. However, the issue of setting the total value of the vertical displacement of the pipeline with taking into account the mechanical compliance of the support elements calls for additional study. Besides, the problem of change in acting loadings for the case of elastically adjustable displacements of supports of overground segment of the main pipeline is not presented completely.

2. Mathematical model of the investigation

For the mathematical representation of the overground segment which rests on supports we use the approach suggested in the works [4, 6]. There, it is accepted to consider a static beam of piece-wise constant rigidity. Since the structure can have different positions of its supports relative to the center, it is assumed that the overground segment necessarily contains left and right underground parts as well as central overground part that rests on point supports (Figure 2). We assume that the origin of the coordinate system xOy is located on the left at the beginning of the overground segment. With this, the positive directions of the axes are the following: Ox – axis is directed to the right along the central axis of the pipeline, and Oy – axis is downward.

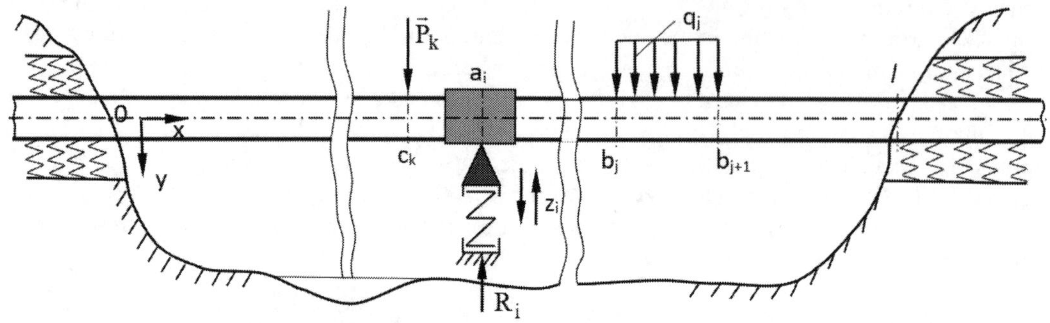

Figure 2. Schematic diagram of the overground segment of the pipeline on supports.

The designations for the overground part of the pipeline are analogical to those in the works [4, 6]. In particular, here it is assumed that the coordinates, coefficients of mechanical compliance, and initial displacements of the pipeline supports are a_i, κ_i, z_i, respectively, $i = \overline{1, N_R}$; $[d_m, d_{m+1})$, $m = \overline{1, N_D}$, are individual segments of constant rigidity of the pipeline for bending. This enables us to take into account the variety of thickness in the overground segment D_m, which is characteristic of its erection. The application of concentrated loads P_k is at the points $x = c_k$, $k = \overline{1, N_p}$, and the distributed loading of constant intercity q_j is applied in the semi-intervals $[b_j, b_{j+1})$, $j = \overline{1, N_q}$, (within a segment of constant wall-thickness). As result, we can more accurately simulate the weight characteristics of the pipeline, those of the transported product, and external influences of diagnostic devices. The reactions of supports are determined by R_i, $i = \overline{1, N_R}$, they are applied at the points whose coordinates are a_i. The sagging $w(x)$ of the overground segment satisfies the differential equation [4, 6].

$$\left(D(x)w''\right)'' = q(x),\ x \in (0, l), \tag{1}$$

Where the external loading is determined as follows

$$q(x) = \sum_{i=1}^{N_R} R_i \delta(x - a_i) + \sum_{k=1}^{N_P} P_k \delta(x - c_k) + \sum_{j=1}^{N_q} q_j \left[\theta(x - b_j) - \theta(x - b_{j+1})\right], \tag{2}$$

$\delta(x - \alpha)$ is the Dirak delta function, $\theta(x - \alpha) = (0,\ x < \alpha) \vee (1,\ x \geq \alpha)$ is the Heviside function.

At each support, there acts a force opposite to the reaction of the support R_i, it causes elastic displacement of the support $U_i = -\kappa_i R_i$, $i = \overline{1, N_R}$. Therefore, the conditions at the supports are of the form:

$$w(a_i) = -\kappa_i R_i + z_i,\ \ i = \overline{1, N_R}. \tag{3}$$

The bending moment and the shearing force is the beam is determined in terms of displacement as follows:

$$M(x) = -D(x)w''(x),\ \ Q(x) = M'(x). \tag{4}$$

Bg means of canchi function [2], the general solution of Equation (1) we write in the form [1]:

$$w(x) = w_0 + \theta_0 x + M_0 \frac{\partial}{\partial \alpha} K(x, 0) + Q_0 K(x, 0) + w^*(x, 0), \tag{5}$$

Where $K(x,\alpha)=\int_{\alpha}^{x}\frac{(x-s)(s-\alpha)}{D(s)}ds$ is the Cauchi function, w_0, θ_0, M_0, Q_0 are the initial parameters,

$$w^*(x)=\sum_{i=1}^{N_R}R_i\theta(x-a_i)K(x,a_i)+\sum_{k=1}^{N_P}P_k\theta(x-c_k)K(x,c_k)+\sum_{j=1}^{N_q}q_jH_j(x),$$

$$H_j(x)=\int_{b_j}^{\varphi_j(x)}K(x,t)\,dt,\qquad \varphi_j(x)=\begin{cases}b_j, & x<b_j\\ x, & b_j\le x<b_{j+1}\\ b_{j+1}, & b_{j+1}\le x\end{cases}.$$

In this case, we assume the anti-bending regidit $D(x)$ to be piece-wise constant. Therefore, it is expedient to represent the expressions for Cauchy function $K(x,\alpha)$ and for the function $H_j(x)$ in the form that is convenient for implementation of numerical calculation [4, 6] the beam's saggings w_1 and w_2 in the underground segments $x\in(-\infty,0)$ and $x\in(l,\infty)$ should satisfy the following equation:

$$(D_i^0 w_i'')''=k_i(f_i-w_i)+q_i^0\,,\quad i=1,2, \tag{6}$$

where D_i^0 – is the anti-bending rigidity of the corresponding segment of the pipeline; q_i^0 – is the reduced coefficient of rigidity of the base; $f_1(x)=\alpha_1 x+\delta_1$, $f_2(x)=\alpha_2(x-l)+\delta_2$ – are the equations determining the shape of the axis lines after ground shift.

As known [12], the underground part of the main pipeline is of constant wall-thickness. Therefore, for the given segments the expressions determining the sagging are the following

$$W_1(x)=A_1e^{\beta_1 x}\sin\beta_1 x+A_2e^{\beta_1 x}\cos\beta_1 x+f_1(x)+q_1^0/k_1\,, \tag{7}$$

$$W_2(x')=B_1e^{-\beta_2 x'}\sin\beta_2 x'+B_2e^{-\beta_2 x'}\cos\beta_2 x'+f_2(x')+q_2^0/k_2\,, \tag{8}$$

where $\beta_i^4=k_i/4D_i^0$; $x'=x-l$ – is the auxiliary coordinate; A_1, A_2, B_1, B_2 – are the unknown constants.

Generally, the solution contains eight unknown constants w_0, θ_0, M_0, Q_0, A_1, A_2, B_1, B_2 and N_R unknown reactions R_i, $i=1,N_R$. To find them, we use the conditions of equality of displacements, tarn-angles, moments, and shearing forces at the points of joining the segments for $x=0$ and $x=l$, as well as conditions at contacts with supports. The calculating program is implemented in Delphi-medium. The testing of the program has been carried out by comparing the results with an alternative solution.

3. Results obtained

The calculations were conducted for a pipeline whose diameter was 0,72 m and the wall-thickness $t=0{,}095$ m. Here, two overground segments are sufficiently long and their length are approximately equal (20,56 m and 19,34 m), and the third segment is somewhat shorter (14,35 m). The material of the pipeline is of K50 – class ($E=2\cdot10^5$ MPa; $\mu=0{,}3$; $\sigma_{0,2}=343$ MPa, $\sigma_B=490$ MPa, coefficient of thermal expansion $\alpha_t=1{,}58\cdot10^{-5\circ}C^{-1}$). The pressure in the pipeline $p=4{,}5$ MPa, the possible temperature drop $\Delta t=-20°C$. The density of the transported oil-product 895 kg/m^3.

To simulate the underground part of the overpass, the following initial data were assumed: density of the ground $2750\,kg/m^3$; depth of the position of the pass $1{,}2\,m$, coefficient of the bed for the initial calculational state of the oil pipeline $10\,MPa/m$, coefficient of mechanical compliance of the support elements 0,1 m/MN [6, 12].

3.1 Calculation of overground segment for rigid state (zero mechanical compliance) of all supports. Such an implementation of design scheme is used mainly in designing calculations. It is basic for choice of necessary materials and elements of the overground segment. However, such a scheme is characteristic of special conditions of erection: stony ground, additional protective shells, pile-type supports, etc.

Here, the maximal sag of the pipeline's axis is observed in the middle of the spans of the overground segment. With this, the maximal value we have for the first span (10,66 mm) at the distance of 10,85 m from the left side. The next span has the sag of 6,64 mm at the distance of 30,92 m, and the last, third span has its sag of 1,96 mm at the distance of 48,28 m. The curves of bending moment (blue lines), distribution of shearing forces (pink lines), and reactions of supports (green lines) are shown in Figure 3. Here, the maximal values of bending moments are observed in the middle of the spans and at support elements. In general, the distribution of bending moments is alternating. The greatest support moment is over the first support, it amounts to 0,144558 MN·m. Considerably less moment is over the second support (0,113860 MN·m), and the least is at the place of the pipeline's entering the ground (0,0514236 MN·m). Analogically, the shearing forces are distributed, they are saw-shaped with maximal gradient and values at sections at supports and at places of entrance and exit of the overground segment of the pipeline into/out of the ground. The value of reaction is greater above the first support, and it amounts to 0,104890 MN, and above the second support 0,0871283 MN.

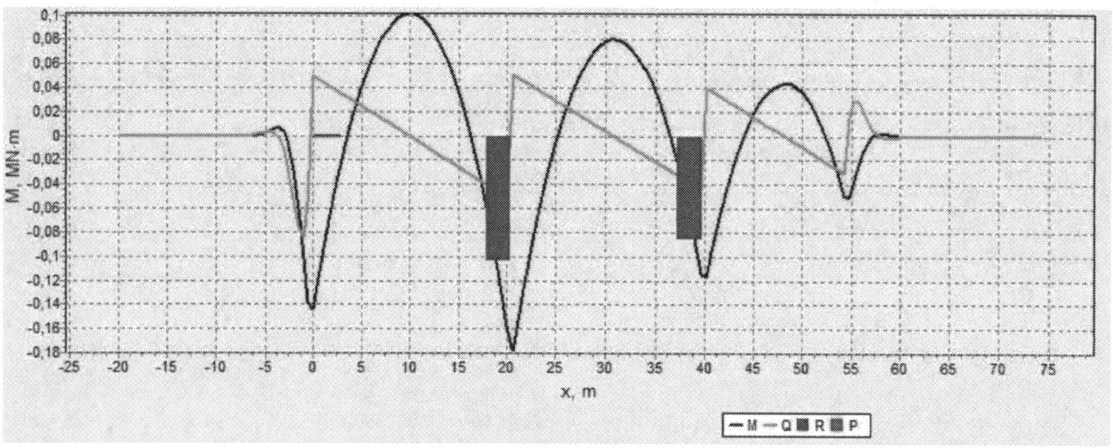

Figure 3. Curves of bending moments, shearing forces, and support reactions for rigid state of overground segment.

The distribution of caused by moment stresses in the pipeline at the lower surface (generetrix) of pipes of the overground segment is shown in Figure 4. If we neglect probable local perturbations in the section where the pipe rests on its support, then the general pattern of stresses at the surface of generatrix will be of asymptote to that shown in Figure 4 character. We see that the stress distribution is also alternating with maximal values in spans and above the supports. With this, maximal tensile stresses at the lower generatrix are observed in the middle of spans, and maximal compressive ones above the supports. Here, the greatest values caused by moment tensile stresses are 26,327 MPa, 20,798 MPa, and 11,367 MPa in the middle of the first, second, and third spans, respectively. Analogically, in the sections above supports the compressive stresses are 46,550 MPa and 30,305 MPa above the first and the second supports, respectively.

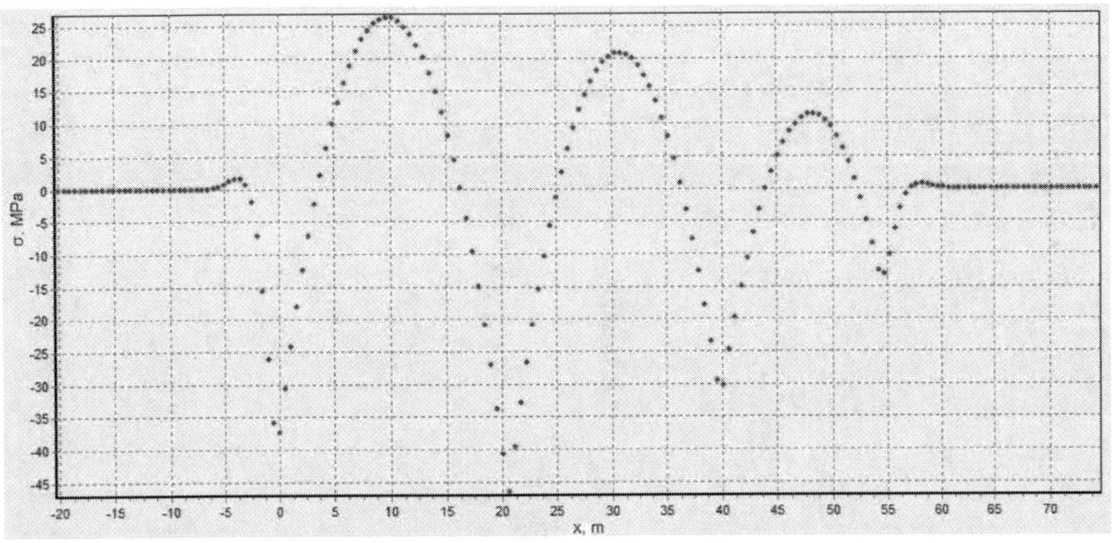

Figure 4. Distribution of caused by moment stresses at lower generatrix.

Thus, the calculation in the case of rigid supports is characterized by rather considerable gradient values. With this, under usual conditions such a state of support surfaces is not simply attainable. More often, there are observed cases when the situation of supports deviates from the design. In this cases periodical inspections and geodetic measurements of overground segments of main pipelines.

3.2 Calculation of overground segment of pipeline in the case of mechanical compliance of ground and support element. Under the action of the weight of the pipeline, there took place a certain shrinkage of support surfaces, but there emerges a greater reverse sagging of the axis of pipes in the vicinity of places of entrance/exit out of ground. The supports shifted downwards depending on the values of the reactions – the first support shifted by 10,88 mm, the second by 8,86 mm. With this, there was observed inconsiderable increase in support reactions to 0,108826 MN and 0,0886608 MN on the first and the second supports, respectively. The calculation of displacements of the overground segment is shown in Figure 5.

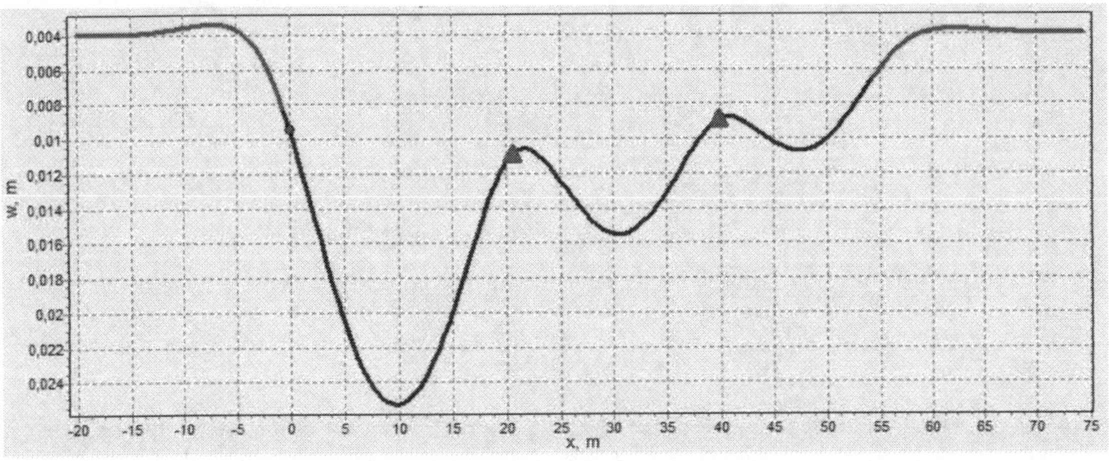

Figure 5. Sagging of pipeline axis in the case of mechanically compliant ground and supports.

The distribution of bending moments, shearing forces, and support reactions is shown in Figure 6. The comparative analysis indicates that taking into account the mechanical compliance of support elements leads to some redistribution of bending moments. With this, for the first span they increase by about 18 %. In underground segments, the change in bending moments is less localized and distributed in a segment 5 m to 10 m long with smooth decrease to zero. Gradients of values of shearing forces in the pipeline above the supports are less in this case, and the maximal values somewhat decrease.

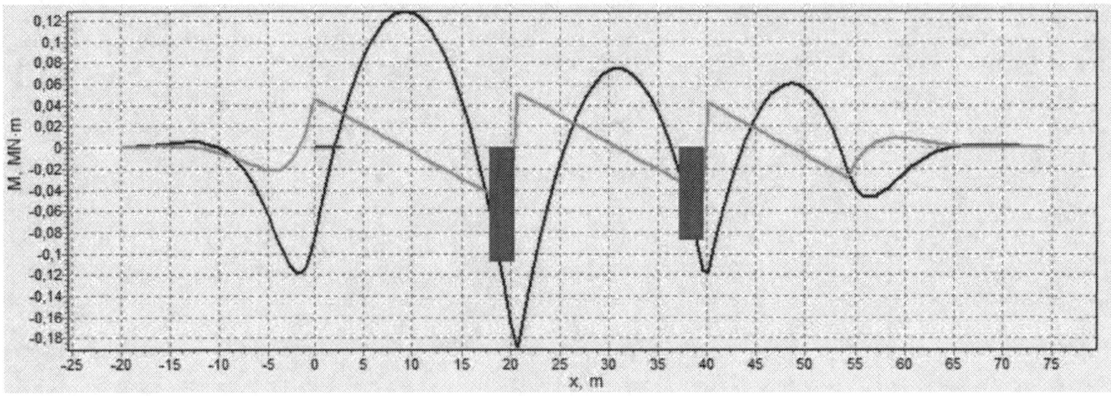

Figure 6. Curves of bending moments, shearing forces, and support reactions of overground segment with mechanically compliant supports.

The distribution of stresses at the lower generatrix of the pipeline for the considered case of compliance of support elements is shown in Figure 7. Comparison of obtained data indicates that the distribution is more smooth. With this, maximal values in the first span increased to 33,301 MPa, in the rest spans they are of approximately the same value. In the first support section, stresses increased to 49,354 MPa. In underground part of the pipeline, there was observed smooth transition from compression at the edge of ground to small tension in segments which are at distances of 5 m to 10 m. However, this alternating character essentially vanishes.

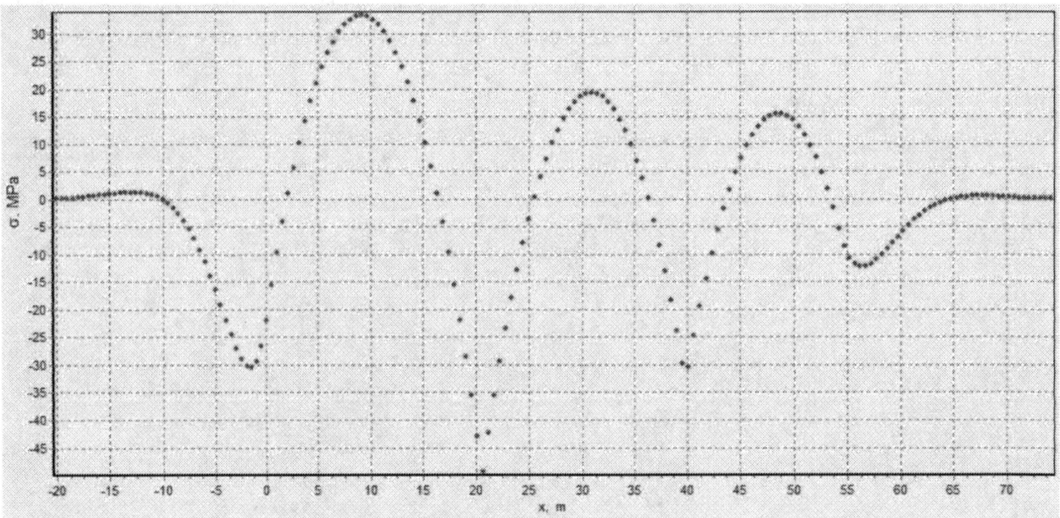

Figure 7. Distribution of bend-stresses at lower generatrix in the case of mechanically compliant supports.

Thus, the consideration of mechanical compliance of supports and ground indicates probable deviations from the design positions. This ensures more smooth distribution of the observed values. However, with this the established maximal values of the observed quantities can locally increase.

Therefore, it is important to carry out a more detailed analysis of the influence of heights of supports upon the strength characteristics of overground segment of the pipeline.

3.3 Influence of altitude of supports on values of acting loads in overground segment. Vertical adjustment of displacement in sections of support elements should be implemented in repairing and restoring works, in adjustment of mutual position of supports, for setting a given position of the pipeline axis, etc. In such a case, it is necessary to take into account the possibility of redistribution of acting loads and stresses in the body of pipeline.

3.3.1 Lifting the first support of overpass. For our investigations we take the elevation value of 250 mm, that is sufficient for most of repair and restoring works [11]. We set the displacement as rigid, but with this we take into account the mechanical compliance of support elements and ground. It is established that at the expense of mechanical compliance of support elements the total elevation is less, and it amounts to 226 mm. Reactions of the supports in the given case amount to 0,235154 MN and 0,0155663 MN for the first and the second support respectively; i. e. comparative analysis indicates that the load at the first support has increased almost 2,5 times (14 %), and at the second support the load essentially decreased. The results of calculation are presented in Table 1. Here are presented most critical segments, that are determined in the course of the carried out calculations: at the edges of ground, above the support, as in the middle of the first (10,28 m), second (30,23 m), and third (47,07 m) spans, respectively.

Analysis of caused by moment stresses indicates that in the give case the maximal values are observed in the place of elevation of the first support, and they amount to 238,97 MPa in magnitude. Here they are compressive at lower generatrix of the pipe. Analogical character (compression) can be observed in the neighborhood of entrance of the pipeline into the ground.

The conclusion can be drawn that the place of the pipeline's exit from ground has experienced elevation relative to its design by 6,37 mm. Further, gradual lifting the checked points to maximal value of 226,67 mm. Then, there becomes decrease with passing neutral axis below the level in the middle of the last span (26,92 mm). The entrance of the overpass into ground takes place below its design position by 11,46 mm.

Table 1. Displacements, bending moments, shearing forces, and moment stresses in overground segment when lifting the first support.

№	x, m	W, m	M, MN·m	Q, MN	σ_M, MPa
1	0	-6,37366E-03	4,18642E-01	-1,40241E-02	1,08234E+02
2	10,28	-1,23380E-01	7,41507E-03	-6,57678E-02	1,91707E+00
3	20,56	-2,26670E-01	-9,24336E-01	1,15642E-01	-2,38975E+02
4	30,23	-9,22985E-02	9,83197E-02	5,57286E-02	2,54192E+01
5	39,90	-4,12952E-03	3,72978E-01	2,03544E-03	9,64283E+01
6	47,07	2,69220E-02	2,62491E-01	-3,33681E-02	6,78635E+01
7	54,25	1,14608E-02	-9,76792E-02	-6,87717E-02	-2,52536E+01

3.3.2 Lifting of the second support by 250 mm. In the given case, the reaction of the second support is 0,276594 MN. The first support has somewhat disloaded, and the reaction became 0,00459908 MN. This indicates its practically completed dislodging. The elevation together with taking into account the mechanical compliance of support surfaces in this case was 233,37 mm for the second support and 0,182 mm for the first support; i. e. the first support has taken its practically initial position relative to its design position. However, the elevation of the second support is less almost by 27 mm because of mechanical compliance of support elements.

In Table 2, main results of calculation of the overpass are presented. The denotations and checked sections of the pipeline are analogical to those in the previous item. In the given case, the values of bending moments and of the acting upon the pipeline stresses are greater. For the first support the magnitude of lend-stresses was 238 MPa.

Table 2. Displacements, bending moments, shearing forces, and axial stresses in overpass when lifting the second support.

№ п/п	x, m	W, m	M, MN·m	Q, MN	σ_M, MPa
1	0	1,44596E-02	-2,07663E-01	7,61732E-02	-5,36884E+01
2	10,28	5,55209E-02	3,52012E-01	1,35362E-02	9,10078E+01
3	20,56	-1,82092E-04	2,96206E-01	-2,27150E-02	7,65800E+01
4	30,23	-1,44758E-01	-1,64947E-01	-7,17353E-02	-4,26449E+01
5	39,90	-2,23376E-01	-1,04001E+00	-1,18032E-01	-2,68881E+02
6	47,07	-1,26507E-01	-6,28129E-02	1,20434E-01	-1,62394E+0
7	54,25	-1,72736E-02	6,61709E-01	8,50309E-02	1,71076E+02

Thus, the lifting of the second support in the given case leads to greater overload of pipeline material.

It is expedient to carry out analysis of the magnitude of elevation of the overground segment above the support with setting the maximal stresses caused by this process. In such a case, we can plan the amount and character of works for repairing the pipeline in a specific section at a support.

Using the aforesaid technique, there have been obtained date on caused by moment stresses and those on reactions of supports for lifting the considered overground part by the set value (Table 3). Taking into consideration the mechanical compliance of support elements and the necessity of access to the inspected part of the pipeline, the initial altitude was taken equal to 50 mm. Further, the calculation of lifting the pipeline was carried out with the step of 50 mm. Maximal stresses and reactions in support sections were studied. Here, stresses and support reactions under the initial position of the pipeline are presented for comparative analysis.

Thus, the analysis of the data of Table 3. indicates that the lifting of the second support by 50 mm leads to greater reaction and stresses as compared to lifting of the first one. With this, the lifting by 50 to 100 mm is characterized by caused by moment stresses amounting to 30 to 50 % of the proportionality limit of material of pipes/ Greater absolute values in support lifting in the given case are not recommended because of essential values of stresses and support reactions in a pipeline.

Table 3. Maximal moment-stresses and support reactions in local lifting of overground segment.

Support		Altitude above the support, mm				
		0	50	100	150	200
I	σ_I^b, MPa	4,93452E+01	8,72710E+01	1,25197E+02	1,63123E+02	2,01049E+02
	R_1, MN	1,08826E-01	1,33692E-01	1,58557E-01	1,83423E-01	2,08288E-01
II	σ_{II}^b, MPa	2,97717E+01	7,75936E+01	1,25415E+02	1,73237E+02	2,21059E+02
	R_2, MN	8,86608E-02	1,26247E-01	1,63834E-01	2,01420E-01	2,39007E-01

4. Discussion and conclusions

The suggested technique of calculation of calculation of overpass provides effective technology when we plane repair and restoring works. Here we can obtain the following data: displacements of support surfaces, acting loads, and stresses in the main pipeline.

Analysis of results of alternating lifting of the supports indicates that in the given case it is necessary to take into account the possibility of mechanical compliance of support elements, that reduces the total elevation relative to the design position. This is important when carring out adjusting displacements to ensure rectlinearity of the overpass. With this, there takes place redistribution of support reactions, moments, and shearing forces. The lifted pipe above a support experiences considerable overloading, and stresses can increase several times.

Adjusting vertical displacements in section over a support enable us to estimate the acting stresses and to plan repairing technology.

5. References

[1] Dzyubyk A R, Palash V M, Nazar I B, Dzyubyk L V, Palash R V 2016 Welding renovation of supporting units in gap-crossing structures. *Naukovyi visnyk NLTU Ukrainy* **26.1** 230 – 238

[2] Bilobran B S, Dziubyk A R, Yanovskyi S R 2013 Influence of installation elastic bending on stress-strain state of pipeline above-ground passages in mountains *Oil & gas industry of Ukraine* **4** 52 – 58

[3] Bilobran B S, Dziubyk A R, Yanovskyi S R 2010 The stress-deformed state of the above-ground beam transition in the casing of the main oil pipeline *Oil & gas industry of Ukraine* **1** 50 – 52

[4] A R Dzyubyk, L V Dzyubyk, G V Pokhmurs'ka, I A Prokopyshyn 2019 Residual Strength of the Overhead Section of a Main Pipeline with Annular Cracks *Mater Sci* **54** 855 – 865 https://doi.org/10.1007/s11003-019-00273-4

[5] A V Andrusyak, A S Velichkovich 2015 Engineering assessment of the strength of an above-ground oil pipeline constructed in a mountainous area *Exploration and development of oil and gas deposits* **1** 89 – 96

[6] Dzyubyk A, Sudakov A, Dzyubyk L, Sudakova D, 2019 Ensuring the specified position of multisupport rotating units when dressing mineral resources *Mining of Mineral Deposits* **13(4)** 91 – 98 DOI: 10.33271/mining13.04.091

[7] Dzyubyk A R, Nykolyshyn T M, Porokhovs'kyi Y V 2016 Influence of Residual Stresses on the Limit Equilibrium of a Pipeline with Internal Crack of Arbitrary Configuration *Mater Sci* **52** 89 – 98 https://doi.org/10.1007/s11003-016-9930-4

[8] Palash V M, Yus'kiv V M, Dzyubyk A R 2000 Stresses in Pipes Joined by a Circular Weld *Mater Sci* **36** 586 – 591 https://doi.org/10.1023/A:1011378508847

[9] Dziubyk A R, Voitovych A A, Student O Z, Dziubyk L V, Khomych I B 2022 Evaluation of the Technical State of Reinforcement of the Concrete-Beam Span of a Bridge Constructed at the End of the Last Century *Mater Sci* **57** 466 – 474 https://doi.org/10.1007/s11003-022-00567-0

[10] Bilobran B, Dzyubyk A. 2007 Calculation of the stressed-strained state of beam transitions of trunk pipelines by numerical methods *Bulletin of the Lviv Polytechnic University "Dynamics, strength and design of machines and devices"* **588** 8 – 12

[11] Bilobran B S, Dzyubyk A R, Pavlyk V P 2006 Stressed – deformed state of an oil pipeline during its local heaving *Bulletin of the Lviv Polytechnic University "Dynamics, strength and design of machines and devices"* **556** 3 – 8

[12] Doroshenko Ya V 2015 *Construction and Repair of Concentrated Objects of Gas and Oil Pipelines: textbook* (Ivano-Frankivsk: IFNTUNG)

Design and technological aspects of functionally oriented technology of manufacturing the three-cone drill bits

A M Slipchuk [1], R S Jakym [2], V M Korendiy [1,3] and Y M Lytvyniak [1]

[1] Department of Robotics and Integrated Mechanical Engineering Technologies, Lviv Polytechnic National University, 12 S. Bandera Street, Lviv, 79013, Ukraine
[2] Department of Technological and Professional Science, Drohobych Ivan Franko State Pedagogical University, 24 I. Franko Street, Drohobych, 82100, Ukraine
[3] E-mail: vitalii.m.korendii@lpnu.ua

Abstract. The quality and efficiency of the drilling process significantly depend on the design parameters and operational peculiarities of the drill bit cutter. The present paper is focused on developing the functionally oriented technology of manufacturing the three-cone drill bits. The research methodology involves the theoretical and experimental studies dedicated to analysing the main reasons for the loss of the cutters working capacity, and the technological factors that affect the strength and durability of the "tooth – cutter" press connection. The major scientific novelty consists in defining the relationships between the design parameters of the cutters, their kinematics, dynamics, and operational conditions. It has been established that the destruction of the rock by the three-cone drill bits is carried out by applying successive impacts of the tipped hard-alloyed teeth upon the rock. The rolling of the cutters (bits) along a face (stope) exerts a dynamic effect due to the alternate change of the contact of the teeth with the rock, and the axial load on the bit ensures that the striking part of the teeth is pressed into the rock. The character of wear and destruction of the tipped teeth also has an important influence. During the drilling process, the rock-destructing parts (teeth), as well as the working areas of the cutters, undergo intensive wear and failure. This leads to a significant decrease in the quality of the rock destruction process and a reduction in the drilling efficiency. The obtained results have a significant practical value and can be effectively used by designers, technologists and researchers while implementing the energy-efficient drilling technologies.

1. Introduction

To ensure the given operational parameters of three-cone drill bits, it is necessary to pre-set the corresponding parameters at the stages of their designing and manufacturing. Therefore, the mentioned processes should be preceded by preliminary in-depth studies of the drill bits' operational peculiarities in terms of providing the specified parameters of the objective function. In this case, the design and technological support being developed should take into account the peculiarities of the drill bit operation during the required time interval. Note that the process of the drill bits operation is characterized by a wide range of different features of their certain sections and elements, such as rock-destructing equipment, supports, cooling systems, etc. During the operation of the chisels, a whole set of static, dynamic, cyclic loads acts on their elements. The bits work in aggressive, chemically active, abrasive-intensive environments. The effect of various factors on the elements of the bit and, in general, on the design of the bit is variable in space and time. In each design element of a bit, there are alternating stresses. In particular, the features of the stress state in the connection "tooth-cone" determine not only the reliability of the fastening of rock-destructing equipment but also its functional and operational indicators. The change in the action of the operational functions causes a decrease in the performance of cone drill bits and, in some cases, sudden failures due to destruction and catastrophic wear of teeth and cutters during the process of drilling the rock.

Numerous scientific papers are devoted to studying the functionally oriented engineering technologies, particularly those related to manufacturing the drill bits and implementing the drilling processes. The influence of various technological parameters on the reliability of the "teeth – cutter" connection has been mathematically modelled and experimentally tested in [1]. The paper [2] is focused on studying the frictional and thermal conditions during the drilling process performed with the help of the three-cone drill bit. The effects of generating and propagating the crack-type defects in the rotary parts depending on their constraints were investigated in [3]. The deformations, fracture, and wear influence on the service life of the drill bits' bearing units was analysed in [4]. The papers [5] and [6] are dedicated to studying the frictional and wear processes during the operation of the roller cone bit inserts, depending on their mechanical and physical properties.

During the drilling process, each element of the drill bit (at its hierarchical level of complexity) performs one or several elementary technical functions for implementing the general target function. The physical meaning of this function is the destruction of the rock and the formation of the walls of the faces (wells, stopes) and the boreholes. At the same time, the limiting conditions ensure the specified safe operation life of the drill bit. All of this is provided by the complex solution of a set of design and technological problems at the main stages of developing the drill bits, which generally forms a system of indicators of the quality of a three-cone drill bit. The issues mentioned above are analyzed in numerous scientific papers. Particularly, in [7], the authors investigated the influence of the drilling parameters and the physical-mechanical properties of the rock being processed on the penetration rate as a factor defining the drilling quality and efficiency. The paper [8] is focused on defining the optimal operational conditions of the three-cone drill bits improving their reliability and the overhaul period. The contact stresses acting in the bearing units of the roller cone drill bits are thoroughly studied in [9]. The papers [10], [11], and [12] present the results of the comprehensive theoretical and experimental research on the enhanced spherical and non-spherical single-cone drill bits performing the drilling processes under various operational conditions.

In order to develop functionally oriented drilling technologies, it is necessary to carefully study and establish correlations between the types of technological operations, their parameters, and the operational and functional properties of three-cone drill bits. Therefore, the design of the bit is considered as an ordered hierarchy of executive fragments and surfaces. Bases on this, the routing technology of the bit manufacturing can be formed. The important and initial foundation for developing the functionally oriented technology is a comprehensive study of the operation features and the reasons for the performance loss of the three-cone drill bits. Based on the obtained results, the structure of operational functions can be formed. As an example, the papers [13] and [14] consider the wear processes and failure cases occurring in the roller cone drill bits under different drilling regimes (penetration rates), rock materials, materials of the bits, etc. In [15], the authors proposed the improved wear model considering the physical and mechanical properties of the rock and the drill bit materials, as well as the drilling regimes. The developed model allowed for describing the drill bit wear processes and its real-time dullness (bluntness). The papers [16] and [17] study the strength parameters of the novel taper-structure teeth of the cone drill bit fixed under different temperatures and pressing conditions taking into account the possible changes in the geometrical parameters of the teeth. Similar research on the reliability and failure analysis of the teeth (inserts) and bearing of the three-cone drill bits during the blasthole drilling is presented in [18].

The present paper is based on the latest investigations dedicated to developing the cutters of the three-cone drill bits and continues the previous research of the authors presented in [19] and [20]. The latter studied the influence of various technological factors of the drilling process and geometrical parameters of the roller cone bit on the tool reliability and molding accuracy. The efficiency of the drilling process performed with the help of the roller-cone drill bits is thoroughly investigated in [21]. The paper [22] is focused on analyzing the dynamics of drilling the laminated metal materials using the double-cone drill and on improving the chips fracture ability. In [23], the authors proposed the novel drill bit design for machining the carbon-fibre reinforced polymers and studied the corresponding drilling process characterized by the decreased thermomechanical loads. The problems of predicting the roller cone bits performance and optimizing the novel single-energizer metal seals are thoroughly investigated in [24]. The comprehensive analysis of numerous scientific papers, particularly those mentioned above, has shown that the problems of implementing the functionally oriented technologies while designing and manufacturing the roller-cone drill bits are currently of significant interest among researchers all over the world.

In this study, the design and technological parameters of the drill bit cutters are considered from the viewpoint of the functionally oriented technology of their manufacturing. The major scientific novelty consists in defining the relationships between the design parameters of the cutters, their kinematics, dynamics, and operational conditions. Based on this, the main reasons for the loss of the cutters working capacity are to be studied, and the technological factors that affect the strength and durability of the "tooth – cutter" press connection are to be defined.

2. Research Methodology

The first stage of developing the functionally oriented technology of manufacturing the three-cone drill bits consists in studying the interaction between the rolling cutter teeth and the rock being drilled. Herewith, it is necessary to consider the distribution of forces and stresses characterizing the rock-breaking process, as well as the stress-strain state of the "hard-alloy tooth – rolling cutter body" connection. The interaction between the rolling cutter tooth and the rock is attended by the pressing process characterized by the force $\vec{R}_t$ taking into account the tooth compound motion (figure 1). The drilling bit is rotating at a certain angular velocity $\vec{\omega}$, and, at the same time, it is rectilinearly moving at a certain velocity $\vec{V}_c$. In the "tooth – rock" contact zone, there also acts the sliding friction force $\vec{F}_s$ directed oppositely to the velocity vector.

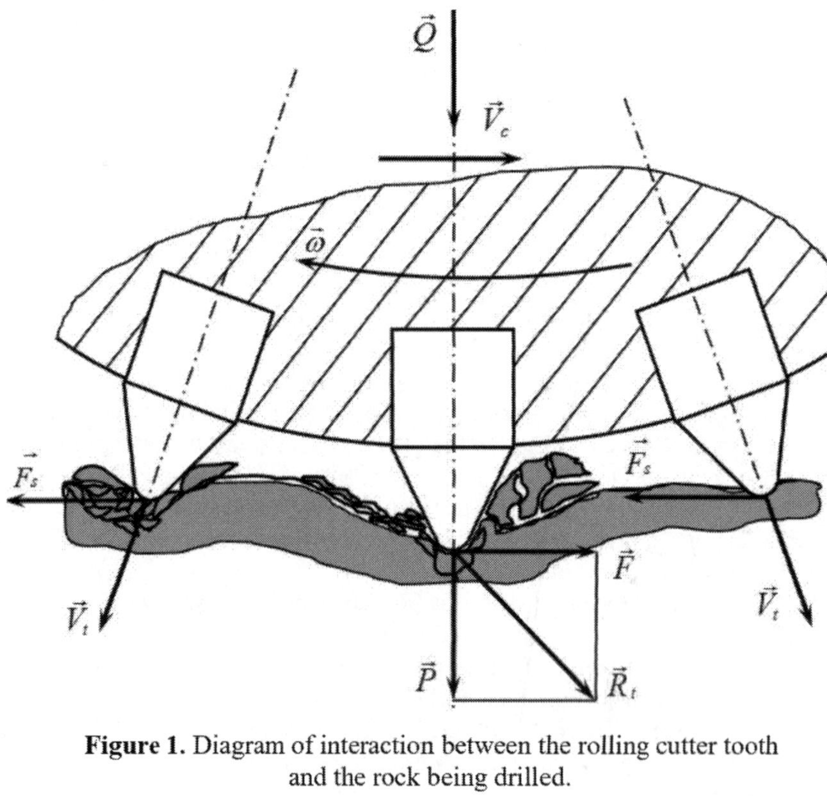

Figure 1. Diagram of interaction between the rolling cutter tooth and the rock being drilled.

Considering the process of pressing the tooth into the rock, the tooth velocity depends on the following factors: the rolling cutter design parameters, the number of teeth, and the angular velocity of the drill bit:

$$\vec{V}_t = \vec{V}_c + (\vec{\omega} r_t) = b \frac{\pi}{60} \frac{n_d r_d}{r_c}, \tag{1}$$

where r_t is the distance between the tooth axis and the instantaneous centre of the cutter rotation; b is the number of teeth on the cutter ring; n_d is the rotation frequency of the drill bit; r_d is the radius of the drill bit; r_c is the radius of the cutter ring.

The contact force acting between the tooth and the rock depends on the number of teeth z_i being simultaneously in contact:

$$\vec{R} = \frac{Q}{z_i}, \tag{2}$$

where Q is the axial pressing (pushing) force acting upon the drill bit.

The maximal horizontal and vertical components of the contact force (figure 2 (a)) can be determined considering the contact angle between the tooth and the rock β :

$$\beta = \arcsin\left(\sin\frac{\pi}{z} \cos\varphi \right), \tag{3}$$

where z is the number of teeth on the cutter ring; φ is the angle between the ring generatrix and the rolling cutter axis.

Analysing the interaction between the teeth and the rock at the beginning of the drilling process, there take place two basic conditions of the teeth loading (figure 2). When two teeth of the rolling cutter are in contact with the rock being drilled, the horizontal component of the contact force takes its maximal value causing the maximal bending moment acting upon the cutter body. In such a case, the sliding friction forces also take their maximal values during the cutter rotation. On the other hand, when the symmetry axis of the tooth coincides with the line of action of the contact force, the sliding friction and the bending moment tend to their minimal values.

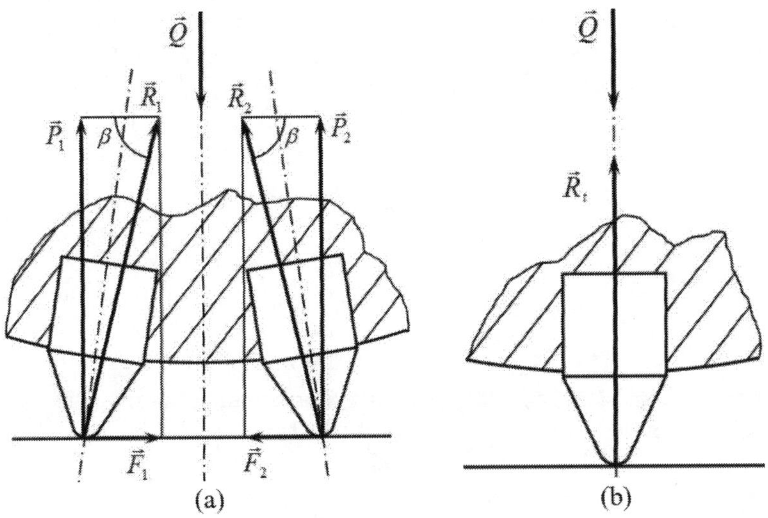

Figure 2. Diagram of forces acting between the teeth of the rolling cutter and the rock: a – double-teeth contact; b – single-tooth contact.

Let us define the law of changing the horizontal component of the contact force (figure 2) acting upon the tooth and causing cyclic stresses in the "tooth – cutter body" contact zone:

$$-R\sin\beta < 0 < +R\sin\beta. \tag{4}$$

The frequency of the tooth disturbance can be determined as follows:

$$N_t = \frac{n_d r_d}{60 r_c}. \tag{5}$$

The time of contact of the tooth with the rock under the conditions of zero sliding friction:

$$t_t = \frac{60}{i n_d z}\frac{r_c}{r_d}, \tag{6}$$

where i is the "drill bit to rolling cutter" transmission ratio $(i \approx 1.5)$.

The intensity of the axial loading exerted upon the cutter tooth:

$$P_t = \frac{Q}{k_o r_d}, \tag{7}$$

where k_o is the overlap factor; Q is the axial load acting upon the drill bit (e.g., for the drill bit of diameter 215.9 mm, $Q_{max} = 260$ kN).

During the drill bit operation, the tooth takes about 2×10^5 loading cycles. The conditions of durable complex cyclic loading or repeatable overloading usually cause the initiation of cracks and their propagation, followed by cutter teeth fracture. When one tooth is damaged, the nearby teeth are subjected to overloading. The process of the cutter rolling becomes more complicated due to the presence of impact loads reducing the drilling speed and causing the intensified wear of the cutter rings and body (figure 3 (a)). In addition, the impact loads cause the crumbling (chopping) of the casehardened (carburized) layers and deformation of the lower metal layers. In such cases, there appear numerous craters, scratches, furrows, and other deformations of the rolling cutter surface.

In the case of chipping the impact surfaces of the teeth out, their shapes change. Some teeth become sharper, while the other ones fail in the zone of contact with the cutter body (figure 3 (b)). The abrasive wear is characterized by micro-scratches when the chipped-out surfaces of the teeth become smooth. The teeth fracture is caused by material fatigue followed by initiation and propagation of cracks. The final wear-out and failure of the rolling cutter take place when most of the teeth are chipped out (figure 3 (c)) or have fallen out of the cutter body (figure 3 (d)).

In order to estimate the strength indicators of the cutter teeth, the following coefficient is usually used:

$$\xi = \frac{d_s}{h_o}, \tag{8}$$

where d_s is the average value of the tooth shank diameter; h_o is the average value of the tooth impact part overhang.

The larger value of the coefficient ξ is calculated, the safer is the rolling cutter operation between the overhaul periods, and the longer is its service life. For example, in order to improve the strength parameters of the teeth located on the cutter head of the 215.9C3 drill bit, their overhang values are reduced. This is caused by the drill bit design peculiarities requiring the use of the teeth with smaller diameters than the ones located in other rings of the rolling cutter. Therefore, considering the main hard-alloy teeth of 12.87 mm diameter, the coefficient $\xi = 1.716$, whereas for the teeth of 11.29 mm diameter, $\xi = 1.737$ (figure 4).

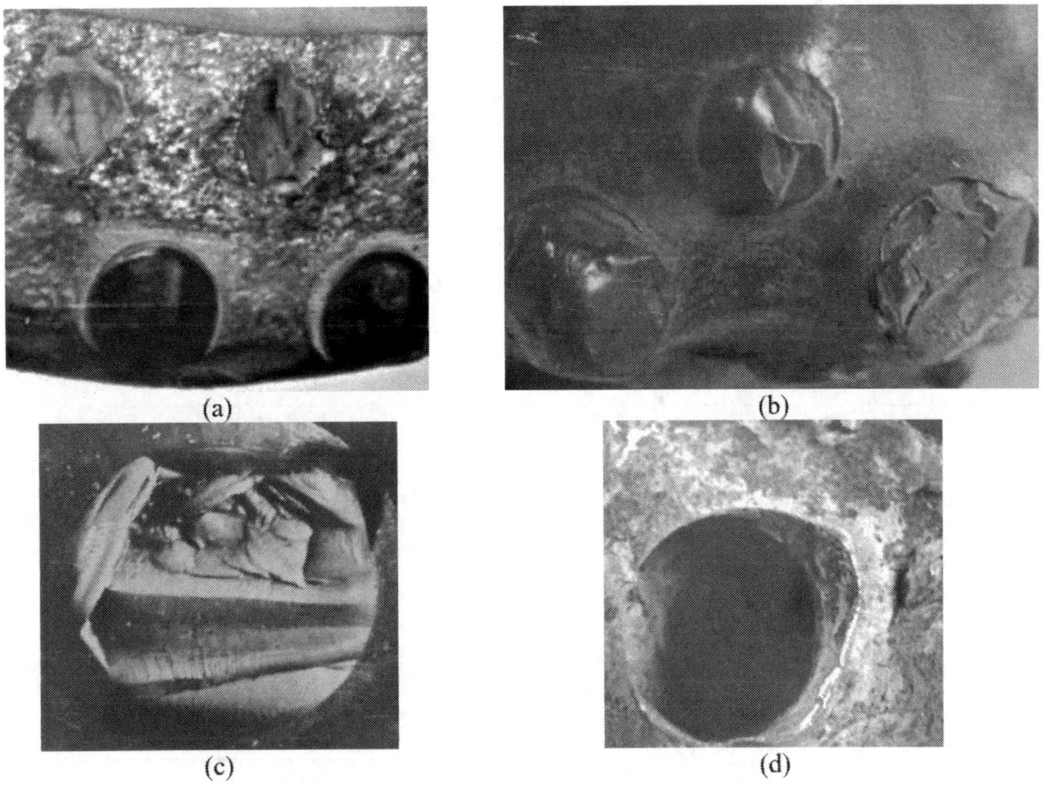

Figure 3. Examples of the teeth fracture and of the cutter working surface failure: a – impact-abrasive wear; b – hydro-abrasive wear; c – teeth cracking and chipping out; d – fatigue failure of the cutter body and falling the teeth out of the mounting seats.

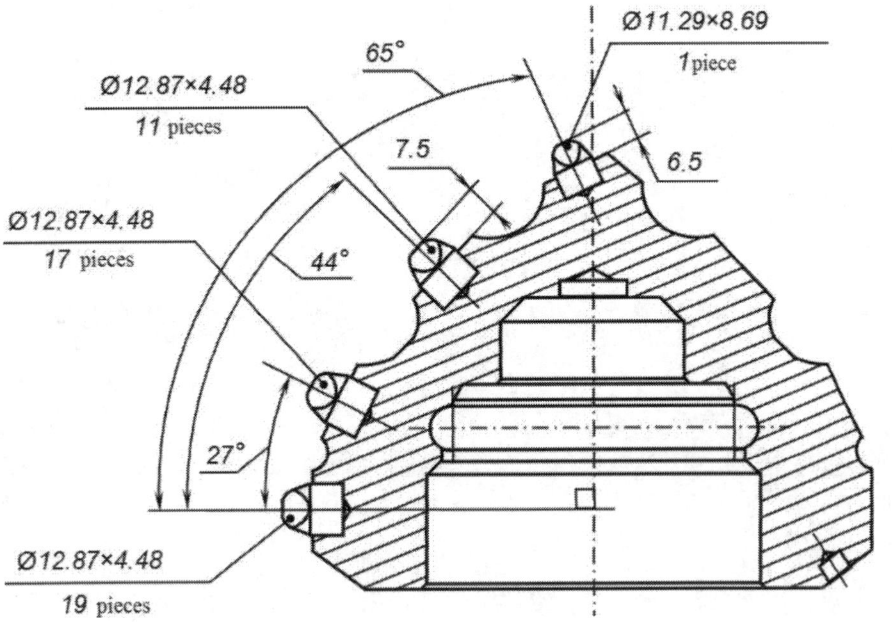

Figure 4. General design of the rolling cutter no. 1 of the three-cone drill bit 215.9C3.

3. Results and Discussion

In order to estimate the operational indicators of the drill bits 215.9C3-ПВ manufactured by Universal Drilling Technique LLC (Drohobych Drill Bits Plant, Ukraine), there were carried out the corresponding performance tests, the results of which are presented in Table 1. While drilling the rocks of the hardness 9…11, the average run distance of the drill bit was equal to 1894 m. The drill bits were subjected to the pressing load of 10…13 tons; the rotation frequency of the bit was in the range of 50…100 rpm; the pressure of the blowing mixture varied from 3.6 to 4.2 atm. Figure 5 presents the photos of the used (spent) three-cone drill bits.

Table 1. Results of performance tests of the three-cone drill bits 215.9C3-ПВ.

No.	Run distance, m	Operation time before failure, h	Average drilling speed, km/h	Reasons and circumstances of failure
1	1792.5	71.7	25	Seizure (cramping) of the third rolling cutter; general wear and failure of teeth
2	1706.5	66.4	25.7	Breaking-down of the third rolling cutter; general wear and failure of teeth; impact-abrasive wear of the rolling cutters' bodies
3	1649.9	58.3	28.3	Seizure (cramping) of the first and the third rolling cutters; general wear and failure of teeth; failure of the cutters' heads
4	2237.2	95.2	23.5	Breaking-down of the third rolling cutter; general wear and failure of teeth; impact-abrasive wear of the rolling cutters' bodies
5	2106.8	86.7	24.3	General wear and failure of teeth located on the peripheral rings; failure of the cutters' heads
6	1871.3	67.8	27.6	Breaking-down of the first and the second rolling cutters; general wear and failure of teeth located on the peripheral rings; impact-abrasive wear of the rolling cutters' bodies

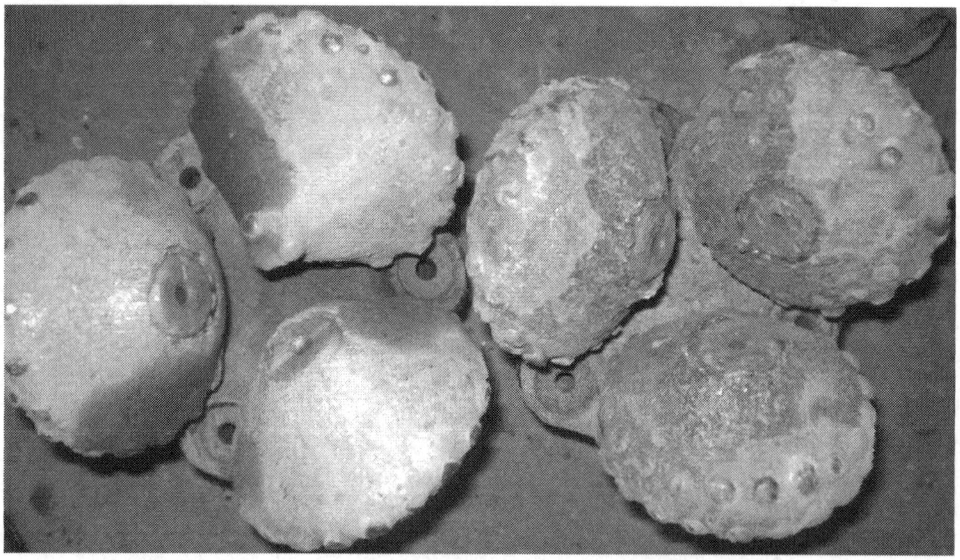

Figure 5. Photos of the used (spent) three-cone drill bits no. 1 and no. 5.

Analysing the obtained results of performance tests, it can be concluded that the largest influence on the drill bit strength and durability is made by the following technological factors: the accuracy of the design parameters of the teeth; the accuracy of the geometric parameters of the holes in the cutter rings; the accuracy and reliability of the compression connection of the teeth with the cutter rings; the heat treatment parameters ensuring optimal distribution of physical and mechanical properties of the cutter rings holes (before pressing) and the "tooth – ring" connection (after pressing); the presence of surfactants on the mated surfaces, etc.

On the other hand, the stress disturbance character between the contacting surfaces of the "tooth – cutter ring" connection is defined by the technological factors described above. Let us calculate the contact pressure taking place during the tooth pressing when maximal tension is equal to $N_{max} = 0.14$ mm. According to [19, 20], the calculations showed us the maximal value of the contact pressure equal to 2317 MPa, whilst under the conditions of the minimal tension of $N_{min} = 0.063$ mm, this pressure equals 1040.5 MPa. As an example, the computer simulation of the stresses character in the 215.9С3-ПВ drill bit cutter has been carried out (see figure 6).

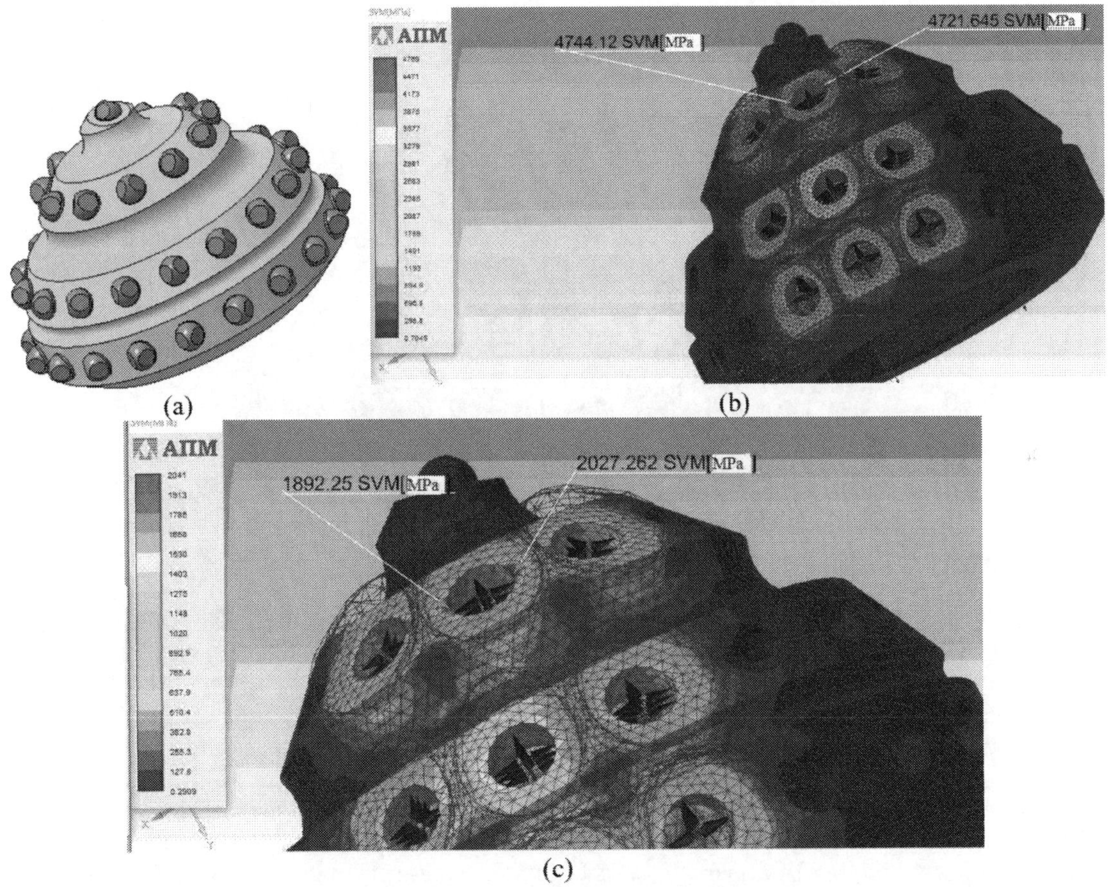

Figure 6. Results of computer simulation of the 215.9С3-ПВ drill bit cutter stress state: a – general design of the drill bit; b – stresses in the contact area under the tension of 140 μm; c – stresses in the contact area under the tension of 63 μm.

Analyzing the obtained results, it can be concluded that the contact pressure on the surfaces being connected is ununiformly distributed. Therefore, the contact stresses are mainly concentrated in the areas where the tooth enters (fits) in the ring seat and at the bottom of the mounting seat. The middle area of the "tooth – cutter ring" contact is significantly weaker loaded.

Considering the case of increasing the contact pressure causing the increased tension during the tooth pressing (see figure 7), the contact stresses increase in direct proportion, while the locations of the most loaded areas (stress concentrators) remain unchanged. It is necessary to mention that the stress distribution in the area of entering (fitting) a tooth into a cutter ring is performed in parallel to the surface of the cutter 3D-model. In the case of increasing the contact pressure, the magnitude of the stress concentration increases, whilst the lines of equal stresses bend towards the unloaded wedge part of the tooth. This fact indicates the occurrence of the axial tensile stresses, which gradually spread to the wedge part of the tooth.

When modelling the working loads on the tooth, the area where the tensile stresses act in the tooth's wedge part is replaced by the sliding area. Herewith, on the side of the working load action, the compressive stresses are much lower than the ones on the free side of the tooth. The concentration of stresses decreases in the area of the tooth entering (fitting) into the cutter body on the side of the working load action; and the stress concentration increases on the side that is not loaded. The magnitude of the change in the stress concentration depends on the absolute value of the working load and the angle of its application to the striking (impact) part of the carbide (hard-alloyed) tipped tooth. Under the conditions of the constant magnitude of the working load, the magnitude of the stress concentration in the area of the tooth entering (fitting) into the cutter body can be increased up to 1.5 times by changing the angle of its application in the range of 20°...60°.

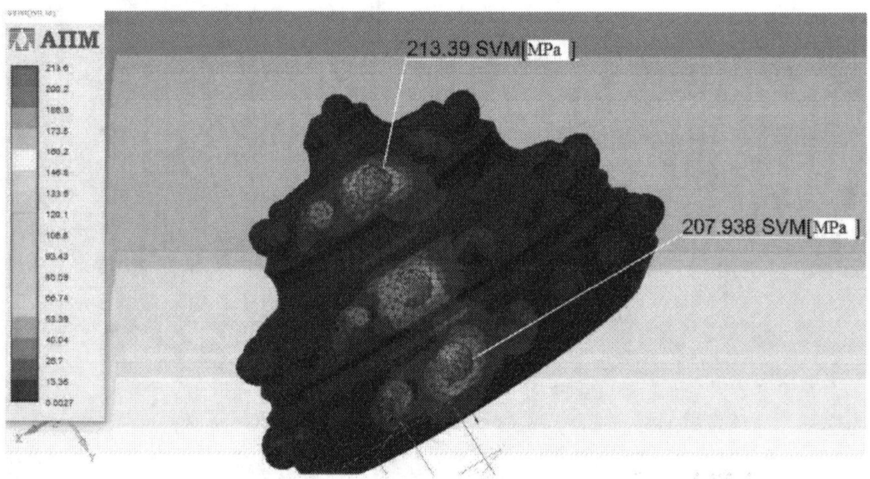

Figure 7. Results of computer simulation of the 215.9С3-ПВ drill bit cutter stress state in the case when the teeth are loaded by the 37 kN force due to their interaction with the rock.

The rolling of the cutter along the face (stope) causes a change in the orientation of the working loads action from + α to – α, where the angle α is determined by the contact angle β, which depends on the geometric dimensions of the drill bit (see figure 2). The change in the orientation of the working load action causes the periodic change in the magnitude of the stresses in the area of the tooth entering (fitting) into the cutter body and the occurrence of variable stresses throughout the wedge-shaped working part of the tooth. The presence of variable stresses causes the fatigue damage and wear of the carbide (hard-alloyed) tipped tooth. This is where the generation of the crack-like defects takes place, as well as the formation and intensive growth (propagation) of cracks. The cracks propagate quickly in the casehardened (carburized) layers and stop when they reach the viscous layers. The intensity of these processes is determined by the magnitude and character of the stress concentration distribution in the area where the tooth enters (fits) into the cutter body and in the area of the bottom of the hole for the carbide (hard-alloyed) tipped tooth.

The neutral lines of relative tension-compression on the wedge working surface of the tooth form possible sliding planes along which the destruction of the carbide tooth takes place due to exceeding the strength parameters of the hard alloy. These planes describe the hyperbolic (higher plane) curves that begin in the area of the working load and end in the area of the stress concentration in the zone where the tooth enters (fits) into the cutter body. The location of the chipping surfaces of the carbide (hard-alloyed) teeth is in good agreement with the orientation of the planes of possible destruction, which is revealed during the analysis of damage and destruction of the teeth of the used (spent) drill bits (see figures 3 and 5).

The magnitude of the stresses concentration in the area where the tooth enters (fits) into the cutter body, as well as in the area of the bottom of the hole, depends not only on the physical and mechanical properties of the conjugated materials but also on the design features of the teeth shanks. A significant influence on the stress concentration distribution is also imposed by the design features of the connection of the carbide (hard-alloyed) tipped teeth with the cutter body. In particular, the high strength of the cutter steel and the too close location of the tipped teeth cause the more intensive formation of cracks. In the cases of non-optimal strength parameters, for example, those assessed according to the DI criterion, the disastrous destruction of the cutters takes place due to the formation of cracks in the standard structures of the cutters with the carbide (hard-alloyed) tipped teeth [19, 20].

In general, the established character of the stress distribution makes it possible to form a system of design and technological problems, the ultimate goal of which is to reduce the magnitude of the stresses concentration in the "tooth – cutter ring" connection, as well as in the cutter's body.

4. Conclusions

The paper is focused on analysing the basic design and technological aspects of functionally oriented technology of manufacturing the three-cone drill bits. The first stage of the present investigations consists in studying the interaction between the rolling cutter teeth and the rock being drilled. At this stage, the distribution of forces and stresses characterizing the rock-breaking process, as well as the stress-strain state of the "hard-alloy tooth – rolling cutter body" connection, are considered. The conditions of durable complex cyclic loading or repeatable overloading usually cause the initiation of cracks and their propagation, followed by cutter teeth fracture. In the case of chipping the impact surfaces of the teeth out, their shapes change. Some teeth become sharper, while the other ones fail in the zone of contact with the cutter body. The abrasive wear is characterized by microscratches when the chipped-out surfaces of the teeth become smooth.

The second stage of the present research is dedicated to estimating the operational indicators of the drill bits 215.9C3-ПВ manufactured by Universal Drilling Technique LLC (Drohobych Drill Bits Plant, Ukraine) on the basis of the corresponding performance tests. The largest influence on the drill bit strength and durability is made by the following technological factors: the accuracy of the design parameters of the teeth; the accuracy of the geometric parameters of the holes in the cutter rings; the accuracy and reliability of the compression connection of the teeth with the cutter rings; the heat treatment parameters ensuring optimal distribution of physical and mechanical properties of the cutter rings holes (before pressing) and the "tooth – ring" connection (after pressing); the presence of surfactants on the mated surfaces.

The established character of the stress distribution makes it possible to form a system of design and technological problems, the ultimate goal of which is to reduce the magnitude of the stress concentration in the "tooth – cutter ring" connection, as well as in the cutter's body. Therefore, the obtained results of the present research can be effectively used by the scientists, designers, and technologists dealing with designing and manufacturing the three-cone drill bits, as well as implementing the improved drilling technologies.

References

[1] Franca L F P 2010 Drilling Action of Roller-Cone Bits: Modeling and Experimental Validation *J. Energy Resour. Technol.* **132** 043101
[2] Suto Y and Takahashi H 2011 Effect of the load condition on frictional heat generation and temperature increase within a tri-cone bit during high-temperature formation drilling

Geothermics **40** 267–74

[3] Ševčík M, Hutař P, Náhlík L and Seitl S 2013 The effect of constraint level on a crack path *Eng. Fail. Anal.* **29** 83–92

[4] Huang Z, Li Q, Zhou Y, Jing S, Ma Y, Hu W and Fan Y 2013 Experimental research on the surface strengthening technology of roller cone bit bearing based on the failure analysis *Eng. Fail. Anal.* **29** 12–26

[5] Yahiaoui M, Paris J-Y, Denape J and Dourfaye A 2015 Wear mechanisms of WC–Co drill bit inserts against alumina counterface under dry friction: Part 1 — WC–Co inserts with homogenous binder phase content *Int. J. Refract. Met. Hard Mater.* **48** 245–56

[6] Yahiaoui M, Paris J-Y, Denape J, Colin C, Ther O and Dourfaye A 2015 Wear mechanisms of WC-Co drill bit inserts against alumina counterface under dry friction: Part 2 — Graded WC-Co inserts *Int. J. Refract. Met. Hard Mater.* **48** 65–73

[7] Deng Y, Chen M, Jin Y, Zhang Y, Zou D and Lu Y 2016 Theoretical and experimental study on the penetration rate for roller cone bits based on the rock dynamic strength and drilling parameters *J. Nat. Gas Sci. Eng.* **36** 117–23

[8] Shigina A A, Shigin A O, Stupina A A, Kurchin G S and Kirsanov A K 2016 Model of rock drilling process in terms of roller cone bit remaining life *Int. J. Appl. Eng. Res.* **11** 9792–9

[9] He W, Chen Y, He J, Xiong W, Tang T and OuYang H 2016 Spherical contact mechanical analysis of roller cone drill bits journal bearing *Petroleum* **2** 208–14

[10] Guo Z, Deng R, Liang Z and Yang S 2018 Non-spherical single cone bit mechanics analysis and tooth load test *J. Pet. Sci. Eng.* **167** 458–65

[11] Guo Z, Deng R and Zhang W 2019 Non-spherical single cone bit tooth failure analysis and tooth load test *Eng. Fail. Anal.* **105** 1–11

[12] Guo Z and Deng R 2021 Research on tooth wear of non-spherical single-cone bit in hard rock formation of deep well *Eng. Fail. Anal.* **125** 105408

[13] Šporin J, Mrvar P, Petrič M, Vižintin G and Vukelić Ž 2019 The characterization of wear in roller cone drill bit by rock material – Sandstone *J. Pet. Sci. Eng.* **173** 1355–67

[14] Šporin J 2020 Characterisation of the Wear of the Roller Cone Drill Bit Caused by Improperly Chosen Drilling Regime *Mater. Geoenvironment* **67** 91–102

[15] Abbas R K, Ghanbarzadeh A and Hassanpour A 2020 A novel method for estimating the real-time dullness of tri-cone oil drill bits *Eng. Fail. Anal.* **109** 104386

[16] Yang Y, Huang K, Song D, Chen X, Chen H and Chen L 2020 The study on tooth failure and fixation strength of cone bits under high temperature *Eng. Fail. Anal.* **110** 104444

[17] Yang Y, Song D, Chen H, Chen X and Chen L 2021 The research on the fixation strength of tapered cone bit teeth *Eng. Fail. Anal.* **124** 105318

[18] Prakash S and Mukhopadhyay A K 2020 Reliability analysis of tricone roller bits with tungsten carbide insert in blasthole drilling *Int. J. Mining, Reclam. Environ.* **34** 101–18

[19] Slipchuk A and Jakym R 2021 Technological Parameters of Hole Shaping in the Cones Rolling-Cutter Row of Roller Cone Bits *Lecture Notes in Mechanical Engineering* pp 123–32

[20] Slipchuk A, Jakym R, Lebedev V and Kurkchi E 2021 The Influence of Technological Factors on the Reliability Connection for Tungsten Carbide Insert Cutter with Cone in the Roller Cone Bits *Lecture Notes in Mechanical Engineering* pp 443–52

[21] de Moura J, Xiao Y, Yang J and Butt S D 2021 An empirical model for the drilling performance prediction for roller-cone drill bits *J. Pet. Sci. Eng.* **204** 108791

[22] Fan L and Wang D 2021 Study on delamination inhibition and chip breakage mechanism in drilling metal laminated materials with double cone drill *J. Manuf. Process.* **64** 81–94

[23] Shu L, Li S, Fang Z, Kizaki T, Kimura K, Arai G, Arai K and Sugita N 2021 Study on dedicated drill bit design for carbon fiber reinforced polymer drilling with improved cutting mechanism *Compos. Part A Appl. Sci. Manuf.* **142** 106259

[24] Ma Y, Yuan Z, Ni Y, Meng X and Peng X 2022 Performance prediction and multi-objective optimization of metal seals in roller cone bits *J. Pet. Sci. Eng.* **208** 109316

IOP Conf. Series: Materials Science and Engineering 1277 (2023) 012016 doi:10.1088/1757-899X/1277/1/012016

Comparative studies of the three-body abrasion wear resistance of hardfacing Fe-Cr-C-B-Ti alloy

B Trembach

Chief Designer Department, Mining and Press-and-Forging Equipment, Novo-Kramatorsky Mashinostroitelny Zavod, Kiev, 04070, Ukraine

btrembach89@gmail.com

Abstract. In this research, the mechanical and tribological properties of two deposited metal of Fe–C–Cr–B–Ti alloying systems. In this study, comparative tests of the deposited metal hardfacing by two self-shielded flux-cored wires electrode were carried out: with an exothermic additive (110Cr4Cu5Ti1MnVB) and without an exothermic additive (140Cr15TiSi1MnVB). Abrasive wear resistance was evaluated using a slightly modified procedure on a standardized tribological machine for abrasive wear, according to the limits established by the ASTM G65. The results of studies of the efficiency of introduction of an exothermic addition CuO-Al system into the core filler of a flux-cored wire electrode were presented.

1. Introdaction

Wearing was and remains an actual problem in the industry. Many scientists are focused at this problem solving [1, 2, 3, 4]. Introduction of new design solutions [5, 6, 7] and new materials [3, 4, 8, 9, 10] provides great benefits to manufacturers and consumers of their products. Metallurgical and mining industries remain important segments of the economy. The cost of worn parts in mining is approximately the same as the cost of maintenance. In addition, wearing of some equipment in mining industry can lead to catastrophic failures and emergency stops, which can have a bad influence to equipment efficiency [11]. Since the greatest number of industrial parts is withdrawn from exploitation due to wear (53% abrasive, 24% adhesive, 10% impact and other kinds of wear) and corrosion (13%) [12]. At that abrasive wear is the most widely spread type of wearing. According to the present-day views of things abrasion resistance to wear can be increased by a suitable selection of abrasion-resistant of the coating material [13]. In most cases, such a material will contain an increased number of alloying elements and will have increased mechanical properties compared to conventional structural materials.

Construction of coatings implies the simulations of melting processes [14, 15], including smart methods of evaluating energy processes [16], thermal calculations [17, 18, 19], and calculations of the stress-strain state of the coating layers under local load [20] and in the crack initiation zone [21].

The development of surfacing technologies is carried out by taking into account theoretical approaches to the selection of the alloy component composition [22, 23, 24], technological heredity [25], and optimization of detail manufacturing technologies [26] to ensure the functioning of products during the life cycle [27, 28].

The hardfacing is considered as economical way to improve the performance of components submitted to severe wear conditions, with wide range of applicable filler materials [29, 30, 31, 32]. Among the methods of applying a wear-resistant layer by the method of hardfacing, the following were most widely used: shielded metal arc or manual metal arc welding (SMAW or MMAW) [33], gas metal arc welding (GMAW), submerged arc welding (SAW), self-shielded flux-cored arc welding (FCAW-S) [34, 35, 36, 37].

Flux-cored wire is one of the most advanced and constantly growing segments of the welding consumables market in the global economy. This growth is conditioned by the inherent advantages of the flux-cored welding process, and is expected to be the first preference for new entrants due to their high deposition rate, efficiency in delivering work, high quality of the deposited metal and arc visibility. Recently, the greatest part of economy sectors (building, automobile construction, industry, etc.) more often use the of flux-cored wires as the main welding material. The worldwide market for flux cored welding wire is valued at 1546.1 million USD in 2020 is expected to reach 2169.2 million USD by the end of 2026. By the end of 2026 it will make ~750 thousand tons of flux-cored wire. Implementation of flux-cored wires and self-shielded flux-cored wires for hardening and restoring worn parts in mining and metallurgical equipment. One of the advanced areas for further development and improvement flux-cored wires is introduction of exothermic addition into the core filler of a flux-cored wire electrode [29, 36, 38, 39]. Niobium microalloying has a positive effect on the wear resistance of stainless steels [40]. Introduction of an exothermic addition into the core filler of a flux-cored wire electrode change metal transfer mode [38] increase the electrode melting rate [41], improve the weld bead morphology [36] and reduce the cooling rate [42] during FCAW process.

Among the whole variety of applied exothermic addition systems, the exothermic addition CuO-Al system is of a great interest, because it allows to perform additional alloying of deposited metal by copper [29]. The positive effect of alloying with copper deposited metal can be an increase in corrosion resistance [11].

The purpose of this work is to make a comparative researches of wear resistance of 110Cr4Cu5TiVBAl alloy having a low chromium content and high copper content with 140Cr15TiSi1MnVB alloy having a high chromium content in two-body abrasive wear conditions.

2. Experimental Section

2.1. Materials

The flux-cored wire of 4 mm diameter was used for investigations. The flux-cored wire were coated with different compositions of the powder mixtures. A marble-fluorspar-rutile was used as the slag basis, the gas slag-forming component. The powder of core filler materials is composed of gas-slag-forming components (fluorite concentrate GOST 4421–73, rutile concentrate GOST 22938–78, calcium carbonate GOST 8252–79), deoxidizers components (ferrosilicon FS-92 GOST 1415–78, ferromanganese FMN-88A GOST 4755–91), alloying components (metal chrome powder X99 GOST 5905–79, boron carbide powder, graphite, ferrovanadium FVd-40 GOST 27130–94, titanium powder PTM–3 TU 14–22-57–92), exothermic addition component (oxide of copper GOST 16539-79, aluminium powder PA1 GOST 6058-73) and iron powder. The difference between filler materials was as follows: an equivalent amount of metal chrome powder was added to the flux-cored wire FCAW-S-140Cr15TiSi1MnVB (with the symbol S) instead of the exothermal addition (CuO-Al) components for a flux-cored wire with an exothermic addition FCAW–SE–110Cr4Cu5TiVBAl (with the symbol SE). The coefficient wire filling (filling factor) of the flux-cored wire electrode is 0.34-0.35.

2.2. Hardfacing process

The hardfacing was carried out by three-layers on plates made from low carbon steel S 235 J2G2 EN 10025-2 with dimensions 10x100x200 mm on reverse polarity by A-874 automatic machine. Thus, hardfacing was performed by two flux-cored wire according to the next hardfacing conditions: wire feed speed WFS = 1,85 m/min; arc voltage U_a = 28 V; torch speed TS = 0,3 m/min; contact tip to work distance CTWD = 45 mm. Morevan, the temperature of preheating is T_p = 250…300 °C. There are 3 layers made during hardfacing. Each layer was formed by sequential deposition of weld bead with a partial overlap of the previous weld bead (1/3). Samples for microstructure analysis, mechanical properties investigation and three-body abrasive wear test where prepared by mechanical cutting from the deposited plates.

2.3. Structure and mechanical properties

The alloys were examined by light optical microscope Neophot. Quantitative metallography was carried out with structural analyzer Epiquant. X-ray diffraction analysis was done to identify the existing phases in produced samples on an X-ray diffractometer DRON-UM-1 with CuKα source.

The physical-mechanical properties of the samples surface were investigated by the depth-sensing indentation (instrumented indentation) method using the universal micro/nano-hardness tester "Micron-Gamma". The measurements were done using a diamond Berkovich indenter. Indenter penetration depth against the specimen surface is recorded with an accuracy of 5 nm. The measurements were carried out with maximum load on the indenter 50 g and a loading rate 5 mN/s.

2.4. Abrasive wear test

The detailed test procedures were discussed elsewhere [43].

The weight loss due to wear was noted down. Volume loss (*WV*) due to wear, the specific wear rate (*SVR*) and relative wear (φ_{abr}) resistance were calculated after the test using the following equations:

$$WV = {}^{\Delta m}/_{\rho} \tag{1}$$

WV – wear volume, mm^3;
Δm – is the mass loss of the specimen, g;
ρ – is the experimental density of the sample, g/ mm^3;

$$SVR = {}^{WV}\!/_{N \cdot L} \tag{2}$$

N – normal load, N;
L – sliding distance, m.

$$\varphi_{abr} = {}^{\Delta m_e}/_{\Delta m} \tag{3}$$

where Δm_e – mean weight loss of etalon sample, g;
Δm – mean weight loss of samples of tested material, g.

3. Results and discussion

3.1. Microstructure of the Hardfacings

In weld metal deposited bt FCAW-S-140Cr15TiSi1MnVB and FCAW-SE-110Cr4Cu5Ti1MnVB, apart from Fe_2B and $Fe_3(B,C)$ borides, the carbides Cr_7C_3 and TiC might be also present according to the reported results through X-ray diffraction analyses, as shown in figure 1. While the matrix is a eutectic of borides, α-Fe и γ-Fe. Whereas, in the high chromium alloy without copper, a large intensity of Cr_2C_3 carbide was observed.

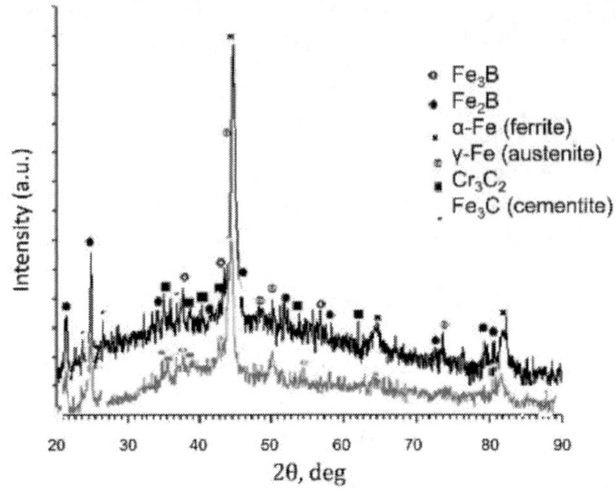

Figure 1. XRD pattern of deposited metal.

Figure 2 presents the SEM image of the surface morphologies after potentiostatic tests at the applied potential of – 1.5 and 0.5 VSCE of the deposited metal sample of hardfacing by FCAW–S–140Cr13Si1TiMnB and the experimental FCAW–SE–110Cr4Cu5TiVBAl with the introduction of exothermic addition (CuO–Al) in the core filler.

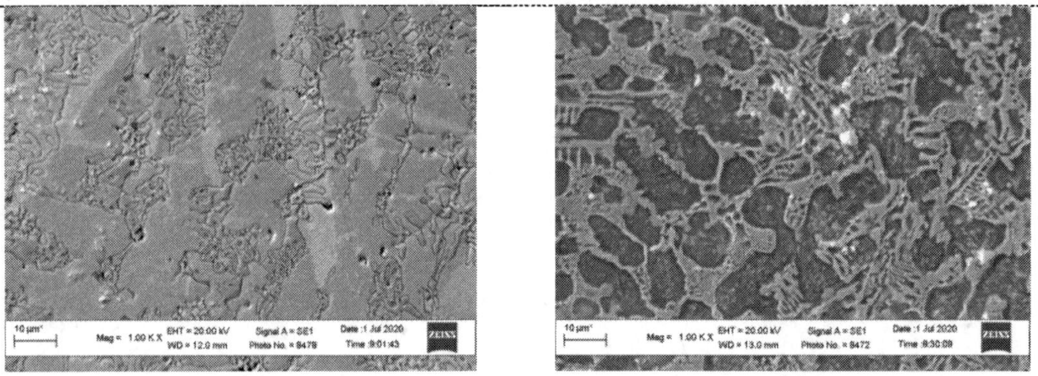

a) b)

Figure 2. SEM image of the deposited metal following the corrosion test hadfacing by: a) FCAW–S–140Cr15Si1MnBTi; b) FCAW–SE–110Cr4Cu7TiVBAl.

Data analysis showed that the introduction of an exothermic addition CuO–Al in the core filler of flux-cored wire electrode had a positive effect on the grain morphology. This can be explain by the formation of a large number of small non-metallic inclusions (NMI), which played the role of grain refiner/modifying agents. Due to the uniform distribution of the components of the exothermic addition (copper oxide and aluminum), as well as their fine fraction (less than 10nm). The manufacturing technology of flux-cored wire consists of preliminary intensive mixing of the core filler components. As a consequence, the components of the exothermic mixture are evenly distributed in the core filler. In addition, they are separated by other core filler components (slag-forming, alloying, iron powder). When an exothermic reaction occurs at the end of the flux-cored wire, copper is reduced from its oxide due to the oxidation of aluminum. Fine inclusions of Al_2O_3 corundum were formed in view of the features of the described above exothermic mixture components distribution.

3.2. Mechanical properties

Analysis and processing of the registered indentation curves allows to obtain mechanical properties of studied samples (instrumented indentation hardness (H_{IT}), modulus of elasticity (E_{TI})). The result of measuring the modulus of elasticity of weld metal deposited with flux-cored wire with an exothermic addition was greater than the modulus of elasticity of deposited metal deposited with flux-cored wire without, despite the higher content of carbon and chromium was shown. (E_{TI}(SE)=187 GPa>E_{TI}(S)=177 GPa). Whereas the instrumented hardness values are almost the same (H_{TI}(SE)=9,94 GPa=H_{TI}(S)=10,08 GPa).

The introduction of the exothermic additive CuO–Al had a positive effect on the mechanical properties. The increase in the hardness can be explained primarily by the strengthening as per the Hall-Petch mechanism associated with the structure dispersion [44]. The positive effect on the elasticity modulus in case of exothermic addition introduction to the core filler can be explained by the increasing of ferrite in the matrix, which have higher modulus of elasticity than austenite.

3.3. Three-bode abrasive wear

After 3-body abrasive wear the weight loss of the sample, hardfaced with flux-cored wire electrode FCAW - S - 140Cr15Si1MnBTi (without exothermic addition), had a value of 0.0867 g. The relative wear resistance (φ) was SVR = 0.1278 mm^3/Nm and relative wear resistance φ=7,8. Whereas in case of using the experimental flux-cored wire electrode FCAW-S-110Cr4Cu5Ti1MnVB with the introduction of an exothermic addition CuO--Al in the core filler, the weight loss was 0.0533 g. At that the wear resistance was SVR = 0.0786 mm^3/Nm and relative wear resistance φ=12,68. Thus, the wear resistance increased 1.58 times.

Morphological features of the friction surfaces due to wear of the 3-body abrasive metal, hardfacing by FCAW-S-140Cr15Si1MnBTi and FCAW-S-110Cr4Cu5Ti1MnVB are showed in figure 3.

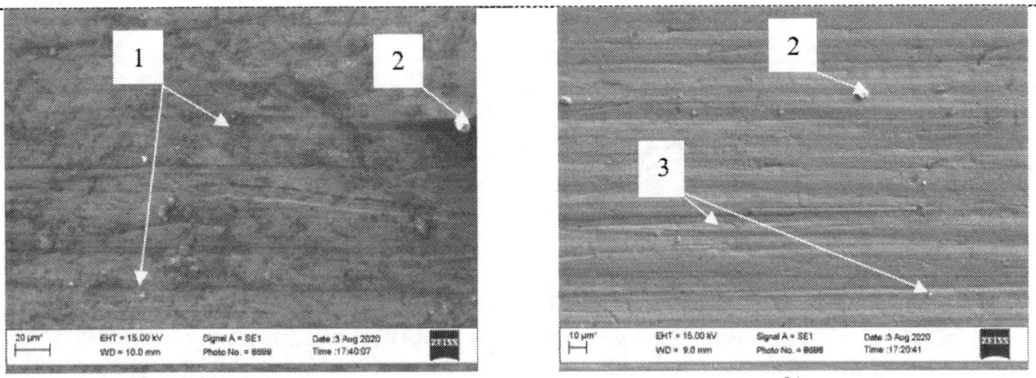

a) b)

Figure 3. Worn surfaces of the hardfacings: (a) FCAW-S-140Cr15TiSi1MnVB; (b) FCAW-SE-110Cr4Cu5Ti1VB with exothermic addition (CuO–Al). 1 - place of high-cycle fatigue failure; 2 - embedded abrasive particle; 3 - ledge of plastically deformed material.

Combination of micro-cutting and microcracing wear mechanisms was observed in reinforced two deposited metals tested under the two-body abrasive wear. Dominant wear pattern of FCAW-S-140Cr15TiSi1MnVB hardfaced surfaces was microcracing as was shoved in figure 3 (a). Sharp tops of solid phase act as a stress concentrators, from which the deposited metal cracking with the further crumbling begins. For the reason that eutectic borides (Fe, Cr)$_2$B and Fe_3(B, C) due to significant stresses after some time they were damaged and separated.

Deposited metal received using proposed flux-cored wire FCAW-SE-110Cr4Cu5Ti1VB had a higher wear-resistance to abrasive wearing. Higher wear- resistance of the reinforcing layer applied by experimental flux-cored wire can be explained by the grain size reducing as well as an increasing of more damage-proof and plastic ferrite phase in the matrix (figure 3 (b)). Increasing of ferrite phase in the matrix and eutectic allows to reduce the intensity of locations concentrations and due to this fact to reduce sensitivity of deposited metal to the stress accumulation.

4. Conclusion

1. Two of flux-cored wire electrode with adjusted proportions of initial components were produced to study the effects of these components on the phase composition, microstructure, hardness, and wear resistance of Fe–C–B-Cr-Ti hardfacings, which were manufactured by self-shielded flux-cored arc welding.

2. Introduction of exothermal addition (CuO-Al) to the core filler alloyed the deposited metal by copper and increased the content of ferrite and austenite phase in the matrix. It's confirmed by more intensity at XRD patterns.

3. Abrasive wear resistance of deposited metal tested under three-body abrasive wear test increased firmly introduction of exothermic addition CuO-Al to the core filler of the flux-cored wire electrode. This is due to an increase in the more elastic ferrite phase as well as a decrease in grain size.

References

[1] Tyczewski P., Nadolny K., Zwierzycki W., Ulbrich D. 2022 *Materials* **15** (19) pp. 6734.

[2] Ulbrich D., Stachowiak A., Kowalczyk J., Wieczorek D., Matysiak W. 2022 *Materials* **15** (11), 3892.

[3] Sukhova O. V. 2020 *Phys. Chem. Solid St.*, 21 (2), 355–360.

[4] Sukhova O.V., Syrovatko Yu.V. 2019 *Metallofiz. Noveishie Tekhnol.* **41** (9) pp. 1171–1185.

[5] Bazaluk O, Slabyi O, Vekeryk V, Velychkovych A, Ropyak L and Lozynskyi V 2021 *Energies* **14** (12) art. no. 3514.

[6] Shatskyi I, Vytvytskyi I, Senyushkovych M and Velychkovych A 2019 *IOP Conference Series: Materials Science and Engineering* **564** (1) art. no. 12073.

[7] Moisyshyn V and Levchuk K 2016 *Mining of Mineral Deposits* **10** (3) pp 65–76.

[8] Ropyak L Y, Pryhorovska T O and Levchuk K H 2020 *Progress in Physics of Metals* **21** (2) pp 274–301.

[9] Prysyazhnyuk P M, Shihab T A and Panchuk V H 2016 *Materials Science* **52** (2) pp 188–193.

[10] Bazaluk O, Velychkovych A, Ropyak L, Pashechko M, Pryhorovska T and Lozynskyi V 2021 *Energies* **14** (14) art. no. 4198.

[11] Trembach B O, Sukov M G, Vynar V A, Trembach I O, Subbotina V V, Rebrov O Yu, Rebrova O M and Zakiev V I 2022 *Metallofiz. Noveishie Tekhnol.* **44** (4) pp 493–515.

[12] Nedeljković B, Lazić V, Aleksandrović S, Krstić B, Mutavdžić M, Milosavljević D and Đorđević M 2010 *Metalurgija–J Metall.* **16** (2) pp 77–90.

[13] Bazaluk O, Dubei O, Ropyak L, Shovkoplias M, Pryhorovska T and Lozynskyi V 2022 *Energies* **15** (1) art. no. 83.

[14] Pawlik J, Cieślik J, Bembenek M, Góral T, Kapayeva S and Kapkenova M 2022 *Materials* 2022 **15** (17) 6019.

[15] Imanbekova U, Hotra O, Koshimbayev S, Popiel P and Tanaś J 2015 *In Proceedings of the Photonics Applications in Astronomy, Communications, Industry, and High-Energy Physics Experiments* 2015 Wilga, Poland 25 May 2015 **9662** art. no. 966246.

[16] Hasanov M, Zakovorotniy A, Leonov S, Khrypunov G, Dmitrienko V and Klochko A 2020 *In Proceedings of the 2020 IEEE 4th International Conference on Intelligent Energy and Power Systems (IEPS)* Istanbul, Turkey 7–11 September 2020 pp 221–224.

[17] Tatsiy R M, Pazen O Y, Vovk S Y, Ropyak L Y and Pryhorovska T O 2019 *Journal of the Serbian Society for Computational Mechanics* **13** (2) pp 36–55.

[18] Tatsii R M, Stasyuk M F and Pazen O Y 2021 *Journal of Engineering Physics and Thermophysics* **94** (2) pp 298–310.
[19] Bandura A and Skaskiv O 2019 *Rocky Mountain Journal of Mathematics* **49** (4) pp 1063–1092.
[20] Ropyak L Ya, Makoviichuk M V, Shatskyi I P, Pritula I M, Gryn L O and Belyakovskyi V O 2020 *Functional Materials* **27** (3) pp 638-642.
[21] Shatskyi I, Makoviichuk M, Perepichka V and Dalyak T 2017 *In 23rd International Conference Engineering Mechanics* May 15–18, 2017 pp 866–869.
[22] Shihab T, Prysyazhnyuk P, Semyanyk I, Anrusyshyn R, Ivanov O and Troshchuk L 2020 *Management Systems in Production Engineering* **28** (2) pp 84–89.
[23] Krivileva S, Zakovorotniy A, Moiseev V, Ponomareva N, Rassokha A and Zinchenko O 2019 *Functional Materials* **26** (2) pp 347–352.
[24] Radchenko T M, Gatsenko O S, Lizunov V V and Tatarenko V A 2020 *Progress in Physics of Metals* **21** (4) pp 580–618.
[25] Kusyi Y and Stupnytskyy V 2020 *Lecture Notes in Mechanical Engineering* pp 276–284.
[26] Dmytriv V T, Dmytriv I V, Horodetskyy I M, Horodniak R V and Dmytriv T V 2021 *INMATEH – Agricultural Engineering* **65** (3) pp 233–240.
[27] Kusyi Y M and Kuk A M 2020 *J. Phys. Conf. Ser.* **1426** (1) art. no. 012034.
[28] Kopei V B Onysko O R and Panchuk V G 2020 *J. Phys. Conf. Ser.* **1426** (1) 012033.
[29] Trembach B, Grin A, Makarenko N, Zharikov S, Trembach I and Markov O 2020 *J. Mater. Res. Technol.* **9** (5) pp 10520-10528.
[30] Dilay Y, Güney B and Özkan A 2022 *Surf. Rev. Lett.* **29** (04) art. no. 2250044.
[31] Sawczuk W, Cañás A M R, Ulbrich D and Kowalczyk J 2021 *Materials* **14** (16) art. no. 4766.
[32] Valenzuela-De la Rosa F, Torres-Sánchez R, Domínguez-Ríos C and Aguilar-Elguézabal A 2021 *Metals* **11** (10) art. no. 1568.
[33] Kocaman E, Kılınç B, Şen Ş, and Şen U 2021 *Arab J Sci Eng.* **7** pp 1-17.
[34] Ivanov O, Prysyazhnyuk P, Lutsak D, Matviienkiv O and Aulin V 2020 *Management Systems in Production Engineering,* **28** (3) pp 178–183.
[35] Trembach B, Grin A, Turchanin M, Makarenko N, Markov O and Trembach I 2021 *Int J Adv Manuf Technol* **114** (3) pp 1099–1118.
[36] Bembenek M, Prysyazhnyuk P, Shihab T, Machnik R, Ivanov O and Ropyak L 2022 *Materials* **15** (14) art. no. 5074.
[37] Kocaman E, Kılınç B, Şen Ş and Şen U 2022 *Journal of Alloys and Compounds* **900** art. no. 163478.
[38] Li H L, Liu D, Guo N, Chen H, Du Y P and Feng J C 2017 *J Mater Proc Technol* **245** pp 149–156.
[39] Rong X, Zhao D, He C, Shi C, Liu E and Zhao N 2021 *Acta Mater* **204** art. no. 116524.
[40] Shyrokov V V, Vasyliv K B, Duryahina Z A, Laz'ko H V and Rats'ka N B 2009 *Materials Science* **45** (4) pp 473–480.
[41] Allen J W, Olson D L and Frost R H 1998 *Weld. J* **77** pp 277–285.
[42] Murzin V V and Russo V L 1994 *Weld Int.* **8** (1) pp 43–46.
[43] Student M M, Hvozdetskyi V M and Dzhoba Y V *Problems Trib* **89** (3) pp 33–41.
[44] Gramajo J, Gualco A and Svoboda H 2019 *Mater Res.* **22** (6) art. no. e20190469.

Development of the material homogeneity criterion for the analysis of processes of technological inheritability for quality parameters of machine parts

Ya M Kusyi[1], O R Onysko[2], A M Kuk[3] , A V Zinko[4]

[1, 3, 4] Department of Robotics and Integrated Mechanical Engineering Technologies, Lviv Polytechnic National University, 12 Stepan Bandera Street, Lviv 79013 Ukraine
[2] Department of Computerized Mechanical Engineering, Ivano-Frankivsk National Technical University of Oil and Gas, 15 Karpatska Street, Ivano-Frankivsk 76019, Ukraine
[1] E-mail: jarkym@ukr.net

Abstract. The material homogeneity criterion and the indicators characterized by this criterion were developed due to the LM-hardness method. The research results were tried out during the manufacturing of steel shaft. The values of the Weibull homogeneity coefficient during shaft's machining increase from 6.12-11.46 (for initial workpiece) to 198.23-344.59 (for finish part) during its manufacturing. Besides the constant of the material A_m in the manufacturing chain "initial workpiece – finish part" changes from 0.814 to 0.966 for different necks of the shaft during its machining by cutting and abrasive methods. The elaborated method can be used when manufacturing and operation of different mechanical engineering parts for analysis of the change of their material homogeneity during machining, control and determining of their residual life.

1. Introduction

The technological processes planning during the parts manufacturing and the assembly of machines obeys the general principles of the technical systems theory [1, 2].

Integration of defined subsystems into a single system of automated design in mechanical engineering allows for the optimization of design and technological preparation of production during the parts manufacturing and assembly of machines [3, 4].

The priority principles of functionally-oriented and object-oriented technological process planning are used in modern integrated systems of automated preparation of mechanical engineering production [3, 5].

The main criterion for the object-oriented technological process planning for part manufacturing is the technological provision of the minimum cost of details manufacturing and assembly of machines in accordance with the regulated indicators of the accuracy for processing or assembly, the quality characteristics of the surface microrelief and macro deviations of the surfaces. The structure of the object-oriented technological process for manufacturing of machine parts is designed in accordance with their annual production program using typical technological processes [3, 4].

Technological provision of operational parameters and reliability characteristics of parts and machines in accordance with defined workloads is implemented in compliance with the bases of functionally-oriented design during their manufacturing and assembly [3, 6], considering the technological inheritability of quality indicators of the mechanical engineering parts [7, 8].

Carriers of inherited information - the processed material and the surface of the part - actively participate in technological processes, passing through various operations and transitions, testing the actions of determining technological factors [7-9].

Technological inheritability is provided by a complex of material properties, surface quality parameters and the surface layer of machine parts during the manufacturing of machine parts, which determines the direction of further research. This approach requires the development of a comprehensive criterion for a step-by-step assessment of the state of the parts material in order to predict failures at priority substages and stages in the Life Cycles of the mechanical engineering parts.

2. Literature Review

The service life of machines is closely related to the physical, chemical and mechanical properties of materials and coatings [10, 11], accuracy indicators [12, 13] and quality characteristics of the surface layers of the functional surfaces of parts [14, 15], including taking into account technological inheritability [7], during the Life Cycles of the mechanical engineering parts [8].

Using the Surface Engineering and Surface Integrity tools makes it possible to form databases related to the technological support of processing accuracy indicators, surface quality parameters, operational parameters and reliability characteristics of the mechanical engineering parts [6, 9, 16, 17].

The technological inheritability of parts parameters determines a set of restrictions for the technological process planning of parts manufacturing from the received initial blank during the implementation of the i-th technological operation ($i \in [1; n]$)). These restrictions provide the requirements of processing accuracy, quality characteristics of the surface microrelief, deviation of the form and mutual arrangement of surfaces and material characteristics in accordance with the technical requirements of design and technological documentation [5].

The methodology of functionally-oriented design is based on the system of parallel engineering - CAPE (Concurrent Art-to-Product Environment) with connections between input and output data of models and software modules.

ABAQUS, LS-Dyna, AdvantEdge, DEFORM cutting process simulation systems are widely used in modern mechanical engineering. The works of V.V. Stupnutskyy [3, 4], J.P. Davim [6, 17], F. Klocke [18, 19], S.K. Lee [20] and others are devoted to the study of the processes of simulation modeling of the processes of primary forming machine parts.

In particular, the optimal structure of functionally-oriented technological processes is formed using an integral qualitative indicator. It systematically characterizes the corrosion resistance, fatigue strength, wear resistance, etc. of the most heavily loaded surfaces of the part. The qualitative indicator is obtained as a result of the formation during the fabrication of the microtopology of the surface layer, residual stresses and deformations [3, 4].

Experimental studies, along with the development of simulation methods for modeling technological processes, are important for confirming theoretical propositions, accumulating databases, and predicting the behavior of products in operational conditions [7, 21].

Technological defects during the operation of products, which are formed at the stage of their development, are transformed into damage under the influence of various force loads, temperature effects, mechanical interactions, electric and magnetic fields, aggressive environments, etc. Parts in products during their operation are, as a rule, exposed to the complex influence of external factors. At the same time, the wear caused by the friction of the conjugate surfaces is a characteristic type of damage to most machines, their mechanisms and parts [7, 8].

It was established [22, 23] that the amount of wear of conjugate surfaces that work under friction conditions directly depends on the hardness of the material of their parts. Therefore, the hardness of the material of the parts is considered as a characteristic for evaluating the processes of material degradation and the regularities of the construction of the structure of the technological routes of the surfaces of the machine parts during studies of the technological inheritability of their quality parameters. This concerns the technological processes planning for the manufacturing of machine parts operating under conditions of force loads and intensive wear.

Hardness, as an important mechanical indicator of a material, characterizes its strength and plasticity and the ability to resist plastic deformation and brittle fracture upon contact with an indenter [22]. At the same time, the traditional method of determining hardness is uninformative due to the weak sensitivity of hardness to many modufications of structure transformation [23] and ineffective due to the ambiguity of the correlation of hardness values with the characteristics of other mechanical properties, in particular yield and strength limits [22, 23].

Increasing the reliability of determining the characteristics of the surface layer of the material is provided by a new approach in materials science research based on the use of the LM-hardness method [7, 8, 22, 23, 24]. According to this method, the degree of homogeneity of the material structure, which is analyzed by the value of the characteristics of the distribution law, is directly related to the degree of dispersion of the indicators of the physical and mechanical properties of the material. This method is most easily implemented in engineering practice when using hardness as a mechanical characteristic, the value of which is used for indirect assessment of properties [17, 18].

Homogeneity is a parameter that fully characterizes the state of the material during the processing of the results of hardness control. Homogeneity is described by the Weibull distribution [22, 23] in constructing statistical theories of strength:

$$P(\sigma)=1-e^{-\left(\frac{\sigma}{k}\right)^{m}}, \tag{1}$$

where the Weibull homogeneity coefficient (m) is defined [7, 8]:

$$m=d(n)/(2{,}30259\cdot S(\ell g(H))), \tag{2}$$

$$S(\ell g(H))=\sqrt{\frac{1}{n-1}\cdot\sum_{i=1}^{n}\left(\ell g(H_i)-\overline{\ell g(H)}\right)^2}, \tag{3}$$

$$\overline{\ell g(H)}=\frac{1}{n}\cdot\sum_{i=1}^{n}\ell g(H_i). \tag{4}$$

Other static parameters, in particular, the coefficient of variation, correspond to the level of dispersion of the studied physical and mechanical indicators, in particular, hardness. It shows how large the dispersion of the values that contain the volume of data under consideration is compared to the average hardness value [22, 23]:

$$\nu=\sqrt{\frac{1}{n-1}\cdot\sum_{i=1}^{n}\left(H_i-\overline{H}\right)^2}\Big/\overline{H}, \tag{5}$$

where $\overline{H}$ is the average value of hardness.

3. Research methods

The analysis of material homogeneity changes provided by various methods at the stage of procurement production and operation of products [5, 7, 8] should be implemented for the surfaces of machine parts during their manufacturing. This makes it possible to check the correct ordering of technological routes for the surfaces machining of machine parts by technical requirements.

Material homogeneity is described by the homogeneity coefficient (m) in the practice of mechanical engineering. The dynamics of changes in the coefficient (m) represent the state of the part surface during its manufacturing or operation for a specific material [7, 8]. The dynamics of changes in the form of the surface will be different for different materials with other physical and mechanical properties.

Therefore, other characteristics must be used to specify the homogeneity of the surface layer of machine parts, in particular, the material constant and technological damageability, establishing their relationship with the Weibull coefficient (m).

The dependence for the use of non-destructive methods of monitoring the state of the material and the calculation of the material constant A_m is established based on the condition of equivalence of the probability of brittle failure according to the Weibull theory [5] for bending and torsion:

$$A_m = \left((m+2)/(4m+4)\right)^{1/m}, \tag{6}$$

where m is the homogeneity coefficient in the Weibull distribution.

The Weibull homogeneity coefficient (m) is determined in formal investigations based on testing samples of different sizes or by Gumbel's formulas based on test data of a batch of identical samples.

Since the material, along with the parameters of the specified surface of the part, serve as carriers of information about technological inheritability during the manufacturing of machine parts, it is good practice to use the parameter of material homogeneity to analyze the technological inheritability of part quality parameters by assessing the degradation of its properties. In particular, it is implemented for the executive surfaces of the object of mechanical engineering based on the dynamics of changes in the homogeneity coefficient (m), material constant A_m and damageability D [7, 8].

The structural heterogeneity of the material by its physical and mechanical characteristics is considered at the sub-micro, micro- and macro levels. The structural heterogeneity of the material is determined by the analysis of physical processes at the macro level in the manufacture of machine parts based on the analysis of its chemical composition, the method of obtaining and the technology of shaping of the blank, the presence of macro defects (inclusions, pores, shells) after procurement operations and technological treatments [5].

Three variants of technological inheritability of material properties at the macro level are implemented in the technological chain "blank – finished part" according to the criterion of material homogeneity in the manufacturing of machine parts [5]: the structurally homogeneous material of the blank after the execution of the technological process of its processing is transformed into the structurally homogeneous material of the finished part; the structurally inhomogeneous material of the blank after the execution of the technological process of its processing provides the structurally inhomogeneous material of the finished part; the structurally inhomogeneous material of the workpiece after the removal of the defective layer due to the implementation of the technological process provides the structurally homogeneous material of the finished part.

According to the provisions of the Kachanov-Rabotnov theory [25, 26]:

$$\psi = d\tilde{A}/dA, \tag{7}$$

$$D = 1 - \psi = \left(dA - d\tilde{A}\right)/dA = dA_D/dA, \tag{8}$$

where ψ, D – integrity and damageability, respectively, dA, dAD, $d\tilde{A}$ – the total area of the studied area of the material, the area of the material without damage and with damage, respectively.

The change of technological damageability D is determined by the dispersion of hardness characteristics according to Brinell, Rockwell, and Vickers using the Weibull coefficient (m) changes according to (8) by means of the LM-hardness method [5]:

$$D = \left(m_{matr.} - m_i\right)/m_{matr.}, \tag{9}$$

where $m_{matr.}$, m_i are the Weibull coefficients (m), characterizing the dispersion of the hardness of the base material and in the i-th section of the part, respectively.

The intensity of growth of technological damageability j_D is determined by [5]:

$$j_D = \left(m_{matr.} - m_i\right)/m_i. \tag{10}$$

Technological damageability D (9) is interpreted as the ratio of the difference in the Weibull coefficients of the base material and in the i-th section of the part to the Weibull coefficient of the base material. At the same time, the intensity of of growth of technological damageability j_D (10) is the ratio of the difference in the Weibull coefficients of the base material and in the i-th cross-section of the part to the Weibull coefficient in the i-th cross-section of the part.

The dynamics of changes in material homogeneity indicators are proposed to be taken into account along with a set of regulated parameters of part quality (processing accuracy, geometric and physical-mechanical quality parameters of the surface layer, etc.) during the analysis of technological inheritability for the principle of functionally-oriented of technological processes planning:

$$\left| \begin{array}{l} m_{x_0} < m_{x_1} < ..m_{x_k} < ... < m_{x_n} \to \infty;\ v_{x_0} > v_{x_1} > ..v_{x_k} > ... > v_{x_n} \to 0 \\ A_{x_0} < A_{x_1} < ...A_{x_k} < ... < A_{x_n} \to 1;\ D_{x_0} > D_{x_1} > ..D_{x_k} > ... > D_{x_n} \to 0; \end{array} \right|, \quad (11)$$

where m_{x_0} .. m_{x_n} ; v_{x_0} .. v_{x_n} ; A_{x_0} … A_{x_n} ; D_{x_0} ... D_{x_n} – the change in the homogeneity coefficient, the coefficient of variation, the material constant, the technological damageability of the part material and the intensity of its growth during the implementation of the rational route of processing the x-th surface; k=0 – blank; k=1,...n – methods of processing the x-th surface, respectively.

The coefficients of Weibull homogeneity (m) and the material constant A_m , the technological damageability of the part material D and the intensity of the growth of the technological damageability j_D associated with (m), are determined by the results of processing test samples (witness samples) or are selected from the database during machining the executive surface according to the rational route of its processing.

The material homogeneity criterion was developed on the basis of conducted investigations (see fig. 1). It is characterized by the homogeneity of the material as a parameter for the analysis of the technological inheritability of the characteristics of the processed surfaces of the part, and is described by the following indicators: the Weibull homogeneity coefficient (m), the coefficient of variation v, the material constant A_m, the technological damageability D and the intensity of its growth j_D.

In the general case, the material homogeneity criterion (MHC) (see fig. 1) is presented for the analysis of the technological inheritability of the quality parameters of parts during their manufacturing by cutting methods, plastic deformation and control of the properties of the surface layer by the LM-hardness method, which is determined by a set of indicators for each technological step in the form:

– for technological step during technological inheritability of properties

$$MHC \in [\, m = f(H_1, H_2......);\ v = f(H_1, H_2......);\ A = f(m);\ D = f(m);\ j_D = f(m)], \quad (12)$$

– for the technological inheritability of quality parameters of the part from the blank to the finished detail

$$\left| \begin{array}{l} m_{x_1} < ..m_{x_k} < ... < m_{x_n} \to \infty;\ v_{x_1} > ..v_{x_k} > ... > v_{x_n} \to 0 \\ A_{x_1} < ...A_{x_k} < ... < A_{x_n} \to 1;\ D_{x_1} > ..D_{x_k} > ... > D_{x_n} \to 0; \end{array} \right| \text{for } T_0 = \sum_{k=1}^{n} t_{ok}, \quad (13)$$

– for the structurally inhomogeneous state of the surface layer for the part material

$$v \to 1(100\,\%);\ m < 10;\ A_m < 0.878;\ D \to 1;\ j_D \to \infty, \quad (14)$$

– for a structurally homogeneous state of the surface layer for the part material

$$v \to 0;\ m \ge 10;\ A_m \ge 0.878;\ D \to 0,\ j_D \to 0 \quad (15)$$

where t_{0k} is the machining time during performing the k-th technological step (processing method); T_0 is the total machining time for processing the specified surface of the part.

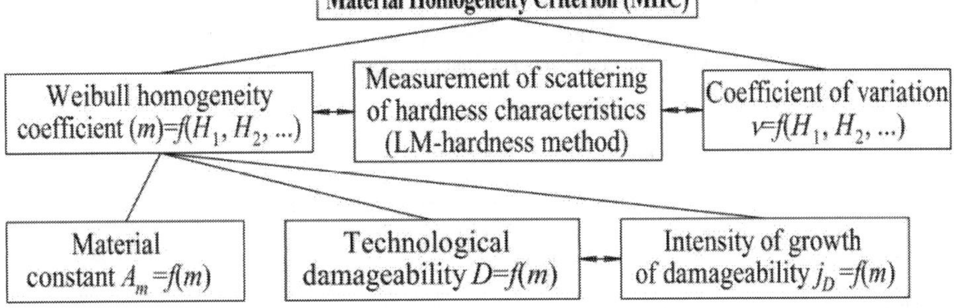

Figure 1. The structure of the material homogeneity criterion with indicators characterized it

The value of the coefficient (m) is determined in dependencies (12)–(15) according to formulas (2)–(4), the coefficient of variation – according to formula (5), the material constant A_m – according to formula (6), the technological damageability D and intensity of its growth j_D - according to formulas (9) and (10), respectively.

The number of used indicators (12)-(15) is determined by the complexity of the performed technological tasks for various design stages to assess the homogeneity of the workpiece material of the part at the specified technological step. The complexity of the performed tasks increases from procurement and rough operations to finishing, finishing and final strengthening operations. In particular, the homogeneity of the material of the blank is analyzed by indicators m, A_m, and the homogeneity of the material of the finished part is analyzed by indicators (m), A_m, D, etc.

4. Results

The testing of the obtained research results was realized during the manufacturing of the shaft 6E4-2717.00.00.01 of the drive drum of the belt conveyor (see fig. 2) with functional surfaces: ∅ 90h7, ∅ $125h7_l$, ∅ $148h9_l$, ∅ 156h14, ∅ $148h9_r$, ∅ $125h7_r$ [5].

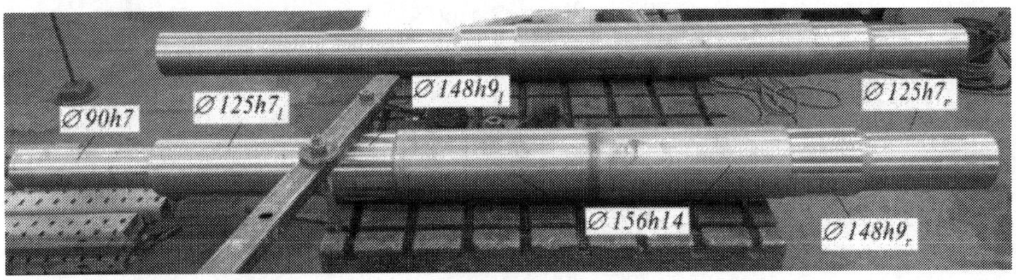

Figure 2. Steel shaft manufactured by the Ukrainian enterprise "Transsystem"

The hardness of the processed surfaces of the shaft was measured by a portable hardness tester TD-42 ("Ultrakon"). The standard method of Brinell hardness measurement after mechanical processing of the workpiece was used. Calculations of the indicators of the material homogeneity criterion were carried out using the integrated mathematical system MathCAD [5, 26].

The results of changes in Weibull homogeneity coefficients (m) and coefficients of variation v for the blank and after each machining method are shown in Table 1. The value of the constant A_m for the blank and after each machining method is shown in Table 2.

Providing the technical requirements for the surfaces of the part according to the principle of functionally-oriented design is carried out using the criterion of material homogeneity. The regulated accuracy and quality parameters of the surface layers of the part are provided by increasing the values of the Weibull coefficients (m) and decreasing the coefficients of variation v after each subsequent technological step in the manufacturing chain "blank – finished part".

Table 1. The results of changes for the indicators of the material homogeneity criterion during the processing of shaft surfaces

Processing methods	Processed surfaces and controlled indicators					
	∅ 90h7		∅ $125h7_l$		∅ $148h9_l$	
	m	ν (%)	m	ν (%)	m	ν (%)
Blank	6.12	18.77	6.13	17.87	9.67	11.98
Rough turning	35.37	3.24	34.02	3.32	40.89	2.74
Finish turning	98.82	1.15	91.93	1.23	100.43	1.13
Rough grinding	239.72	0.49	215.62	0.53	206.46	0.55
Finish grinding	333.89	0.34	344.59	0.33		

Processing methods	Processed surfaces and controlled indicators					
	∅ 156h14		∅ $148h9_r$		∅ $125h7_r$	
	m	ν (%)	m	ν (%)	m	ν (%)
Blank	11.46	9.63	11.14	10.10	6.10	18.80
Rough turning	32.06	3.59	37.69	2.96	33.81	3.34
Finish turning			93.54	1.22	88.03	1.28
Rough grinding			198.23	0.57	218.55	0.52
Finish grinding					336.86	0.34

Table 2. Change of the constant of the part material for processing routes of a shaft surfaces in compliance with technological inheritability

Processing methods	∅ 90h7	∅ $125h7_l$	∅ $148h9_l$	∅ 156h14	∅ $148h9_r$	∅ $125h7_r$
Blank	0.815	0.815	0.874	0.892	0.889	0.814
Rough turning	0.962	0.961	0.967	0.959	0.965	0.961
Finish turning	0.986	0.985	0.986		0.987	0.985
Rough grinding	0.994	0.994	0.993		0.993	0.994
Finish grinding	0.996	0.996				0.996

This behavior is characterized by an increase in the homogeneity of the material structure of the steel shaft and reduces the tendency to its technological damageability from the blank to the finished part (see Table 1).

On the other hand, improving the material constant A_m involves using the principle of technological inheritability of shaft parameters after each technological step during machining. At the same time, the dispersion of values of the material constant A_m is greater for blanks and rough technological treatments with stabilization of values for finishing ones. This regularity indicates greater uniformity of the structure of the material for finishing treatments compared to rough (preliminary) methods of machining (see table 2).

5. Conclusions

The importnat conclusions have been formed based on the investigation results.

1. Further development of the principle of functionally-oriented technologies for the machining of machine parts is systematically provided using the approaches of theoretical, experimental investigations and computer modelling of technological processes, taking into account the technological inheritability of characteristics from the blank to the finished part.

2. Technological inheritability of material properties of steel parts during their machining by the principle of functionally-oriented design ensures the transformation of structurally inhomogeneous material of workpieces into structurally homogeneous material of finished products. It is verified by the increasing values of the Weibull homogeneity coefficients (*m*) from 6.123-11.457 (for the initial workpiece) to 198.227-344.588 (after different grinding methods).

3. The constant of the part material A_m in the manufacturing chain "blank – finished part" increases from 0.814 (for the blank) to 0.966 (after grinding) for the shaft 6E4-2717.00.00.01. Besides, the scattering of the values of material constant A_m is greater for the workpiece of the part and rough technological processing with stabilization of values of material parameters for finishing ones. This regularity indicates the greater homogeneity of the material structure after finishing compared to the rough methods of surface treatment and gives practical recommendations for the rational technological routes planning of the functional surfaces of the parts.

4. The developed method can be used to analyze the change in the homogeneity of mechanical engineering parts during their machining, inspection and operation.

5. Further research should be performed for other materials and different types of mechanical engineering parts as well as for technological processes to realisation the developed scientific method into engineering practice.

References

[1] Hubka V and Eder W E 1988 *Theory of Technical Systems. A Total Concept Theory for Engineering Design* (Berlin Heidelberg: Springer-Verlag)

[2] Haken H 2006 *Information and Self–Organization. A Macroscopic Approach to Complex Systems* (Berlin: Springer)

[3] Stupnytskyy V 2013 Features of Functionally-Oriented Engineering Technologies in Concurrent Environment *International Journal of Engineering Research & Technology* **2(9)** 1181–1186

[4] Stupnytskyy V 2013 Computer aided machine- building technological process planning by the methods of concurrent engineering *Europaische Fachhochschule: Wissenschaftliche Zeitschrift ORT Publishing* **3** 346–354

[5] Kusyi Y 2021 *Scientific and Applied Basics of the Technological Inheritability of the Quality Parameters for Providing of the Products Operation Characteristics*, Lviv, Lviv Polytechnic National University, Ukraine, Doctoral Thesis (in Ukrainian)

[6] Davim J P 2010 *Surface Integrity in Machining of Hard Materials* (London:Springer-Verlag)

[7] Kusyi Y and Stupnytskyy V 2020 Optimization of the Technological Process Based on Analysis of Technological Damageability of Casting *Lecture Notes in Mechanical Engineering* **1** 276–284 https://doi.org/10.1007/978-3-030-50794-7_27

[8] Kusyi Ya M and Kuk A M 2020 Investigation of the technological damageability of castings at the stage of design and technological preparation of the machine Life Cycle *Journal of Physics: Conference Series* **1426(1)** Retrieved from: https://iopscience.iop.org/article/10.1088/1742-6596/1426/1/012034/meta

[9] Rajmohan T, Palanikumar K and Davim J P 2012 Analysis of surface integrity in drilling metal matrix and hybrid metal matrix composites *Journal of Materials Science & Technology* **28** 761–768 Retrieved from: https://www.jmst.org/EN/Y2012/V28/I8/761

[10] Shihab T, Prysyazhnyuk P, Semyanyk I, Anrusyshyn R, Ivanov O and Troshchuk L 2020 Thermodynamic Approach to the Development and Selection of Hardfacing Materials in Energy Industry *Management Systems in Production Engineering* **28(2)** 84-89 doi:10.2478/mspe-2020-0013

[11] Bembenek M, Prysyazhnyuk, P, Shihab T, Machnik R, Ivanov O and Ropyak L 2022 Microstructure and Wear Characterization of the Fe-Mo-B-C—Based Hardfacing Alloys Deposited by

Flux-Cored Arc Welding *Materials* **15(14): 5074** doi: 10.3390/ma15145074

[12] Bazaluk O, Velychkovych A, Ropyak L, Pashechko M, Pryhorovska T and Lozynskyi V 2021 Influence of heavy weight drill pipe material and drill bit manufacturing errors on stress state of steel blades *Energies* **14(14) 4198** https://doi.org/10.3390/en14144198

[13] Pryhorovska T and Ropyak L 2019 Machining Error Influnce on Stress State of Conical Thread Joint Details *Proc. of the 8th Int. Conf. on Advanced Optoelectronics and Lasers (CAOL)* **9019544** 493-497 doi:10.1109/CAOL46282.2019.9019544

[14] Bazaluk O, Dubei O, Ropyak L, Shovkoplias M, Pryhorovska T and Lozynskyi V 2022 Strategy of compatible use of jet and plunger pump with chrome parts in oil well *Energies* **15(1) 83** https://doi.org/10.3390/en15010083

[15] Ropyak L Ya, Makoviichuk M V, Shatskyi I P, Pritula I M, Gryn L O and Belyakovskyi V O 2020 Stressed state of laminated interference-absorption filter under local loading *Functional Materials* **27(3)** 638-642 http://dx.doi.org/10.15407/fm27.03

[16] Gupta K 2020 *Surface Engineering of Modern Materials* (Switzerland: Springer)

[17] Davim J P 2012 *Materials and Surface Engineering: Research and Development* (Cambridge: Elsevier)

[18] Klocke F 2011 *Manufacturing Processes 1: Cutting* (Berlin: Springer-Verlag)

[19] Klocke F, Beck T, Hoppe S, Krieg T, Müller N, Nöthe T, Raedt H-W and Sweeney K 2002 Examples of FEM application in manufacturing technology *Mat. Proc. Technol.* **120(1-3)** 450–457 doi:10.1016/S0924-0136(01)01210-9

[20] Lee B Y, Tarng Y S and Ma S C 1995 Modeling of the process damping force in chatter vibration *Journal of Materials Processing Technology* **35(7)** 951–962

[21] Stotsko Z A., Kuzin O A, Kuzin M O and Mechnik V A 2022 Estimation of strength properties of functionally graded structures with elliptical stress concentrators *Archives of Materials Science and Engineering* **113(1)** 35–41 doi: 10.5604/01.3001.0015.6973

[22] Lebedev A A, Makovetskii I V, Muzyka N R, Volchek and Shvets V P 2006 Assessment of damage level in materials by the scatter of elastic characteristics and static strength *Strength of Materials* **38** 109-116

[23] Muzyka N R, Shvets V P and Boiko A V 2020 Procedure and Instruments for the Material Damage Assessment by the LM-Hardness Method on the In-Service Scratching of Structure Element Surfaces *Strength of Materials* **52** 432-439 https://doi.org/10.1007/s11223-020-00195-6

[24] Kopei V B, Onysko O R, Panchuk V G 2020 Principles of development of product lifecycle management system for threaded connections based on the Python programming language *Journal of Physics: Conference Series* **1426(1)** Retrieved from: https://iopscience.iop.org/article/10.1088/1742-6596/1426/1/012033

[25] Kachanov L M 1999 Rupture time under creep conditions *International Journal of Fracture* **97(1-4)** 11–18 doi:10.1023/A:1018671022008

[26] Murakami S 2012 Continuum Damage Mechanics – A Continuum Mechanics Approach to the Analysis of Damage and Fracture (Dordrecht/Heidelberg/ London/New York: Springer)

Acknowledgments

This research has been conducted as part of ongoing project "Comprehensive system of functional-oriented planning of machining difficult-to-cut materials for the military-industrial complex (Komplekssys)" and has been funded by the Research Council of Lithuania and the Ministry of Education and Science of Ukraine under the Lithuanian–Ukrainian Cooperation Programme in the Fields of Research and Technologies (Grant No. S-REP-21-5).

Accuracy of Thread Profile Depends on the Lathe Tool Angle of Inclination. Theoretical Investigation

O R Onysko[1], **Ya M Kusyi**[2], **V B Kopei**[1], **A P Dzhus**[3]

[1]Department of Computerized Mechanical Engineering, Ivano-Frankivsk National Technical University of Oil and Gas, Ukraine
[2]Department of Robotics and Integrated Mechanical Engineering Technologies, Lviv Polytechnic National University, Ukraine
[3]Department of Oil and Gas Field Machinery and Equipment, Ivano-Frankivsk National Technical University of Oil and Gas, Ukraine

onysko.oleg@gmail.com

Abstract..Among threaded connections of large sizes, tapered thread pipe connections are especially often used, which are a very important part of drill strings. The efficiency of the drill strings largely depends on the accuracy of the tool-joint tapered thread. The production of such joints is implemented on lathes, with the help of tools that have a carbid insert. To ensure high performance, such inserts are recommended to be installed not parallel to the axis of the thread, but at the lead angle of thread. However, the profile of the insert itself is equal to the profile of the thread, and therefore it is important to have a theoretical predictive calculation of the probable influence of the angular setting of the carbide insert on the accuracy of the thread. Based on a detailed consideration of the geometry of the mutual placement of the plate and the tapered thread and the kinematic features of the process, an algorithm for predictive calculation of the accuracy of the thread is created. The result shows that only one of the parameters of the cutting edge really depends on the angle of inclination of the cutting edge - it is its profile. The deviation can reach 7% of the tolerance on the semi-profile angle.

Keywords: Pitch, half angle of thread, tolerance,tool-joint, lathe tool, carbid insert.

1. Introduction

The most common process for manufacturing tapered threads of large diameters from 30 mm to 500 mm is turning. Such fittings include pins and boxs of pipes of the oil and gas assortments and elements of drill strings. Characteristic for grooving is the use of cutters with a special cutting profile, as well as special geometric parameters, including the angle of inclination of the cutting edge.

An important aspect of the thread machining process is compliance with the requirements for the accuracy of the obtained thread. In this regard, it is good to know which of the tool parameters, turning modes or kinematic features are the most influential on predicting the obtained threading accuracy of parts with large diameters.

Therefore, the increase in the production of hydrocarbons as a whole largely depends on the modernization of approaches in the production of thread surfaces, increasing their life cycle and increasing the wear-stability and efficiency of cutting tools.

Thus, research is based on the consideration of the influence of machining methods, their modes and parameters on the accuracy of the obtained tapered drill-pipe threads. The topic of research requires a detailed consideration of the theory of shaping and its calculation and experimental confirmation.

2. Literature Review

Increasing of the volume of hydrocarbon production requires the use of new equipment designsand tools with improved operational characteristics [1, 2, 3]. A characteristic feature drilling and oil and gas equipment is used in its composition of a large number of threaded ones connections. Reliability and durability of threaded connections is achieved through a rational approach selection of materials and coatings [4, 5, 6], precision parameters of mechanical processing [7, 8, 9] and indicators of the quality of the working surfaces of the nozzles [10, 11, 12], including taking into account the technological heredity [13], and lubrication [14] to ensure the operability of products throughout their lifetime cycle [15, 16].

All threads for connecting of drill string elements are tapered . According to the API 7 standard, there are 27 types of triangular rounded thread, which are distinguished by several parameters, one of which is the diameter of the crest at the smallest base of the cone. Among the most used of those threads according to the parameter of pitch P and height H are those made in the form of V-0.038R (Fig. 1) [10].

Modern manufacturing of tool-joint tapered thread is oriented towards the use of turning processes and occasionally milling processes [17]. The theoretical study of deviations from the specified dimensions of the cutting due to the deformation of the workpiece during the turning process is presented in [18], however, the work does not cover the topic of the accuracy of the thread pitch. In an experimental study [19], rather than a theoretical study, the effect of the geometric parameters of the cutter insert on its durability was proven. There is no mention of the accuracy of the thread obtained by turning. The results of the investigations of experiments in which the influence of turning modes on the accuracy of the thread are presented in the publication [20], but there are no experimental or theoretical data of the effect on the thread pitch. The distribution of pressure along the casing pipe joint depending on the size of the pitch is presented at the models developed in studies [21]. However, this work has nothing to do with the process of making the thread. The process of turning an external thread is worked out in [22], which deals with the effects of the feed method on the regimes of turning forces, and not on the accuracy of the obtained threaded surface. Considerable theoretical materials on the research of bolted joints are carried out in [23], but we are talking about the distribution of axisymmetric loads, and not about accuracy. The influence of the geometric parameters of the thread cutter on the strength, but not the accuracy of the obtained part, is studied in [24]. Research [25] is related to the influence of parameters not of turning, but of grinding, on the accuracy of the profile of the obtained thread. The process of turning hard-working steels is studied in the article [26], although among the parameters, there is no mention of the parameters of the cutting tool. The use of MTP to create segmented chips requires in threading operations is shown in [27]. The paper [28] presents the main dynamics and stability of thread turning operations with custom multipoint inserts. Modeling of milling processes (rather than turning) of asymmetric threaded profiles is demonstrated in work [29]. The influence of the geometric parameters of the turning tool on the cutting performance is carried out in the study [30]. The paper [31] covers the study of thread processing with a multithreaded cutter, but does not touch on the issue of the accuracy of the thread obtained. The study of turning processes and its power parameters is presented in articles [32, 33], which do not contain the results of the study of cutting accuracy.

The problem of modern conical drill thread connections is their certain asymmetry, because according to the standard [34], such threads have a longer straight section, for example, segment *AB* in Fig. 1 and a shorter one - for example, segment *AD* in **Fig. 1**.

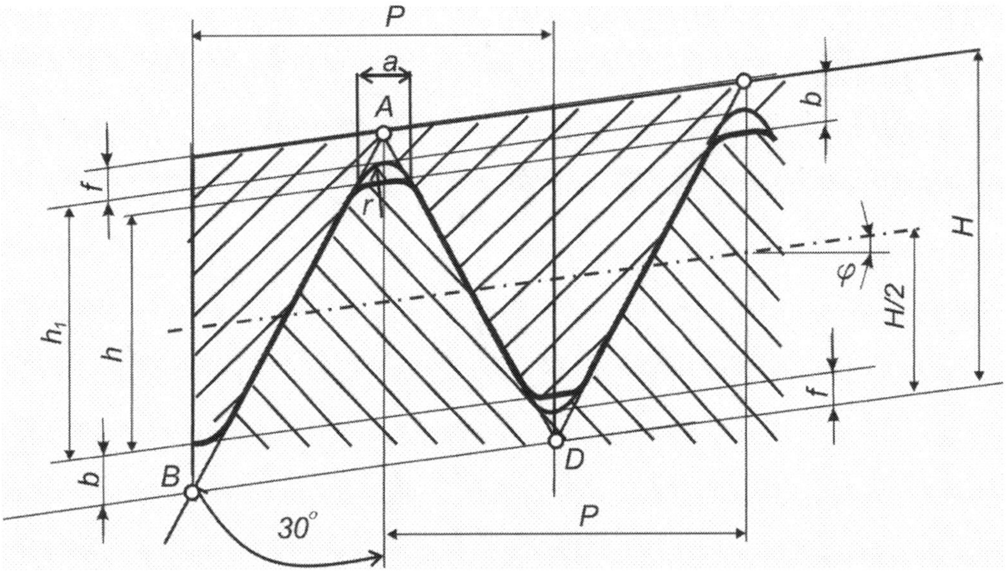

Fig. 1. Scheme of the profile of the tapered thread according to the standard API 7

Most modern tool-makers recommend using an inclined installation of the carbid insert to ensure an equal distribution of cutting forces on the left and right perpendicular sections of the cutting edge [35]. As a rule, this angle of inclination of the cutting edge λ is equal to the lead angle of the thread, which is indicated by the symbol φ in **Fig. 2**.

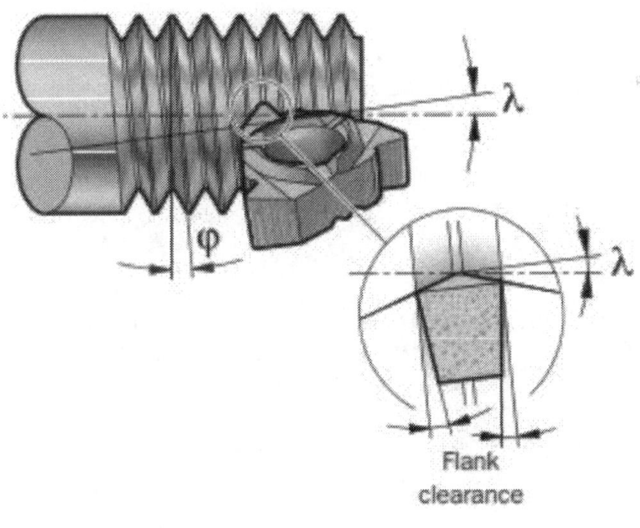

Fig. 2. Scheme of installation of the carbide insert at an angle λ

3. Research methods

The main parameters that regulate the accuracy of the thread include pitch accuracy, profile accuracy, and pitch diameter accuracy. The latter is evaluated only in the connection: bolt-nut, box-pin and others. The profile mainly depends on the precision of the cutting edge and the set-up errors of the cutter and the blank. The pitch is considered to depend on the accuracy of the kinematics of the machine. However, based on the fact that the tapered thread can be asymmetrical (Fig. 1), the ratio of the greater and lesser part of the distance between the top points of the thread B and D relative to the perpendicular to the axis, which is drawn through the point A (**Fig. 3**), should be determined.

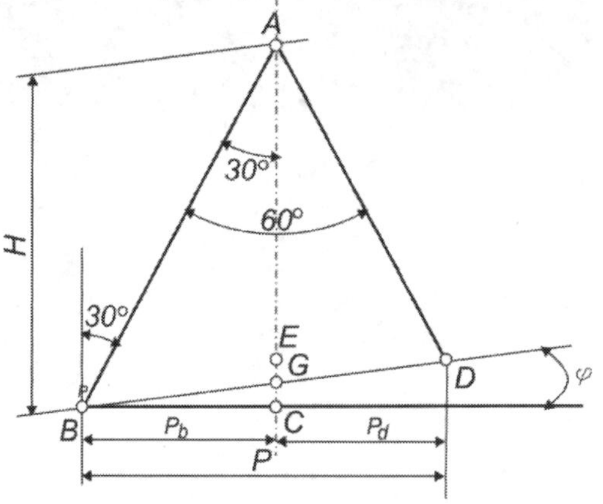

Fig. 3. Scheme for calculating the components of the pitch *Pd* and *Pb* of the tool-joint thread

Therefore, the pitch P of the tapered thread consists of two unequal sections: segment *Pb* along the larger section of the profile of the thread *AB*, and segment *Pd* - along the smaller section of the profile of the thread *AD*. The sum of both segments is the size of the pitch:

$$P = P_b + P_d$$

From the triangle *ABG*, according to the theorem of sines, the relation is valid::

$$\frac{|AB|}{sin(\widehat{AGB})} = \frac{H}{sin(\widehat{ABG})} \quad (1)$$

From the triangle *AGD*, according to the theorem of sines, the relation is valid:

$$\frac{|AD|}{sin(\widehat{AGD})} = \frac{H}{sin(\widehat{ADG})} \quad (2)$$

where:

angle $\widehat{AGB} = \frac{\pi}{2} + \varphi$

angle $\widehat{AGD} = \frac{\pi}{2} - 30° - \varphi$

In this way, using equations (1) and (2), we obtain expressions for determining the parts of the pitch, which are the components of the pitch *P* for the asymmetric version of the profile of driil-string tapered thread.

$$P_b = \frac{H \cdot cos\,\phi}{2\,sin(60° - \phi)} \quad (3)$$

$$P_d = \frac{H \cdot cos\,\phi}{2\,sin(60° + \phi)} \quad (4)$$

However, in order to determine the accuracy of the pitch, taking into account the indicated two unequal components of it, one should also take into account the manufacturers' recommendations [35], that is, apply the scheme from **Fig. 2** for calculation. As can be seen from Fig. 4, the angle of inclination of the cutting edgeequalls lead angle of thread $\lambda = \psi$.

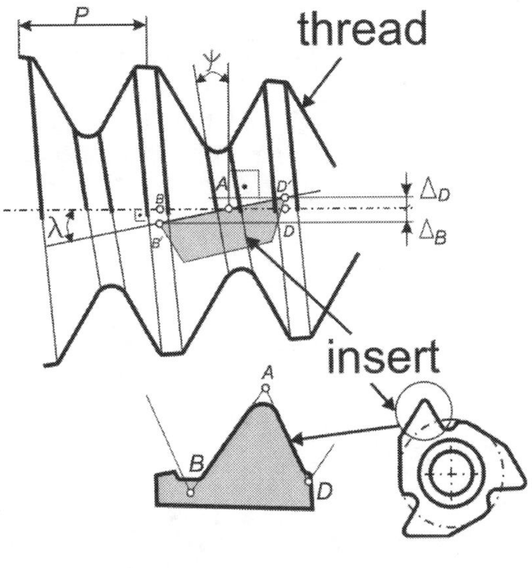

Fig. 4. Scheme of installation of the carbide insert at an angle λ

The illustration in Fig. 4 emphasizes the visualization of tangential deviations of points *D* and *B* of the carbide insert: Δ_D and Δ_B, respectively. These deviations can cause changes in the components of the pitch, because in the process of turning, the specified points do not work simultaneously: first, the workpiece, making a rotational movement, contacts point *D*, and later with point *B* (**Fig. 5**). All intermediate points between *A* and *B*, as well as between *A* and *D*, will also be initialized in the form of profile deviations, which means pitch. The tangential deviations Δ_D and Δ_B indicate the following: point *D* of the theoretic profile of the thread due to the axial movement of the cutter will be performed as point *D'*, i.e. faster than point *A*; point *B* of the theoretic profile of the thread due to the axial movement of the cutter will be made as point *B'*, i.e. later than point *A*. Thus, due to the axial feed of the cutter, there is an axial displacement of the ridge points of the thread *D* and *B* by δ_D and δ_B, respectively. At the same time, point A does not move either in the axial or tangential directions, since it is located in the axial plane of the threaded part. All other points of the profile of the obtained thread can be placed on the line segments A_1D_1' A_1B_1', which approximately reflect the axial profile of the obtained thread.

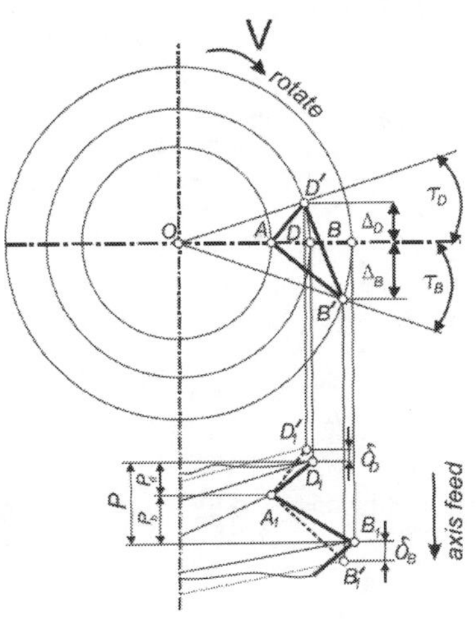

Fig. 5..Inaccuracy of the thread profile due to the use of the angle λ

IOP Conf. Series: Materials Science and Engineering 1277 (2023) 012018 doi:10.1088/1757-899X/1277/1/012018

Under such circumstances, the size of the pithc *P* will also undergo certain changes, namely, it will shift by the value $\delta_B - \delta_D$ compared to the standard (fig. 6). This happens as a result of trimming the width of the crest of each turn first from the right and then from the left. At the same time, the width of the crest a - as can be seen from the figure - will decrease quite significantly and, as can be seen from figure 6, it will be called a':

$$a' = a - \delta_D - \delta_B, \tag{5}$$

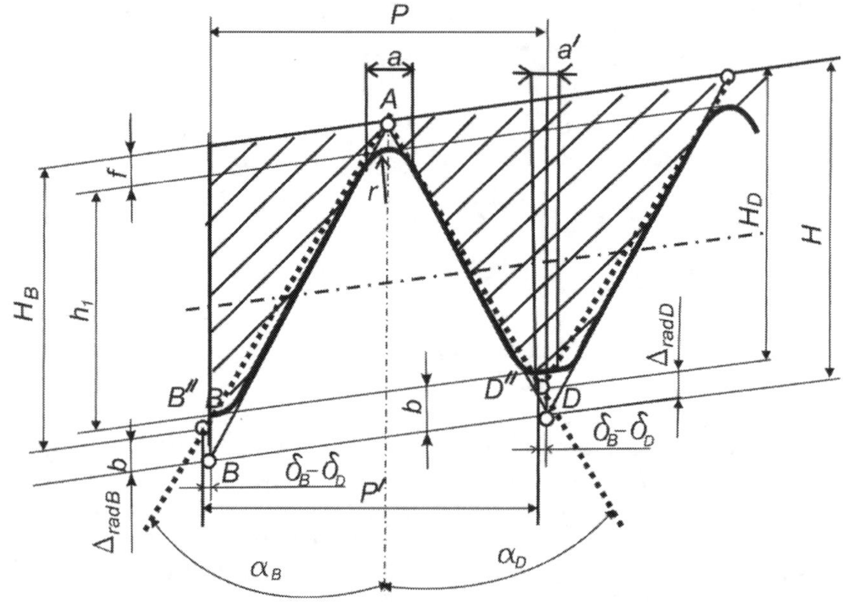

Fig. 6. Change in the width of the crest of the profile of the driil-string thread due to the using of the angle λ

Using the diagram in **Fig. 5**, we have the formula for δ_D and δ_B :

$$\delta_B = \frac{P \cdot \tau_B}{2\pi}; \tag{6}$$

$$\delta_D = \frac{P \cdot \tau_D}{2\pi}, \tag{7}$$

where:

$\tau_B = arcsin\frac{\Delta_B}{|OB'|};$

$\tau_D = arcsin\frac{\Delta_D}{|OD'|},$

where:

$\Delta_B = P_B.tg(\lambda);$

$\Delta_D = P_D.tg(\lambda);$

$|OB'| = |OB| = r_B;$

$|OD'| = |OD| = r_D,$

where:

r_B i r_D – radii of poits D and B. Using the standard [33] and fig. 7 obtained formulas for their determination:

$$r_D = \frac{d_3}{2} + b; \quad (8)$$

$$r_B = \frac{d_3}{2} + P \cdot tg(\varphi) + b, \quad (9)$$

where d_3 – the outer diameter of the conical keyhole at its smallest turn. b – crest truncation.

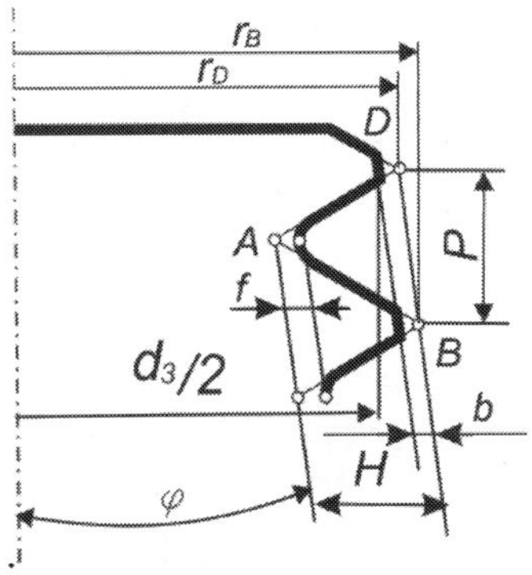

Fig. 7. Scheme for determining the radii of points B and D

3.1. Effect on the value of the angle of the profile of the thread

According to the standard [34], the half angle of the profile is 30 ° (fig.1). The tolerance of this angle according to [34] is ±0.75°. For predictive calculation of the deviations of the angles α_B and α_D (fig. 6) depending on the angle λ, іт іс поссібле то apply **fig. 3**, **fig. 5**, fig. 6 apply the following algorithm:

$$\alpha_B = arctg\left(\frac{P_b+\delta_B}{H_B+P_b tg\varphi}\right), \quad (10)$$

$$\alpha_D = arctg\left(\frac{P_d+\delta_D}{H_D-P_d tg\varphi}\right), \quad (11)$$

where:

$H_B = \mathrm{H} - \Delta_{radB},$

$H_D = \mathrm{H} - \Delta_{radD},$

where:

$\Delta_{radB} = r_b - r_b cos(\tau_B),$

$\Delta_{radD} = r_d - r_d cos(\tau_D),$

4. Results

The application of the algorithm (1-9) makes it possible to analyze the possible influence of the angle of inclination of the cutting edge on the displacement of the pitch and the reduction of the width of the cutting crest. For the example, the smallest diameter tapered tool-joint thread NS-23 is used, which according to formulas (6,7) can lead to the largest values of axial deviations.

Determination of the lead angle along the first turn can be done according to the formula 8 (**fig.4**) [35]:

$$\psi = arctg\frac{P}{d_3 \cdot \pi} \quad (12)$$

For the smallest of the drill - pipe threads, this value gets the largest result. According to the data in Table 1, the value ψ=2.21° is obtained.

Table 1

Data of rotary shouldered thread connection NC-23 according API 7

Pitch P, mm	Tape angle φ, °	Root flat width a, mm	Root truncation b, mm	Major diameter d_3, mm	Thread Height, non truncated H mm
6,35	4,763	1,651	1,427	52,43	5,487

The obtained results are listed in **Table 2**.

Table 2

Calculation results

P_d, mm	P_b, mm	δ_D, mm	δ_B, mm	δ_B - δ_D, mm	a*, mm	α_B, °	α_D, °
3,02	3,33	0,0041	0,0048	0,0007	1,642	30,052	30,013

The tolerance for pitch deviation according to the standard [33] is ±0.05 mm, accordingly, the predicted axial displacement of the pitch δ_B - δ_D = 0.0007 (mm) is only 1.4% of the specified tolerance. The reduction in the width of the crest a - a ' is 0.009 mm, while the tolerance on the width of the crest is not a regulated standard [34]. Deviations from the nominal value by half angle α are: 0.052° and 0.013°, which are 6.9% and 1.7%, respectively.

5. Conclusions

The main conclusions have been drawn based on the research results.

1. According to theoretical prediction , the largest values of axial deviations for conical locking threads when applying the recommendations for the inclination of the cutting insert do not exceed 0.005 mm.
2. The probable decrease in the width of the crest does not exceed 0.01 mm.
3. The projected axial displacement of the pitch is only 0.0007 mm, which is 1.4% of the tolerance for its length.
4. The specified values do not have a noticeable effect on the accuracy of the tool-joint tapered threads
5. The deviation from the nominal by thread half angle α can reach the value of 0.052°, which is almost 7% of the tolerance for the tolerance deviation of 0.75° and can obviously be an important component of the real deviation from the nominal.

Acknowledgments

The authors are grateful to the Ministry of Science and Education of Ukraine for the grant to implement the project D-2-22-P (RK 0122U002082). The team of authors express their gratitude to the reviewers for valuable recommendations that have been taken into account to improve signifi-cantly the quality of this paper.

References

[1] Bazaluk O, Slabyi O, Vekeryk V, Velychkovych A, Ropyak L. and Lozynskyi V 2021 A technology of hydrocarbon fluid production intensification by productive stratum drainage zone reaming *Energies* 14 (12) art. no 3514. DOI: 10.3390/en14123514
[2] Moisyshyn V and Levchuk K 2016 The impact of vibration mechanism' installation place on the

process of retrieving stuck drill pipe *Mining of Mineral Deposits* 10 (3) pp 65-76 DOI: 10.15407/mining10.03.065

[3] Shatskyi I, Vytvytskyi I, Senyushkovych M and Velychkovych A 2019 Modelling and improvement of the design of hinged centralizer for casing *IOP Conference Series: Materials Science and Engineering* 564 (1) art. no. 12073. DOI: 10.1088/1757-899X/564/1/012073

[4] Ropyak L Y, Pryhorovska T O, Levchuk K H 2020 Analysis of materials and modern technologies for PDC drill bit manufacturing *Progress in Physics of Metals* 21 (2) pp 274-301 DOI: 10.15407/ufm.21.02.274

[5] Bazaluk O, Dubei O, Ropyak L, Shovkoplias M, Pryhorovska T and Lozynskyi V 2022 Strategy of/ compatible use of jet and plunger pump with chrome parts in oil well *Energies* 15 (1) art. no. 83 DOI: 10.3390/en15010083

[6] Bembenek M, Prysyazhnyuk, P, Shihab T, Machnik R, Ivanov O and Ropyak L 2022 Microstructure and Wear Characterization of the Fe-Mo-B-C—Based Hardfacing Alloys Deposited by Flux-Cored Arc *Welding Materials* 15 (14) art. no. 5074. DOI: 10.3390/ma15145074

[7] Onysko O, Kopei V, Medvid I, Pituley L and Lukan T 2020 Influence of the Thread Profile Accuracy on Contact Pressure in Oil and Gas Pipes Connectors *Lecture Notes in Mechanical Engineering* pp 432-441 DOI: 10.1007/978-3-030-50794-7_42

[8] Pryhorovska T and Ropyak L 2019 Machining Error Influnce on Stress State of Conical Thread Joint Details Proceedings of the International Conference on Advanced Optoelectronics and Lasers, *CAOL, 2019*- September art. no. 9019544 pp 493-497 DOI: 10.1109/CAOL46282.2019.9019544

[9] Bazaluk O, Velychkovych A, Ropyak L, Pashechko M, Pryhorovska T and Lozynskyi V 2021 Influence of heavy weight drill pipe material and drill bit manufacturing errors on stress state of steel blades *Energies* 14 (14) art. no. 4198. DOI: 10.3390/en14144198

[10] Kopei V, Onysko O, Panchuk V, Pituley L and Schuliar I 2021 Influence of Working Height of a Thread Profile on Quality Indicators of the Drill-String Tool-Joint *Lecture Notes in Mechanical Engineering in book Interpartner 2021: Advances Manufacturing Processes III* pp 395-404. DOI: 10.1007/978-3-030-91327-4_39

[11] Medvid I, Onysko O, Panchuk V, Pituley L and Schuliar I 2021 Kinematics of the Tapered Thread Machining by Lathe: Analytical Study *Lecture Notes in Mechanical Engineering in book Interpartner 2020: Advances Manufacturing Processes II* pp 555-565. DOI:10.1007/978-3-030-68014-5_54

[12] Kopei V B, Onysko O R and Panchuk V G 2019 Computerized system based on FreeCAD for geometric simulation of the oil and gas equipment thread turning *IOP Conference Series: Materials Science and Engineering,* 477 (1) art. no. 012032. DOI: 10.1088/1757-899X/477/1/012032

[13] Kusyi Y and Stupnytskyy V 2020 Optimization of the Technological Process Based on Analysis of Technological Damageability of Castings *Lecture Notes in Mechanical Engineering* pp 276-284. DOI: 10.1007/978-3-030-50794-7_27

[14] Kotsyubynsky V, Shyyko L, Shihab T, Prysyazhnyuk P, Aulin V and Boichuk V 2019 Multilayered MoS 2 /C nanospheres as high performance additives to lubricating oils *Materials Today: Proceedings* 35 pp 538-541. DOI: 10.1016/j.matpr.2019.10.021

[15] Kusyi Y M and Kuk A M 2020 Investigation of the technological damageability of castings at the stage of design and technological preparation of the machine Life Cycle *Journal of Physics:* Conf. Ser. 1426 art. no. 012034

[16] Kopei V B, Onysko O R and Panchuk V G 2020 Principles of development of product lifecycle management system for threaded connections based on the Python programming language *Journal of Physics: Conference Series* 1426 (1) art. no. 012033. DOI: 10.1088/1742-6596/1426/1/012033

[17] Fromentin G, Döbbeler B, Lung D 2015 Computerized Simulation of Interference in Thread Milling of Non-Symmetric Thread Profiles *15th CIRP Conference on Modelling of Machining Operations Procedia CIRP 2015* 31 pp 496–501 DOI:10.1016/j.procir.2015.03.018

[18] Koleva S, Enchev M, Szecsi T 2017 Compensation of the deviations caused by mechanical deformations during machining of threads *Procedia Manufacturing*, vol 13 pp

480–486 DOI:10.1016/j.promfg.2017.09.066

[19] Wang J, Philip Mathew, Xiaoping Li, Chuanzhen Huang, Hongtao Zhu 2010 Experimental Study on Cutting Characteristics for Buttress Thread Turning of 13%Cr Stainless Steel, *Ad-vances in Materials Processing IX in Key Engineering Materials* (Volume 443) pp 262–267

[20] LI Kang, JIANG Li, CHEN Ming 2014 Experimental study on the influence of cutting parameters on thread precision of external thread turning *Materials Science Forum* Vol. 770 pp 7-12 DOI:10.4028/www.scientific.net/MSF.770.7

[21] Liangliang, Dong, Jing Wang, Xiaohua Zhu 2019 Design and mechanical behavior study of ultrahigh-torque variable pitch casing joint *Advances in Mechanical Engineering* **1**(11) art.no.168781401881408 DOI: 10.1177/1687814018814084

[22] Mustafa GÜNAY 2011 Investigation of the Effects on Screw Thread of Infeed Angle during External Threading *Gazi University Journal of Science* 24(1) pp 153–160

[23] Jianhua Liu a Huajiang Ouyang b Lijun Ma c Chaoqian Zhang d Minhao Zhu 2015 Numerical and theoretical studies of bolted joints under harmonic shear displacement *Latin American Journal of Solids and Structures* 12 pp 115-132

[24] Gunay M, Aslan E, Korkut I, Seker U 2004 Investigation of the effect of rake angle on main cutting force, *Int J Mach Tools Manuf* 44 pp 953–959

[25] V H Nguyen, S L Ko 2013 Determination of workpiece profile and influence of singular point in helical grooving *CIRP Annals* Volume 62 Issue 1 pp 323-326 DOI: 10.1016/j.cirp.2013.03.009

[26] Luis Wilfredo Hernández González, Yassmin Seid Ahmed, Roberto Pérez Rodríguez, Patricia Del Carmen Zambrano Robledo, Martha Patricia Guerrero Matav 2018 Selection of Machining Parameters Using a Correlative Study of Cutting Tool Wear in High-Speed Turning of AISI 1045 Steel, *Journal of Manufacturing and Material Processes* 2(4), art. 66 DOI: 10.3390/jmmp2040066

[27] Berglind L, J Ziegert J 2015 Modulated Tool Path (MTP) Machining for Threading Applications *Procedia Manufacturing* Vol. 1 pp 546–555 DOI:10.1016/j.promfg.2015.09.029

[28] Mohammad R. Khoshdarregi, Yusuf Altintas 2018 Dynamics of Multipoint Thread Turning—Part I: General Formulation. *J. Manuf. Sci. Eng.* Jun 140(6) art.no. 061003

[29] Fromentin G, Döbbeler B, Lung D 2015 Computerized Simulation of Interference in Thread Milling of Non-Symmetric Thread Profiles *15th CIRP Conference on Modelling of Machining Operations Procedia CIRP* **31** pp 496–501 DOI:10.1016/j.procir.2015.03.018

[30] Baizeau T, Campocasso S, Fromentin G, Rossi F, Poulachon G 2015 Effect of rake angle on strain field during orthogonal cutting of hardened steel with c-BN tools *15th CIRP Conference on Modelling of Machining Operations Procedia CIRP* 31 pp 166–171

[31] Mohammad R. Khoshdarregi, Yusuf Altintas 2015 Generalized modeling of chip geometry and cutting forces in multi-point thread turning *International Journal of Machine Tools and Manufacture.* Vol.98 November pp 21–32 DOI:10.1016/j.ijmachtools.2015.08.005.

[32] Zhang G, Guo C 2016 Modeling Flank Wear Progression Based on Cutting Force and Energy Prediction in Turning Process. *Procedia Manufacturing* 5 pp 536–545 DOI:10.1016/j.promfg.2016.08.044

[33] Dongmei Zhang, Shiqiao Gao, Shaohua Niu, Haipeng Liu 2018 Prediction method for load distribution in threaded connections *Journal of theoretical and applied mechanics* 56 1 pp 157-168 DOI: 10.15632/jtam-pl.56.1.157

[34] API SREC 7-2. Specification for Threading and Gauging of Rotary Shouldered Thread Connection. [Effective Date June 2011] First Edition p 125

[35] https://www.sandvik.coromant.com/en-gb/pages/default.aspx (last visit 31.07.22)

Analytical study of heat transfer through friction overlay of band-pad brakes

O Y Pazen[1], R M Tatsiy[2], L Ya Ropyak[3] and V S Vytvytskyi[4,5]

[1] Departament of Supervision-preventive Activity and Fire Automatics , Lviv State University of Life Safety, 079007, 35 Kleparivska Street, Lviv, Ukraine
[2] Departament of Applied Mathematics and Mechanics , Lviv State University of Life Safety, 079007, 35 Kleparivska Street, Lviv, Ukraine
[3] Department of Computerized Engineering, Ivano-Frankivsk National Technical University of Oil and Gas, 076019, 15 Karpatska Street, Ivano-Frankivsk, Ukraine
[4] Department of Engineering and Computer Graphics, Ivano-Frankivsk National Technical University of Oil and Gas, 076019, 15 Karpatska Street, Ivano-Frankivsk, Ukraine
[5] vytvytskyi.v.s@gmail.com

Abstract. The paper analyses the operating conditions of band-pad brakes of drilling winch, materials and coatings for strengthening the metal elements of friction pairs. Methods of researching the thermal and stress-strain state of elements of brake friction units are considered. The problem of the development and theoretical research of the properties of new polymer materials with gradient properties for the production of friction overlays of band-pad brakes with improved operational characteristics is singled out. A mathematical model heat transfer through friction overlays of band-pad brakes of drilling winches with a variable coefficient of thermal conductivity of the material based on the thickness of the specified overlays has been developed. According to the developed mathematical model, an analytical study of the process of heat transfer through the friction overlays of band-pad with variable coefficient of thermal conductivity of their polymer material according to the thickness of the friction overlays was carried out according to different laws. It was established that the highest intensity of heat removal from the friction zone of the brakes is ensured when the thermal conductivity coefficient changes according to the law described by a second-order polynomial.
Keywords: polymer brake overlay, thermal conductivity coefficient, wear, functionally gradient material, coating, mathematical model.

1. Introduction

To increase the volume of oil, gas and coal production, it is necessary to increase the volume of well drilling, and this, in turn, requires the development of high-performance drilling equipment and tools with improved operational characteristics [1, 2, 3]. One of the ways to increase the productivity of the well drilling process is to reduce the time of lowering and lifting operations for replacing worn drill bits and increasing the length of the drill pipe column, which requires the use of a fast-acting and reliable braking system of drilling rig winches equipped with band-pad brakes.

The friction process in the metal-polymer friction pairs "metal pulley – friction polymer overlay" of the band-pad brakes is a high-energy phenomenon. To ensure stable thermodynamic conditions for the performance of the tribo assembly, the elements of its friction pairs must dissipate the generated thermal energy during braking in such a way that the material of the friction polymer overlays do not reach the maximum permissible surface temperature. This thermodynamic state of overlays allows to stabilize the process of friction and wear.

During lowering and lifting operations of drill strings with a high speed of movement, the cooling time of the elements of the friction pairs is sharply reduced. It can lead to overheating of the friction polymer overlays and their increased wear, and as a result, premature failure of the braking system of the drilling winch and the occurrence emergency situations. When developing brake structures, researchers focus on the rational choice of materials, the study of temperature and stress distribution, wear tests [4], including taking into account the electrochemical indicators of tribocorrosion [5] and tribocurrents [6], as well as the study of floor images [7].

Steel, cast iron, aluminum, titanium alloys and composite metal and non-metal polymer materials are used as materials for the manufacture of brake parts. Brake discs, drums, and pulleys are usually made from cast or rolled billets and 3D printed. During the manufacture of machine parts, considerable attention is paid to achieving accuracy [8, 9, 10] and the quality of operational surfaces [11, 12, 13], optimization of technologies [14], iincluding taking into account technological heredity [15], to ensure long-term operation of engineering products during their life cycle [16, 17].

To increase the wear resistance of the operational surfaces of steel discs and brake drums, friction hardening [18], laser processing [19] and coating are formed in various ways, for example, by vacuum-arc deposition [20], electric spark deposition [21, 22], as well as surfacing [23, 24]. Electrochemical chrome coatings [25] have high wear resistance, but are not used due to the low coefficient of friction in a pair with a polymer brake overlay.

Theoretical approaches to the development of the composition of new composite materials and strengthening technologies are proposed in [26, 27]. Researchers [28] developed a method of thermodynamic prediction of the phase composition of high-entropy alloys of transition metals for parts of friction nodes, including tungsten-free ones [29]. It is promising to use composite oxide coatings formed by plasma electrolytic oxidation in an electrolyte at high voltages [30, 31] to strengthen the operational surfaces of disks and drums made of aluminum and titanium alloys.

Polymeric materials are widely used in various friction pairs [32]. Friction linings are made by casting, pressing, as well as 3D printing, which is a prerequisite for creating functionally gradient materials for brakes.

To ensure trouble-free operation of braking systems, it is necessary to study the temperature and stress-strain state of the friction parts. The works [33, 34, 35] developed an analytical method for studying heat transfer and calculating non-stationary temperature fields in multilayered bodies of simple shapes. The works [36, 37] present a method of calculating the stress-strain state of layered structures during the application of forces, respectively, perpendicular to the surface of the layers and parallel to them and near defects [38].

Research of complex objects is carried out with the help of computer modeling. Using the finite element method [39], the force interaction in the friction pair of metallic elements – non-metallic rock was investigated without taking temperature factors into account. Researchers [40] performed a simulation of the frictional contact of a disc brake and established the nature of the temperature distribution.

Traditional friction polymer materials do not fully satisfy modern growing requirements for ensuring the stable performance of parts under extreme operating conditions of the drill winch brake. The results of tribological studies of materials with constant thermal conductivity are usually described in the literature, but there is practically no information on the behaviour of materials with a variable coefficient of thermal conductivity in friction pairs.

In paper [41], the process of wear and heating of the composite overlay with a constant coefficient of thermal conductivity was experimentally verified. The work [42] investigated the temperature distribution in the elements of a friction pair consisting of functionally gradient and homogeneous materials, but did not consider various options for changing the thermal conductivity of materials.

Thus, there is a need to develop and theoretically study the new polymer materials with gradient properties for friction overlays of band-pad brakes, as well as the creation of analytical methods for calculating and analysing the thermal state of friction brake pair elements.

This study aims to carry out an analytical study of heat transfer through the friction overlay of band-pad brakes of drilling winches with a variable coefficient of thermal conductivity of the material.

To achieve the goal, the following tasks were set:

- to develop a mathematical model of heat transfer through the friction overlays of band-pad brakes of drilling winches with a variable coefficient of thermal conductivity of the material;
- investigate the process of heat transfer through the friction overlays of the band-pad brake with a variable thermal conductivity coefficient of their polymer material and choose a rational law of thermal conductivity change.

2. Materials and Methods

2.1. Studied brake unit, materials of its friction elements

The band-pad brake of drilling winches (U2-5-5) include a metal brake pulley with side flanges (Steel 35KhNL State standard GOST 4543-71), wrapped around the bang with a set of friction polymer overlays attached to it. The run-up end of the metal band is connected to the adjusting tension screw, and the run-up end of the band is hinged to the mechanical brake control actuator.

Analytical studies of heat transfer through the friction overlays of band-pad brakes of drilling winches have been carried out (figure 1). Retinax of the A brand (FK-16L) State standard GOST 10851-94 and polymer material with variable thermal conductivity coefficient depending on the thickness of the overlay were chosen as the main material of the overlay. To increase the efficiency of heat removal from the friction zone through the friction pads, it was proposed to use a functionally gradient material with the thermal conductivity coefficient varying by thickness in the radial direction of this pad according to various dependencies (Fig. 2). The radius of the pulley was $r_0 = 0.725$ m, and the thickness of the polymer overlay was 0.02 m.

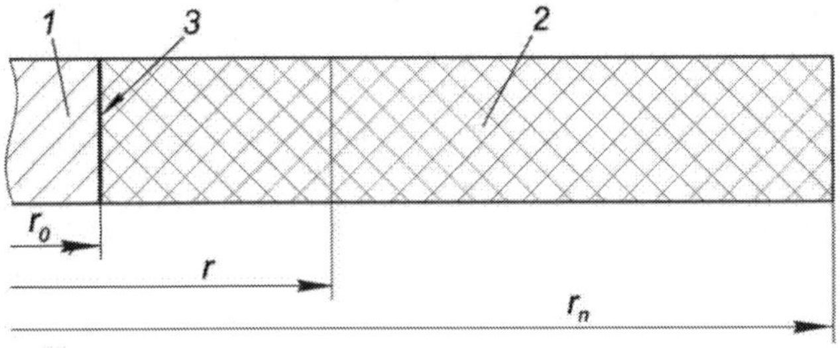

Figure 1. Calculation diagram of heat transfer study through friction overlays with variable thermal conductivity depending on the overlay thickness (in the radial direction) of the band-pad brake of drilling winches: 1 – pulley; 2 – overlay; 3 – friction zone.

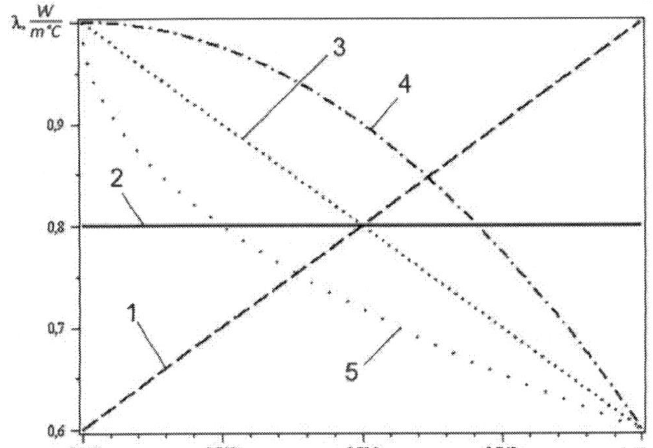

Figure 2. Laws of change in the coefficient of thermal conductivity, which varies depending on the thickness of the overlay in the radial direction:

$1 - \lambda(r) = const = 0,8 \ \frac{W}{m \cdot {}^0C}$;

$2 - \lambda(r) = 20r - 13,9$;

$3 - \lambda(r) = -20r + 15,5$;

$4 - \lambda(r) = 1 - 0,2\sqrt{-145 + 200r}$;

$5 - \lambda(r) = -1000r^2 + 1450r - 254,625$

By the time of braking, the ambient temperature is +20 °C. The surface temperature of the friction between the overlay and the disk is +1100 °C. On the other hand, there is a process of heat exchange between the overlay and the surrounding environment (we neglect the metal tape located on the outer surface of the polymer overlay).

2.2. Analytical research methodology and problem formulation

The process of heat exchange in the overlay can be described by a mathematical model, which includes the differential equation of heat conduction in a cylindrical system [43]

$$c\rho\frac{\partial t(r,\tau)}{\partial \tau}=\frac{1}{r}\frac{\partial}{\partial r}\left(r\lambda\frac{\partial t(r,\tau)}{\partial r}\right), \tag{1}$$

with a system of boundary conditions of the third kind

$$\begin{cases}\alpha_0 r_0 t(r_0,\tau)-t^{[1]}(r_0,\tau)=\alpha_0 r_0 \psi_0(\tau),\\ \alpha_n r_n t(r_n,\tau)+t^{[1]}(r_n,\tau)=\alpha_n r_n t_c(\tau),\end{cases} \tag{2}$$

and the initial condition as

$$t(r,0)=\varphi(r). \tag{3}$$

Here $t(r,\tau)$ – the distribution of the temperature field, $r_0<r_1<...<r_n$ – is an arbitrary division of the interval $[r_0,r_n]$ of real axis into n parts, $t^{[1]}\overset{df}{=}r\lambda t'$ – is the quasi-derivative (the physical essence – is the heat flux density q multiplied by the radius r), c – the specific heat capacity, ρ – the density, α_0 – is the heat exchange coefficient between the friction zone and the inner surface of the brake overlay, α_n – the heat exchange coefficient between the outer surface of the brake overlay and the environment, $\psi_0(\tau)$ – the law of temperature change in the friction zone [44].

3. Results

To confirm the assumptions, an analytical study was conducted to determine the amount of heat that leaves the brake overlay over a certain time interval.

The amount of heat removed from the working surface of the brake overlay is determined by the formul [45]

$$Q_{surf}=2\pi r_{surf}q\tau, \tag{4}$$

where Q_{surf} – where, is the amount of heat dissipated from the working surface of the brake overlay, r_{surf} – is the coordinate of the outer side of the overlay, q – is the heat flux density at point r_{surf}, τ – time.

The heat flux density is determined as

$$q=-\lambda\frac{\partial t(r_{surf},\tau)}{\partial r}, \tag{5}$$

where $t(r_{surf},\tau)$ – is the distribution of the temperature field on the surface of the brake overlay, λ – is the thermal conductivity coefficient.

In order to determine the amount of heat that leaves the surface of the overlay, it is necessary to solve the problem (3) – (5), we will carry out according to the following scheme, which is studied in detail and described in the work [44]:

1. The solution $t(r,\tau)$ is sought by the method of reduction in the form

$$t(r,\tau)=u(r,\tau)+v(r,\tau). \tag{6}$$

2. A quasi-stationary boundary value problem is solved for one of the functions (for example, for $u(r,\tau)$ the function)

$$\frac{1}{r}\frac{d}{dr}\left(r\lambda\frac{du(r,\tau)}{dr}\right)=0, \tag{7}$$

with the boundary conditions (4) for the function $u(r,\tau)$, i.e.

$$\begin{cases}\alpha_0 r_0 u(r_0,\tau)-u^{[1]}(r_0,\tau)=\alpha_0 r_0\psi_0(\tau),\\ \alpha_n r_n u(r_n,\tau)+u^{[1]}(r_n,\tau)=\alpha_n r_n\psi_n(\tau).\end{cases} \tag{8}$$

The solution structure of problem (7) – (8) is described in detail in the work [44].
On each of the intervals $[r_i,r_{i+1})$, the solution of this problem has the form

$$\mathbf{u_i}(r,\tau)=B_i(r,r_i)\cdot B(r_i,r_0)\cdot\mathbf{P_0}(\tau), \tag{9}$$

$$\mathbf{P_0}(\tau)=\left(P+Q\cdot B(r_n,r_0)\right)^{-1}\cdot\mathbf{\Gamma}(\tau)=$$

$$=\left[\begin{pmatrix}\alpha_0 r_0 & -1\\ 0 & 0\end{pmatrix}+\begin{pmatrix}0 & 0\\ \alpha_n r_n & 1\end{pmatrix}\begin{pmatrix}1 & \sum\limits_{i=0}^{n-1}\dfrac{\ln r_{i+1}-\ln r_i}{\lambda_i}\\ 0 & 1\end{pmatrix}\right]^{-1}\begin{pmatrix}\alpha_0 r_0\psi_0(\tau)\\ \alpha_n r_n t_c(\tau)\end{pmatrix}=$$

$$=\begin{pmatrix}\dfrac{(\alpha_n r_n\sigma_n+1)\alpha_0 r_0\psi_0(\tau)+\alpha_n r_n t_c(\tau)}{\alpha_n r_n+1+\alpha_0 r_0(\alpha_n r_n\sigma_n+1)}\\ \dfrac{\alpha_0\alpha_n r_0 r_n(t_c(\tau)-\psi_0(\tau))}{\alpha_n r_n+1+\alpha_0 r_0(\alpha_n r_n\sigma_n+1)}\end{pmatrix}.$$

The expression (9) allows writing the solution of the problem on the interval $[r_0,r_n]$ using characteristic functions θ_i in the form as

$$\mathbf{u}(r,\tau)=\sum_{i=0}^{n-1}\mathbf{u_i}(r,\tau)\theta_i. \tag{10}$$

where $\theta_i(r)=\begin{cases}1, r\in[r_i,r_{i+1}),\\ 0, r\notin[r_i,r_{i+1}), i=\overline{0,n-1}.\end{cases}$

3. We get a mixed heterogeneous problem for the function $v(r,\tau)$

$$c\rho\frac{\partial v(r,\tau)}{\partial\tau}=\frac{1}{r}\frac{\partial}{\partial r}\left(r\lambda\frac{\partial v(r,\tau)}{\partial r}\right)-c\rho\frac{\partial u(r,\tau)}{\partial\tau}. \tag{11}$$

with zero boundary conditions for the function $v(r,\tau)$,

$$\begin{cases}\alpha_0 r_0 v(r_0,\tau)-v^{[1]}(r_0,\tau)=0,\\ \alpha_n r_n v(r_n,\tau)+v^{[1]}(r_n,\tau)=0,\end{cases} \tag{12}$$

and the initial condition

$$v(r,0)=f(r)=\varphi(r)-u(r,0). \tag{13}$$

The solution of the mixed problem (11) – (13) is obtained in the form of a series [44]

$$v(r,\tau)=\sum_{k=1}^{\infty}\left[f_k\cdot e^{-\omega_k\tau}-\int_0^{\tau}e^{-\omega_k(\tau-s)}u_k(s)ds\right]\cdot R_k(r,\omega_k)=\sum_{i=0}^{n-1}v_i(r,\tau)\cdot\theta_i, \tag{14}$$

where, ω_k – the roots of the characteristic equation of the eigenvalue problem, $R_k(r,\omega_k)$ – the eigenfunctions of the corresponding eigenvalue problem, f_k – the Fourier coefficients of the expansion of the initial condition (13), $u_k(s)$ – the Fourier coefficients of the expansion of the function $\frac{\partial u(r,\tau)}{\partial\tau}$ eigenfunctions.

Considering the definition (6) and formulas (10) and (14), we obtain the solution of problem (3) – (5) in the form

$$t(r,\tau)=\sum_{i=0}^{n-1}\left[u_i(r,\tau)+v_i(r,\tau)\right]\cdot\theta_i. \tag{15}$$

Using the above-described method, the amount of heat that leaves the surface of the brake overlay was investigated depending on the law of change in the thermal conductivity coefficient (figure 3).

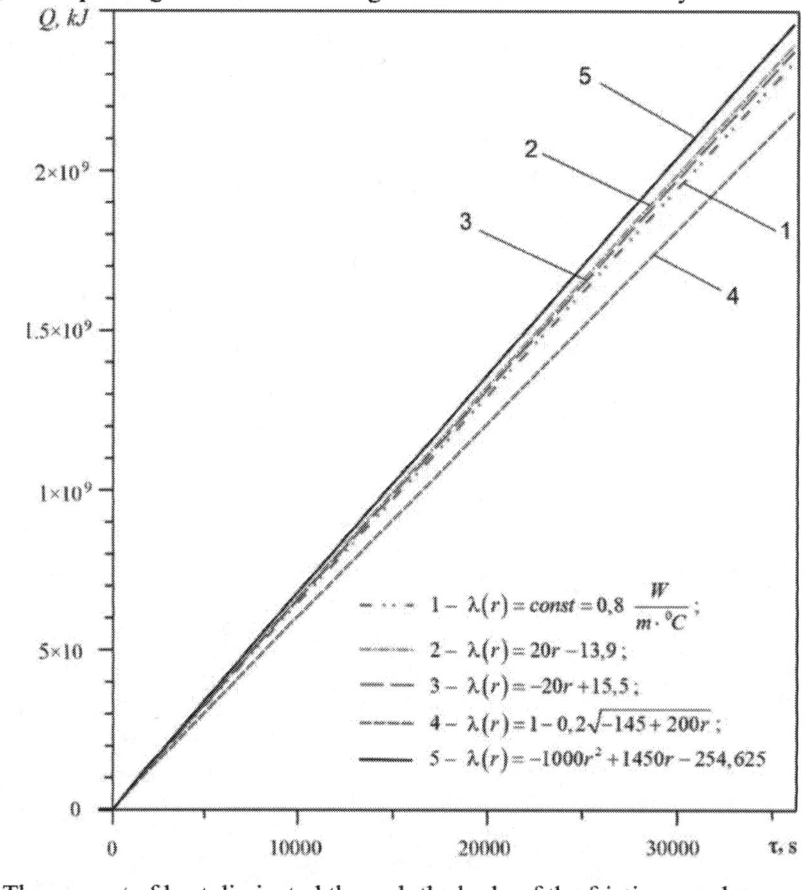

Figure 3. The amount of heat dissipated through the body of the friction overlay, as a function of time under different laws of the thermal conductivity coefficient change (1 – 5).

Figure 3 shows the dependence of the amount of heat dissipated through the body of the polymer overlay a certain time interval under different laws of change in the thermal conductivity coefficient depending on the thickness of the overlay:

- $\lambda(r) = const = 0{,}8\ \frac{W}{m \cdot {}^0C}$ (figure 3, line 1);
- $\lambda(r) = 20r - 13{,}9$ (figure 3, line 2);
- $\lambda(r) = -20r + 15{,}5$ (figure 3, line 3);
- $\lambda(r) = 1 - 0{,}2\sqrt{-145 + 200r}$ (figure 3, line 4);
- $\lambda(r) = -1000r^2 + 1450r - 254{,}625$ (figure 3, line 5).

The analysis of the calculations shows that the coefficient of thermal conductivity, at which the least amount of heat is removed from the friction zone through the body of the polymer overlay, varies according to the law $\lambda(r) = 1 - 0{,}2\sqrt{-145 + 200r}$, and the most according to the law $\lambda(r) = -1000r^2 + 1450r - 254{,}625$.

Thus, the developed mathematical model of the process of heat transfer through frictional polymer overlay s (with a variable coefficient of thermal conductivity depending on its thickness) of band-pad brakes of drilling winches allow the use of materials with gradient properties during the design and manufacture of brake overlay s. This will allow provide modern requirements for achieving stable performance of the parts of the brake unit of the drilling winch under extreme operating conditions.

4. Conclusions

The mathematical model of heat transfer through friction overlay s made of polymer material with the variable thickness in the radial direction of the overlays thermal conductivity coefficient, of a band-pad brake of a drilling winch, has been developed. This model makes it possible to analytically estimate the amount of heat that is released during the corresponding time interval through the overlay, depending on the law of change of the thermal conductivity coefficient.

Five different laws (according to the type of overlay) of the variation of the coefficient of thermal conductivity by thickness in the radial direction of the friction overlay were studied. It was established that the most heat leaves the surface of the polymer overlay when the coefficient of thermal conductivity changes according to the law described by the dependence $\lambda(r) = -1000r^2 + 1450r - 254{,}625$.

Acknowledgments

The authors are grateful to the Ministry of Science and Education of Ukraine for the grant to implement the project 0121U109591 and 0122U002082.

References

[1] Bazaluk O, Slabyi O, Vekeryk V, Velychkovych A, Ropyak L. and Lozynskyi V 2021 *Energies* **14** (12) art. no. 3514. DOI: 10.3390/en14123514

[2] Falshtynskyi V, Lozynskyi V, Saik P, Dychkovskyi R and Tabachenko M 2016 *Min. Miner. Deposits* **10** (1) pp 16–24. DOI: 10.15407/mining10.01.016

[3] Shatskyi I, Vytvytskyi I, Senyushkovych M and Velychkovych A 2019 *IOP Conference Series: Materials Science and Engineering* **564** (1) art. no. 12073. DOI: 10.1088/1757-899X/564/1/012073

[4] Bembenek M, Krawczyk J and Pańcikiewicz K 2020 *Materials* **13** (24) art. no. 5782. DOI: 10.3390/ma13245782

[5] Vynar V A; Dovhunyk V M and Student M M 2011 *Materials Science* **46** (5) pp 633–639. DOI: 10.1007/s11003-011-9334-4

[6] Kryshtopa S I, Prun'ko I B, Dolishnii B V, Panchuk M V, Bogatchuk I M, and Mel'nyk V M 2019 *Materials Science* **55** (2) pp 187–194. DOI: 10.1007/s11003-019-00288-x

[7] Vorobel R, Ivasenko I, Berehulyak O and Mandzii T 2021 *Automation in Construction* **123** art. no. 103515. DOI: 10.1016/j.autcon.2020.103515
[8] Pryhorovska T and Ropyak L 2019 *Proceedings of the International Conference on Advanced Optoelectronics and Lasers, CAOL, 2019-September* art. no. 9019544 pp 493–497. DOI: 10.1109/CAOL46282.2019.9019544
[9] Onysko O, Kopei V, Medvid I, Pituley L and Lukan T 2020 *Lecture Notes in Mechanical Engineering* pp 432–441. DOI: 10.1007/978-3-030-50794-7_42
[10] Bazaluk O, Velychkovych A, Ropyak L, Pashechko M, Pryhorovska T and Lozynskyi V 2021 *Energies* **14** (14) art. no. 4198. DOI: 10.3390/en14144198
[11] Kopei V, Onysko O, Panchuk V, Pituley L and Schuliar I 2022 *Lecture Notes in Mechanical Engineering* pp 395–404. DOI: 10.1007/978-3-030-91327-4_39
[12] Medvid I, Onysko O, Panchuk V, Pituley L and Schuliar I 2021 *Lecture Notes in Mechanical Engineering* pp 555–565. DOI: 10.1007/978-3-030-68014-5_54
[13] Kopei V B, Onysko O R and Panchuk V G 2019 *IOP Conference Series: Materials Science and Engineering*, **477** (1) art. no. 012032. DOI: 10.1088/1757-899X/477/1/012032
[14] Dmytriv V T, Dmytriv I V, Horodetskyy I M, Horodniak R V and Dmytriv T V 2021 *INMATEH – Agricultural Engineering* **65** (3) pp 233–240.
[15] Kusyi Y and Stupnytskyy V 2020 *Lecture Notes in Mechanical Engineering* pp 276–284. DOI: 10.1007/978-3-030-50794-7_27
[16] Kopei V B, Onysko O R and Panchuk V G 2020 *Journal of Physics: Conference Series* **1426** (1) art. no. 012033. DOI: 10.1088/1742-6596/1426/1/012033
[17] Kusyi Y M and Kuk A M 2020 *Journal of Physics: Conference Series* **1426** (1) art. no. 012034. DOI: 10.1088/1742-6596/1426/1/012034
[18] Gurey V and Hurey I 2020 *Lecture Notes in Mechanical Engineering* pp 63–72. DOI: 10.1007/978-3-030-40724-7_7
[19] Duriagina Z, Kulyk V, Kovbasiuk T, Vasyliv B and Kostryzhev A 2021 *Metals* **11** (3) art. no. 434. DOI: 10.3390/met11030434
[20] Beresnev V M, Klimenko S A, Sobol' O V, Grankin S S, Stolbovoi V A, Turbin P V, Novikov V Y, Meilekhov A A, Litovchenko S V and Malikova L V 2016 *Journal of Superhard Materials* **38** (2) pp 114–122. DOI: 10.3103/S1063457616020052
[21] Kryshtopa S I, Petryna D Y, Bogatchuk I M, Prun'ko I B and Mel'nyk V M 2017 *Materials Science* **53** (3) pp 351–358. DOI: 10.1007/s11003-017-0082-y
[22] Tarelnyk V B, Paustovskii A V, Tkachenko Y G, Konoplianchenko E V, Martsynkovskyi V S and Antoszewski B 2017 *Powder Metallurgy and Metal Ceramics* **55** (9-10) pp 585–595. DOI: 10.1007/s11106-017-9843-2
[23] Bembenek M, Prysyazhnyuk P, Shihab T, Machnik R, Ivanov O and Ropyak L 2022 *Materials* **15** (14) art. no. 5074. DOI: 10.3390/ma15145074
[24] Ivanov O, Prysyazhnyuk P, Lutsak D, Matviienkiv O and Aulin V 2020 *Management Systems in Production Engineering* **28** (3) pp 178–183. DOI: 10.2478/mspe-2020-0026
[25] Bazaluk O, Dubei O, Ropyak L, Shovkoplias M, Pryhorovska T and Lozynskyi V 2022 *Energies* **15** (1) art. no. 83. DOI: 10.3390/en15010083
[26] Radchenko T M, Gatsenko O S, Lizunov V V and Tatarenko V A 2020 Model *Progress in Physics of Metals* **21** (4) pp 580–618. DOI: 10.15407/ufm.21.04.580
[27] Krivileva S, Zakovorotniy A, Moiseev V, Ponomareva N, Rassokha A and Zinchenko O 2019 *Functional Materials* **26** (2) pp 347–352.
[28] Shihab T, Prysyazhnyuk P, Semyanyk I, Anrusyshyn R, Ivanov O and Troshchuk L 2020 *Management Systems in Production Engineering* **28** (2) pp 84–89. DOI: 10.2478/mspe-2020-0013
[29] Shihab T A, Shlapak L S, Namer N S, Prysyazhnyuk P M, Ivanov O O and Burda M J 2021 *Journal of Physics: Conference Series* **1741** (1) art. no. 012031. DOI: 10.1088/1742-6596/1741/1/012031
[30] Sakhnenko N D, Ved' M V, Androshchuk D S and Korniy S A 2016 *Surface Engineering and*

Applied Electrochemistry **52** (2) pp 145–151. DOI: 10.3103/S1068375516020113
[31] Student M M, Ivasenko I B, Posuvailo V M, Veselivs'ka H H, Pokhmurs'kyi A Y, Sirak Y Y and Yus'kiv V M 2019 *Materials Science* **54** (6) pp 899–906. DOI: 10.1007/s11003-019-00278-z
[32] Berladir K V, Hovorun T P, Bilous O A and Baranova S V 2018 *Functional Materials* **25** (2) pp 342–347. DOI: 10.15407/fm25.02.342
[33] Tatsii R M and Pazen O Y 2018 *Journal of Engineering Physics and Thermophysics* **91** (6) pp 1373–1384. DOI: 10.1007/s10891-018-1871-3
[34] Bandura A and Skaskiv O 2019 *Demonstratio Mathematica* **52** (1) pp 82–87. DOI: 10.1515/dema-2019-0008
[35] Tatsii R M, Stasyuk M F and Pazen O Y 2021 *Journal of Engineering Physics and Thermophysics* **94** (2) pp 298–310. DOI: 10.1007/s10891-021-02302-z
[36] Ropyak L Ya, Makoviichuk M V, Shatskyi I P, Pritula I M, Gryn L O and Belyakovskyi V O 2020 *Functional Materials* **27** (3) pp 638–642. DOI: 10.15407/fm27.03.638
[37] Velychkovych A, Ropyak L and Dubei O 2021 *Advances in Materials Science and Engineering* **2021** art. no. 4517657. DOI: 10.1155/2021/4517657
[38] Shatskyi I P, Makoviichuk M V and Shcherbii A B 2020 *Mater. Sci.* **55** pp 484–491. DOI: 10.1007/s11003-020-00329-w
[39] Pryhorovska T O 2017 *Machining Science and Technology* **21** (1) pp 37–66. DOI: 10.1080/10910344.2016.1260429
[40] Belhocine A and Omar W Z W 2018 *International Journal on Interactive Design and Manufacturing* **12** (1) pp 63–80. DOI: 10.1007/s12008-017-0376-1
[41] Yusubov F F and Hurey I V 2021 *Materials Science* **57** (2) pp 221–227. DOI: 10.1007/s11003-021-00535-0
[42] Yevtushenko A, Kuciej M, Topczewska K and Zamojski P 2022 *Materials* **15** (10) art. no. 3600. DOI: 10.3390/ma15103600
[43] Lykov A V 1967 Teoriya teploprovodnosti (Moskva: Vyisshaya shkola)
[44] Tatsii R M, Stasyuk M F, Vlasii O O and Pazen O Y 2017 *Informatsiyni tehnologii ta komp'yuterne modelyuvannya: mater. Mizhnar. nauk.-prakt. konf. (Ivano-Frankivsk 15-20 travnya)* pp 436–440. (date of access: 17.082022). https://sci.ldubgd.edu.ua/handle/123456789/3717
[45] Velichko L D, Lozinskiy R Ya and Semerak M M 2011 *Termodinamika ta teploperedacha v pozhezhniy spravi* (Lviv: "Soplom")

To the question of mathematical modeling of the crushed wood particles movement in pneumatic transport and aspiration systems

A V Liashenyk [1], O M Udovytskyi [2, 4], L M Dorundiak [1], M V Udovytska [3]

[1] Kolomyia Polytechnic College, Lviv Polytechnic National University, Ukraine
[2] Department of Applied Mechanics and Engineering Technology, Ukrainian National Forestry University, Ukraine
[3] Technological College, Ukrainian National Forestry University, Ukraine
[4] E-mail: o.udvytskyi@nltu.edu.ua

Abstract. Transport pneumatic systems (TPS) are widely used in various branches of industries. The transportation of crushed material in such systems is carried out due to the air flow. In the woodworking industry, there are technological processes that are often accompanied by the formation of dust and other small air pollutants. Studying the movement of particles in polluted air makes it possible to transport or separate such particles, while spending less energy. In this paper, the mathematical model of the movement of dusty air containing separate solid particles (production waste) was worked out, and based on it, solution will allow to separate and capture crushed wood particles and to reduce the cost of consumed electricity for transportation is proposed.

1. Introduction

At industrial enterprises of the woodworking industry transport pneumatic systems (TPS) are used to remove air contaminated with small wood waste, followed by its aspiration with the help of cyclones [1, 2]. The transportation of crushed material with the help of TPS is carried out due to the flow of air moving through the pipelines. TPS make it possible to transport crushed material along a complex trajectory, collect material from various loading devices and hard-to-reach places, protect the environment from gust pollution during transportation. TPS are characterized by simplicity of construction and operation, ease of management, the possibility of automation and remote control. They take up little production space for their placement.

The main disadvantage of TPS is the high specific consumption of electricity, which is determined by the speed of the air flow in the pipeline, the need to separate the material from the air at the end of the transportation route [1, 2].

In this article, an attempt is made to describe the main concepts of the pneumatic transportation process using mathematical modeling and CFD systems.

2. Methodology

The basic concepts in the theory of pneumatic transport are transportation speed, the speed of floating (soaring) of a part and the starting speed of movement of a material particle.

The transportation speed of the flow is the minimum speed of the air flow at which a reliable and stable movement of the material is achieved.

The speed of floating (soaring) of a particle is the speed of the upward internal flow in a vertical pipe, at which the particle is in a soaring state.

The starting speed of movement of a particle is the air speed averaged over the cross-section of a pneumatic transport pipeline, at which a solid particle begins to move along the bottom of the pipeline by sliding.

When analyzing the movement of dust particles along the TPS pipeline, the following main sections should be distinguished:

- movement along a horizontal section;
- movement along the vertical section;
- movement in the elbow (ell).

During the analysis, we will study the value of the pressure decrease in the certain element of the transport pneumatic system, the nature of the movement of air flows and dust particles, and the value of the pressure decrease on this hydraulic resistance.

Movement along a vertical pipeline (**Figure 1**). It is assumed that a dust particle in a vertical pipeline will move if the value of the air flow speed exceeds the soaring speed of the particle. At the same time, the weight ***G*** of the particle will exceed the force ***F*** of capturing the particle by the flow.

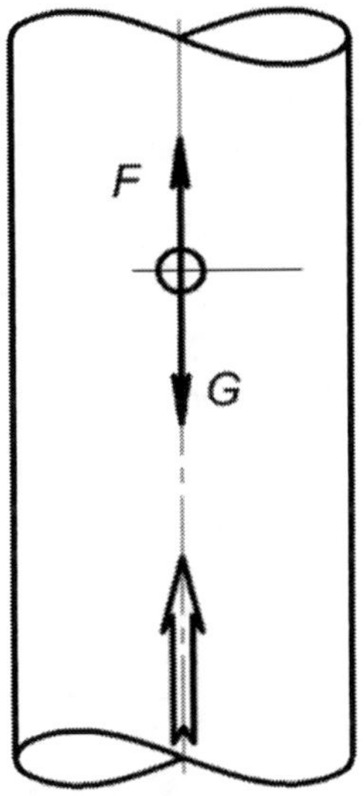

Figure 1. Movement of a particle in a vertical pipeline

The force ***F*** with which the flow acts on the particle is determined by the dependence:

$$F = C_F \cdot S \cdot \frac{\rho_n}{2} \cdot V^2 \quad (1)$$

where C_F – the resistance coefficient, which depends on the shape of the particle; S – the area of the projection of the particle on the plane that is perpendicular to the flow direction; ρ_n – air density; V - the speed of air flow.

The value of the coefficient C_F depends on the nature of the movement of the air flow in the pipeline and can be determined by the formula $C_F = 24 / R_e$ for the values $R_e < 2$; $C_F = 18{,}5 / R_e^{0,6}$ for values $2 < R_e < 500$; $C_F = 0{,}44$ for values $500 < R_e < 200000$ [1].

$$R_e = \frac{V d \rho_n}{\mu}, \quad (2)$$

where R_e – the Reynolds criterion when the particle is flown around by the air stream; μ – the coefficient of dynamic viscosity; V – the speed of the flow around the particle;
d – the diameter of the equivalent sphere.

The weight of the particle:

$$G = m_p \cdot g, \tag{3}$$

where m_p – the mass of the particle; g – the acceleration of free fall.

In this way we will find:

$$V = \sqrt{\frac{2m_p g}{C_F S \rho_n}}, \tag{4}$$

In the case when the speed of the flow will be greater for this value, the particle will move in the desired direction.

In addition to the described influences, the particle in the flow will be affected by the force of the added weight, which arises as a result of the acceleration of one of the phases, the Basset force [2], the resistance force of the medium, the Magnus force, the force caused by the collision of solid phase particles with each other, and other forces. It is customary [2] not to take these forces into account due to their insignificant effect.

Movement along a horizontal pipeline (**Figure 2**). In order for a particle to move in a horizontal pipeline, the force with which the air flow acts on it must be greater than the sum of all other forces acting on the particle. It is believed [2, 3] that reliable transportation is possible in the case when the flow rate in the pipeline is greater than the starting speed of movement of the particle.

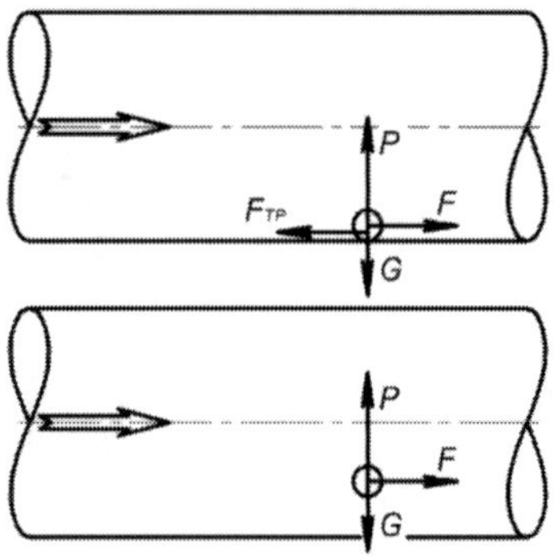

Figure 2. Movement of a particle in a horizontal pipeline

A particle lying at the bottom of the pipeline is affected by the force F of being captured by the flow, the force of gravity G, the force of friction F_{TP} and the lifting force P, which are determined by the following dependencies:

$$P = p_d \, \xi S, \tag{5}$$

where p_d – the dynamic pressure; ξ– the aerodynamic coefficient.

$$F_{TP} = Gf, \tag{6}$$

where f – the coefficient of friction of the wood against the pipeline wall.

In the work [2], it is proposed to determine the speed of movement according to the following formula:

$$V_m = \sqrt{\frac{2G}{\rho_n S\xi}}, \tag{7}$$

When the speed of the air flow begins to exceed the value of the starting speed of movement, the particle begins to rise. When the particle is placed above the axis of the pipeline, the force P will change its direction. Under the influence of weight and lifting force, the particle will periodically rise and fall. The nature of particle movement in a horizontal pipeline depends on such parameters as the mode of movement and speed of air flow, weight and shape of the particle. Possible trajectories are presented in **Figure 3**. With a further increase in the value of the flow rate and a decrease in the size of the dust particle, the character of its movement will be rectilinear.

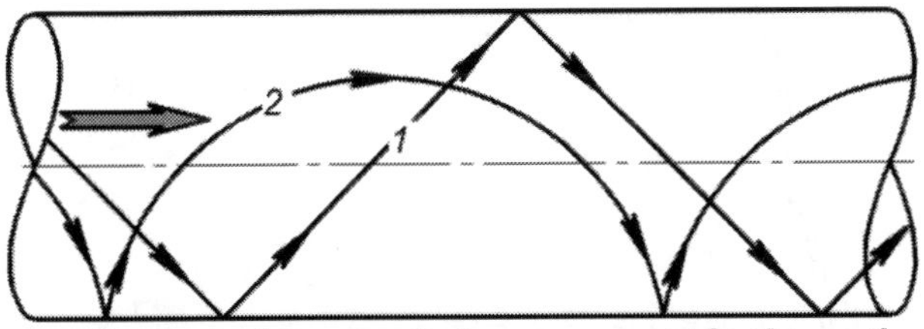

Figure 3. The nature of the movement of dust particles in a horizontal pipeline: 1 - movement along an indirect line; 2 - movement along a curvilinear trajectory

Pressure decrease during movement in the vertical and horizontal pipeline.

The pressure decrease in a round pipeline is simply determined by the following formula [4]:

$$\Delta p_{TP} = \lambda \cdot \frac{L}{d} \cdot \frac{\rho_n V^2}{2}, \tag{8}$$

where λ — the resistance coefficient; L - the length of the pipeline; d – pipeline diameter.

A more detailed analysis of the causes of pressure decrease during pneumatic transportation is made in the work [5] (**Figure 4**), where the following main components are highlighted. Pressure decrease, which is caused by friction of the flow against the walls of the pipeline Δp_L. The value of the pressure decrease increases in proportion to the square of the flow rate. The decrease Δp_R, which is associated with the slowing down of the movement of particles near the walls of the pipeline and the need for their further acceleration. The decrease Δp_F, which is caused by the interaction of particles with each other in the air flow. The total pressure decrease dependence curve in the pipeline is shown by the Δp_{tot} curve. Its form largely depends on the characteristics of the material being transported, its concentration in the flow, the configuration of the pipeline, however, the parabolic nature of the dependence is characteristic of all transport pneumatic systems. The power of the TPS fan is determined by the value of the pressure decrease at the required air flow rate.

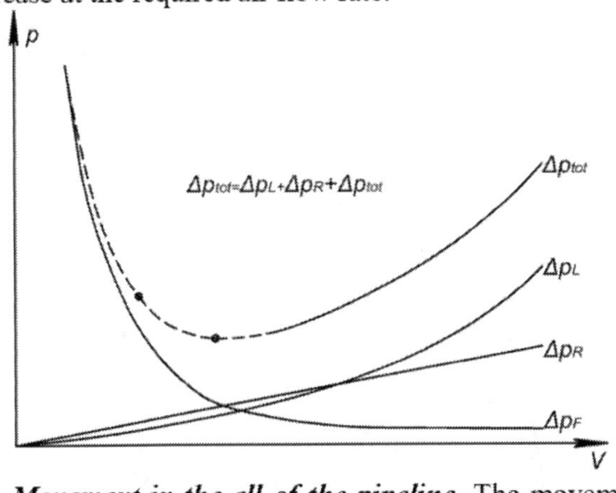

Figure 4. Graphs of pressure decrease for components of total pressure

Movement in the ell of the pipeline. The movement of dust and air flow in the ell is the most complicated, so let's consider it in more detail. It is advisable [6] to use applied CFD packages.

We will analyze the flow of dusty air through a pipeline with a 90° ell. The diameter of the pipeline is 240 mm; the length of the section before the ell is 1500 mm, and after the ell – 3000 mm. The radius of the ell is 500 mm. Numerical analysis of the solution of the problem was carried out at the following parameter values: air density ρ_{Π}=1.3 kg/m^3, solid particle density ρ_P=700 kg/m^3; air flow temperature T_0 = 20°C; static pressure at the entrance to the pipeline is equal to atmospheric pressure; speed of the air flow at the exit from the pipeline – 20 m/s.

When the flow of air moves along a curved channel, the velocity of the flow particles decreases with an increase in the radius of curvature and, as a result, the pressure near the inner wall is lower than near the outer wall. It follows from this that at the beginning of the curvilinear flow near the inner wall there is a confusing section, and near the outer one - a diffusive section (**Figure 5**).

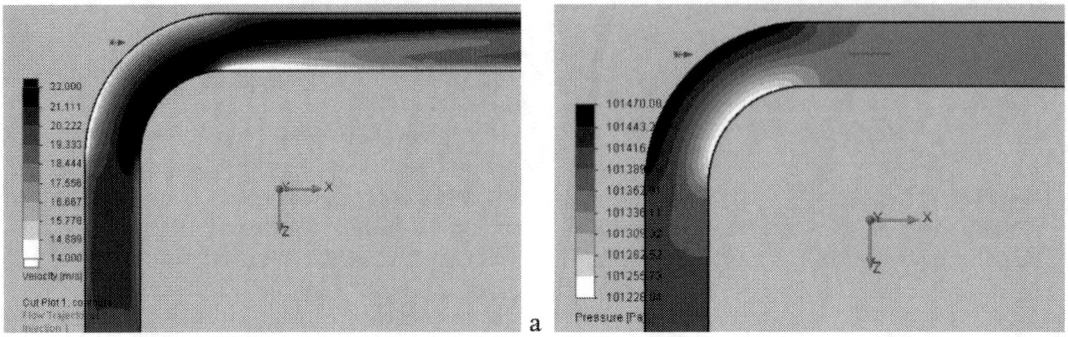

Figure 5. Distribution of air flow speed (a) and air flow static pressure (b) in the ell

In the diffuser zone, the near-surface layer grows very intensively, it is unstable and can easily break away from the wall [7]. This phenomenon is one of the main causes of air flow turbulence. It is observed in the diffuser zones of the curvilinear flow.

Figure 6 shows the distribution of air flow velocity in the pipeline after the ell in the section, which is distant equal to one (curve I), two (curve II), four (curve III) and six (curve IV) diameters of the pipeline. The distance from the bottom of the pipeline is plotted along the abscissa axis, and the value of the flow velocity along the ordinate axis. We see that even in section IV, which is distant from the ell by a distance equal to six diameters of the pipeline, the flow rate in the upper part will be higher than in the lower part. In cross-section I, the speed difference is about 9 m/s at a distance of about 200 mm. Such an uneven nature of the distribution of the flow velocity value will lead to an increase in the influence of such forces as aerodynamic, pushing out, and Magnus force, which are usually not taken into account when modeling the pneumatic transportation process [2].

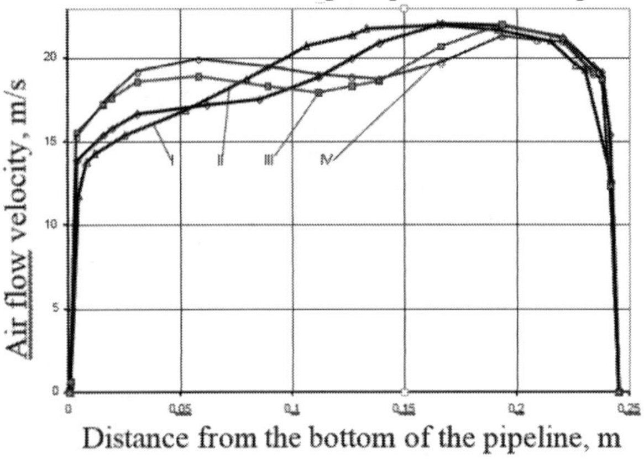

Figure 6. Distribution of air flow velocity in the pipeline

Due to the fact that the pressure near the outer wall is greater than near the inner wall, along the side walls where the velocity is close to zero, the flow will flow from the outer wall to the inner wall, due to which a pair vortex is formed. As the pressure gradient increases, the pair vortex will strengthen, which is the cause of the radial flow.

3. Results and discussion

These phenomena determine the nature of losses in a curvilinear flow, which consist of frictional losses; losses associated with the occurrence of paired vortices; and losses caused by the presence of local flow separations. The last ones have the largest relative value, and friction losses make up the smallest share of total losses.

The results of the study of the movement of dust particles with a diameter of 10^{-6}, 10^{-4}, 10^{-2} m are presented in **Figure 7** a-c.

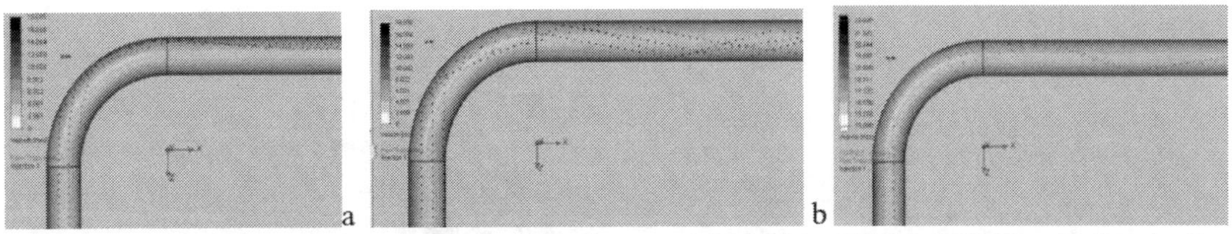

Figure 7. Movement of dust particles with a diameter of 10^{-6} (a), 10^{-4} (b), 10^{-2} (c) m in the pipeline

At first, dust particles in the ell try to maintain a straight trajectory. Especially small (10^{-6} m) particles are captured by the flow and their trajectories are distorted. Larger particles continue to move in a straight line, strike the wall of the ell, as a result of which they lose speed to a large extent, then are captured by the air flow again, which leads to an increase in pressure losses in the ell. With a significant concentration of crushed wood particles in the stream moving in the ell, a significant layer of particles is formed near the outer wall (**Figure 8**). The section through which the flow moves in the ell will be significantly reduced. Those confusive and diffusive areas will be observed in it, secondary flows will arise. The flow rate in such a section will increase sharply, which will lead to a significant increase in hydraulic resistance.

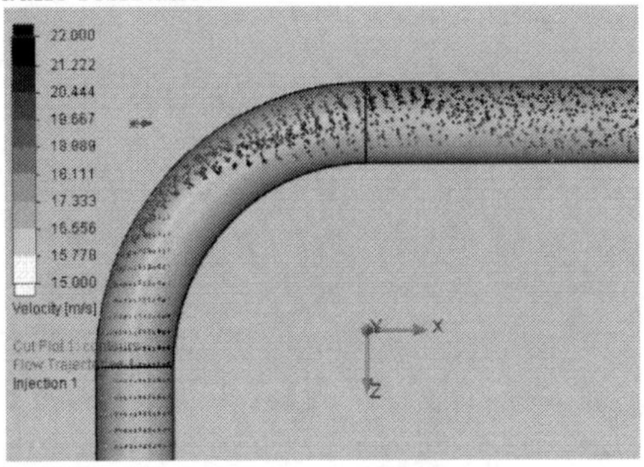

Figure 8. Movement of dust particles with a diameter of 10^{-3} m in the pipeline

According to the results of the study, which are described in [3], with a pneumatic transportation productivity of 10-30 kg/s in an ell, the radius of which is equal to two diameters of the pipeline, the equivalent length of the straight pipeline is 18 meters. Therefore, the presence of an ell is one of the significant causes of pressure decrease. That is why it is necessary to try to reduce their number.

When trying to improve transport pneumatic systems, we are dealing with the following contradiction, which must be resolved. For reliable transportation of crushed wood, it is necessary to increase the value of the air flow speed, which in turn leads to a sharp increase in the hydraulic resistance of the pipeline. A decrease in the value of the flow rate leads to a decrease in energy consumption, but at the same time to an increase in the danger of clogging the pipeline with the material being transported.

One of the possible solutions to this contradiction is the combination of an ell with a capture for large particles. **Figure 9** shows the proposed design.

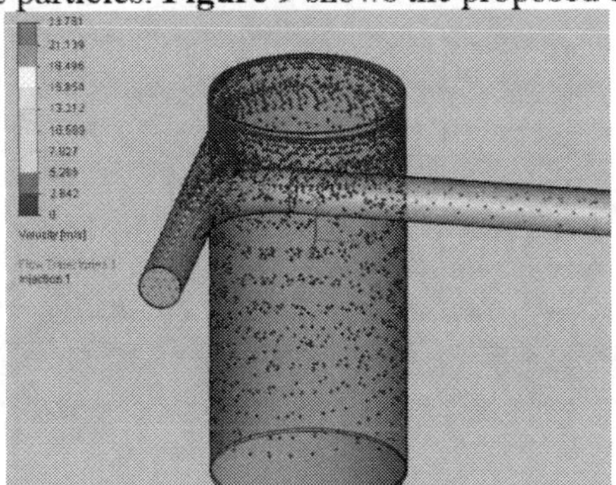

Figure 9. The movement of dust particles in a pipeline with an ell, which is combined with a dust settling chamber

The ell is replaced by a tee. A straight section of the pipeline enters a cylindrical chamber, in which the flow rate is low. The static pressure in the chamber will be close to the maximum value in the ell (**Figure 10**).

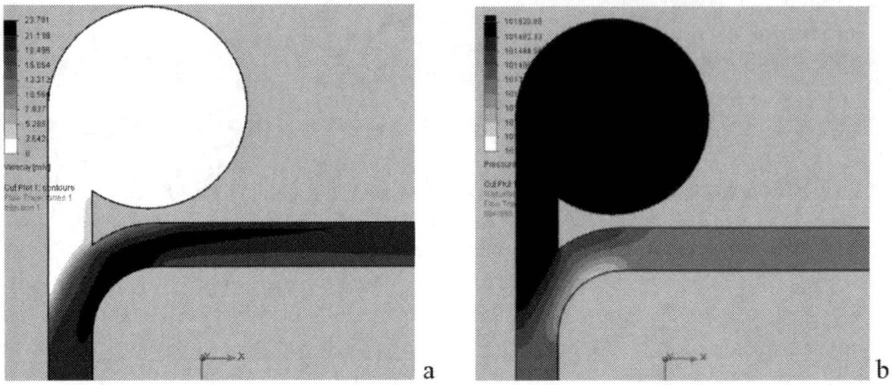

Figure 10. Distribution of air flow velocity (a) and air flow static pressure (b) in an ell with a capture for large wood particles

Particles of wood whose trajectories are close to rectilinear will continue their movement and enter the cylindrical chamber, where they will lose speed and sink to the bottom. Small particles whose trajectories are close to the flow lines will move further along the pipeline. In the ell, the diffusive and confusive areas are significantly reduced.

The proposed solution makes it possible to significantly reduce the transport speed of transportation beyond the ell, which will make it possible to reduce the hydraulic resistance of the transport pneumatic system. Reducing the speed can be achieved by increasing the diameter of the pipeline.

4. Conclusion

The proposed design will make it possible to achieve the following results.

1. Separation and capture of large particles of crushed wood, which makes it possible to reduce the transport speed in the pipeline.
2. Elimination of the phenomenon of a decrease in the "live" section of the ell, which will lead to a decrease in its hydraulic resistance.
3. Reduction of flow rate difference and static pressure in the cross-section, which is perpendicular to the axis of the pipeline outside the ell.
4. Reducing the concentration of crushed wood particles in the pipeline after the ell, which will make it possible to reduce the cost of electricity for transportation.

5. References

[1] Alexandrov A N and Kozoriz G F 1988 *Pneumatic transport and dust collection facilities of woodworking enterprises* Handbook Moscow Forest industry 248

[2] Svyatkov S N 1963 Some issues of pneumatic transport of chopped wood *Pneumatic transport of woodworking waste* Moscow Forest industry 12-20

[3] Lyashenyk A V, Tysovsky L O, Dorundyak L M and Dadak Yu R 2012 Study of the bunker influence on the cyclone aerodynamics by modeling processes using Cosmos Floworks *Scientific Bulletin of the UNFU: Collection of science and technology works Lviv: RNV UNFU* 22-01 113-119

[4] Ivantsov V V 1962 *Simplified calculation of intrashop systems of pneumatic transport of wood waste* Moscow Giprodrevprom 107

[5] S ZOUAOUI, H DJEBOURI, A MENGUELTI, K MOHAMMEDI and A OUAKED 2014 Analytical Approach Modelling of the Motion of a Solid Particle in a Fluid Flow inside a Pipe *Conference Paper, Conference: ACMA* April, At: Marrakech, Morrocco

[6] Electronic resource: http://neuero.de. *Introduction to the Theoretical and Practical Principles of Pneumatic Conveying* SCOTTNEIDIGH, Neuero Corporation, West Chicago, IL, USA

[7] David C Wilcox 2006 *Turbulence Modeling for CFD* (Third Edition) - Dcw Industries, Incorporated

[8] Schlichting G 1974 *Theory of the boundary layer* / Translation from German - Moscow: Science 712

Composite wheat-amaranth flour baking properties and bread freshness

S Yr Mykolenko, E B Aliiev* and V Yr Dudin

Dnipro State Agrarian and Economic University, 25, Voroshylova str., Dnipro, 49027, Ukraine

*aliev@meta.ua

Abstract. The study has reviled the benefits of amaranth groat flour application into wheat bread making towards enhancing the technological efficiency and the quality of the bread, prolonging the freshness of the product without artificial improvers involved in the formulations. Paying attention to textural characteristics of the wheat-amaranth bread during its shelf-life, amaranth groat flour as a derived product of amaranth grain processing presented a positive impact on the change of textural properties of the bread crumb during wheat-amaranth bread shelf-life. The effect of ranging 5–30% substitution of common wheat high-grade flour by the amaranth flour derived from partially defatted amaranth grain was investigated and the complex organoleptic quality of the bread made of the composite flour reached the highest level at 15% of the amaranth flour in the formulation. The technological parameters of bread making such as the duration of proofing, shape stability, specific volume, crumb porosity, baking and drying losses, when wheat high-grade flour was partially replaced by the amaranth grain-derived milling product showed better performance for the bread enriched by the amaranth flour at higher substitution level approving the wider application of the pseudocereal flour in the food industry. The freshness of the bread made of the composite flour is related to the textural characteristics of the crumb which tend to change drastically due to styling. The amaranth groat flour application into the wheat bread formulation led to changes in compressive force and tension of the crumb, the stress value at relative strain $\varepsilon = 25\%$, the elastic modulus, and the statical elastic hysteresis index. The resulting application of the amaranth groat flour into bread formulation allowed to ensure a 2-2.5-fold increase in the freshness of wheat bread with a 15% amaranth groat flour content when providing higher comprehensive quality of the product.

1. Introduction

Bread is a staple food all around the world. The nutritional value of bread and bakery products due to their daily consumption has a significant influence on human health, being the cornerstone to posing a risk to diseases of affluence, such as diabetes, obesity and atherosclerosis. Recently, global trends of a healthy diet have been faced with high consumer expectations regarding the sensory properties of food products [1]. Moreover, sustainability in food chains requires particular attention to seeking alternative and more environmentally friendly sources, processes and food systems. Climate change in recent decades has prompted to look for non-traditional food resources, e.g. cereal and pseudocereal crops, especially drought-resistant and undemanding soil quality, which can replace wheat, rye, and corn [2–4]. Under the influence of global trends and state policy, agribusiness and the food industry are increasingly inclined to the need of providing manufacturers with a sustainable approach to processing food and reducing waste along the food chain [5]. It is known that the majority of waste from bakery products is related to the last stage of the food chain. Styling of bread is the main reason to accumulate waste due to losing the freshness of the product and decreasing sustainability of the food chain [6]. Therefore, the use of improvers to prolong the freshness of bread is a widespread practice, which often goes beyond the principles of healthy food production.

Amaranth is a promising crop, which, despite its long cultivation history in Polynesian cultures, has not yet received wide use as a source in the food industry of the XXI century [7]. This is mainly associated with the lack of agricultural techniques, insufficient awareness of the population about its nutritional value and limited research and developments on the effective use of amaranth processing products in the food industry. Amaranth grain among cereals and pseudocereals is unique due to its high content of squalene (4–8 %), proteins (13–18 %) and its precious amino acid composition close to the ideal protein requirements, explicitly high content of lysine, dietary fibers (8–18 %), which are closer rather to vegetables and fruits than to cereals [8]. Amaranth grain-derived products are a unique plant-based source of squalene, which is triterpene and precursor of steroids, exhibiting anti-carcinogenic, immunomodulating, and antioxidant biological activity. This way, the introduction of amaranth grain-derived products to food might be extremely beneficial to the nutritional value of the product. The main fraction of amaranth protein is presented by globulins and albumins, which are abundant in essential amino acids, but the absence of gluten in amaranth grain-derived products also affects technological qualities, for instance, related to baking properties of the flour, triggering deterioration of rheological characteristics of the dough, which should be optimal to obtain the product of high quality. Moreover, an additional influential factor contributing to the change of technological behaviour of the dough including amaranth grain-derived products is small-sized amaranth starch and the presence of alfa-amylase inhibitor [9].

Due to its complete absence of gluten, Amaranth flour can only be used as a partial substitute of wheat flour in traditional wheat bread formulations [10]. The baking qualities of composite wheat-amaranth mixtures depend significantly on both the grain processing and the varietal characteristics of amaranth grain [11]. However, most of the research works that investigated composite mixtures with amaranth grain-derived products in bread making were devoted to the use of whole grain flour [12]. This approach does not address sustainability and efficiency along the food chain, because the fatty acid composition of amaranth grain includes linolenic acid, which can quickly oxidize and lead to high product rancidity, turning them into food waste. Moreover, the fat content in amaranth grain overpasses cereals [13]. Dehulling and cold-pressing of amaranth grain allow an additional concentration of squalene in the amaranth oil. Therefore, defatted amaranth grain-derived products as baking ingredients are more promising to use in composite flour formulations. This applies to amaranth flour, mainly consisting of amaranth grain perisperm, which does not contain fat. Studies on the baking properties of composite wheat-amaranth flour made from amaranth groats are scarce [14], and the effect of such baking ingredients on the freshness and staling of bread remains unexplored. Therefore, the study aimed to determine the impact of amaranth groat-derived product as an ingredient of wheat-amaranth composite flour on the quality emphasising the changes of the bread crumb textural properties during storage.

2. Materials and methods

2.1. Materials

Wheat high-grade flour was purchased from the market (Dnipro, Ukraine), with the following characteristics: moisture content of 14 %, wet gluten content of 26 %, and gluten deformation index (GDI) of 65 units. Amaranth flour was provided by the Association of Amaranth Grain Producers and Processors (Dnipro, Ukraine). This flour was produced from debranned amaranth flour as a milled product of polished amaranth grain with germ and pericarp removed (Karlivka, Ukraine). During the research, the flour was stored at a temperature of +4–+10 °C for 2–3 months. Individual indicators of the quality of wheat and amaranth flour and the composite flours are given in Table. 1. Compressed fresh yeast (Lviv, Ukraine) and salt (Artemivsk, Ukraine) were also used for trials.

Table 1. Quality characteristics of wheat, amaranth flours and composite mixtures.

Flour (%)	Moisture content, %	Whiteness, units	Falling number, s	Titratable acidity, mL 0.1N NaOH
Wheat high-grade flour (100)	14.7±0.14	59.1±0.02	503±2	3.0±0.28
Amaranth groat flour (100)	11.7±0.14	12.9±0.03	61±1	9.5±0.42
Wheat-amaranth composite flour (95:5)	14.7±0.14	54.3±0.03	502±18	2.8±0.28
Wheat-amaranth composite flour (90:10)	14.7±0.14	49.3±0.04	516±8	3.3±0.14
Wheat-amaranth composite flour (85:15)	14.4±0.06	44.3±0.04	516±4	3.1±0.14
Wheat-amaranth composite flour (80:20)	14.4±0.57	38.7±0.03	499±12	3.3±0.14
Wheat-amaranth composite flour (75:25)	14.1±0.14	36.8±0.04	475±6	3.9±0.14
Wheat-amaranth composite flour (70:30)	13.7±0.42	40.4±0.04	453±9	4.3±0.14

2.2. Baking trial tests

Laboratory bread baking tests were performed according to the standard method GOST 27669 when kneading the dough with 200 g of flour. Wheat-amaranth composite flours were prepared as mixtures of the wheat and the amaranth flour at a ratio of 100:0 (control), 95:5, 90:10, 85:15, 80:20, 75:25, and 70:30. Thus, amaranth flour content (α) varied in the following concentration range of 0–30% with a step of 5%. The amount of water needed for the formation of the dough was determined based on the moisture content of the flour. Compressed baker's yeast (5 g) and salt (1.5 g) were added to the formulation. The fermentation of the dough was for 170 minutes at 31±1°C. The loaves were baked after proofing, the duration of which varied depending on the test sample (T_e, min), in a laboratory oven with the humidification of the baking chamber at 220–230 °C for 30 minutes. After 4–6 hours, the organoleptic, and physicochemical quality characteristics of the bread were determined.

2.3. Bread quality characteristics

Baking loss (ω) was determined as the ratio of the weight of hot bread, determined immediately after removing the loaf from the oven, to the mass of cooled bread after 4 hours of storage at room temperature, expressed in %.

Drying loss (φ) was determined as the ratio of the weight of cooled bread after 4 hours of storage at room temperature to the weight of bread after 24 hours of storage at room temperature, expressed in %.

The total loss of bread weight was determined as baking and drying losses together, expressed in %.

The specific volume of bread (ρ) was measured according to the AACC Method 10-05.01 as the volume of displaced rapeseed about the weight of the bread, g/ml.

Shape retention (H:B) was defined as the ratio of the height to the width of the central part of the bread [12].

The porosity of the bread crumb (δ) was determined according to the common method [15] using Zhuravlev's device, and expressed in %.

The moisture content of the crumb (W) was determined by the air-oven method as a loss in weight of 5 g of crushed bread crumb taken from the central part of the loaf according to AACC Method 44-15.02 and expressed in %.

To measure the titratable acidity (TA, °), 25 g of the crushed bread crumb taken from the central part of the loaf was placed in a flask having a capacity of 500 ml, where 250 ml of water of 20 °C was added. The flask was closed with a plug and vigorously shaken for 2 min, settled for 10 min, shaken for 2 min and settled again for 8 min. Then the solution was filtered, 50 ml of the filtrate were taken with a pipette and transferred to two 150 ml conical flasks, 2-3 drops of phenolphthalein solution were added, and titration was carried out with 0.1 N solution of sodium hydroxide till the appearance of light pink colour was observed for 1 min.

The titratable acidity was calculated with the use of the following formula:

$$TA=2V \tag{1}$$

where V is the volume of 0.1 N solution of sodium hydroxide used for the titration.

The complex quality of the loaf was determined by taking into account sensory characteristics, particularly crust colour, crust surface condition, crumb colour, crumb porosity structure, crumb rheological properties, flavour, taste, and crumb chewability, expressed on a five-point scale [15] taking into account the significance of the criteria. The organoleptic evaluation was carried out by a panel group of 7 employees of the Faculty of Engineering and Technology, Dnipro State Agrarian and Economic University.

The trials were performed in 3–5 repetitions. Statistical analyses were performed in MS Excel.

The textural characteristics of the bread were determined after 24 and 48 hours of storage of the samples at room temperature according to the following indicators: the dynamics of changes in compressive force and tension, the stress value at relative strain ε = 25%, the elastic modulus, the static elastic hysteresis index.

The elastic characteristics of the product were determined according to the developed methodology, which includes the following stages:

1. A loaf of bread is cut crosswise into parts 25 mm thick.
2. Parts of bread products are alternately placed on a flat horizontal surface (work table) in the area under the cylindrical indenter (diameter of 21 mm, area of the working surface $S = 3.464 \cdot 10^{-4}$ m^2), which is fixed on a strain gauge that moves in a vertical plane. The strain gauge with the indenter attached to it must be calibrated and correspond to a force of 0 N in a state of rest.
3. The maximum value of the compressive force F_{max} (H) is set, and the largest value of the relative deformation ε_{max} (%) is required.
4. Next, the indenter starts to descend uniformly straight at a speed of 10 mm/s from the upper point. When a part of the bread crumb is reached, the force on the strain gauge begins to increase from 0 N. This moment is recorded as the beginning of the measurement process. At this point, the value of the layer thickness of the part of the bread product X (mm) is determined on the condition that the work table corresponds to the origin of coordinates 0 mm.
5. During the measurement process, the value of time t (ms), absolute deformation Δx (mm) and compression force F (N) are determined.
6. Upon reaching the value of absolute deformation $\Delta x \geq \varepsilon_{max} \cdot X/100$ (mm) or force $F \geq F_{max}$ (H), the indenter stops.
7. Next, the indenter starts to move upwards uniformly and in a straight line at a speed of 10 mm/s. At the same time, the determination of the time value t (ms), the absolute deformation Δx (mm) and the compressive force F (N) continues.
8. All received data are stored in the database.
9. Further, the textural properties of the bread crumb are calculated based on the obtained data. Thus, the compressive force F (H) is converted into tension σ (kPa) according to the formula

$$\sigma = \frac{F}{1000 \cdot S} = 2.887 \cdot F \tag{2}$$

and stored in the database.

10. From the database, the stress value at relative strain $\varepsilon = 25\%$ is rotated and fixed (Fig. 2) as the textural characteristic of the product σ.
11. The elastic modulus of the product E (kPa) is determined by the formula (Fig. 1):

$$E = \frac{100 \cdot \sigma(\varepsilon = 25\%)}{25} = 4 \cdot \sigma(\varepsilon = 25\%). \tag{3}$$

12. Based on the obtained dependences σ (ε), which are presented in the form of elastic hysteresis, their area (static elastic hysteresis index) S (kPa·%) is determined by the formula (Fig. 1)

$$S = \sum_{i=1}^{\frac{\varepsilon_{max}}{\Delta\varepsilon_i}} \left(\sigma_{up\,i} - \sigma_{down\,i}\right)\Delta\varepsilon_i = \sum_{i=1}^{10\varepsilon_{max}} 0.1\left(\sigma_{up\,i} - \sigma_{down\,i}\right). \tag{4}$$

IOP Conf. Series: Materials Science and Engineering 1277 (2023) 012021 doi:10.1088/1757-899X/1277/1/012021

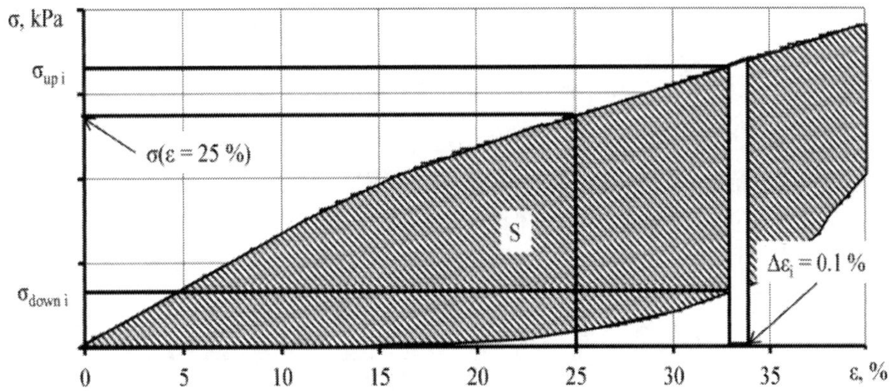

Figure 1. Bread crumb tension and relative deformation.

To implement the abovementioned methodology and to measure the textural characteristics of bakery products, a device for automatic determination was constructed. The structural and technological scheme and general view of the device for automatic measurement of the food textural properties are shown in Figure 2.

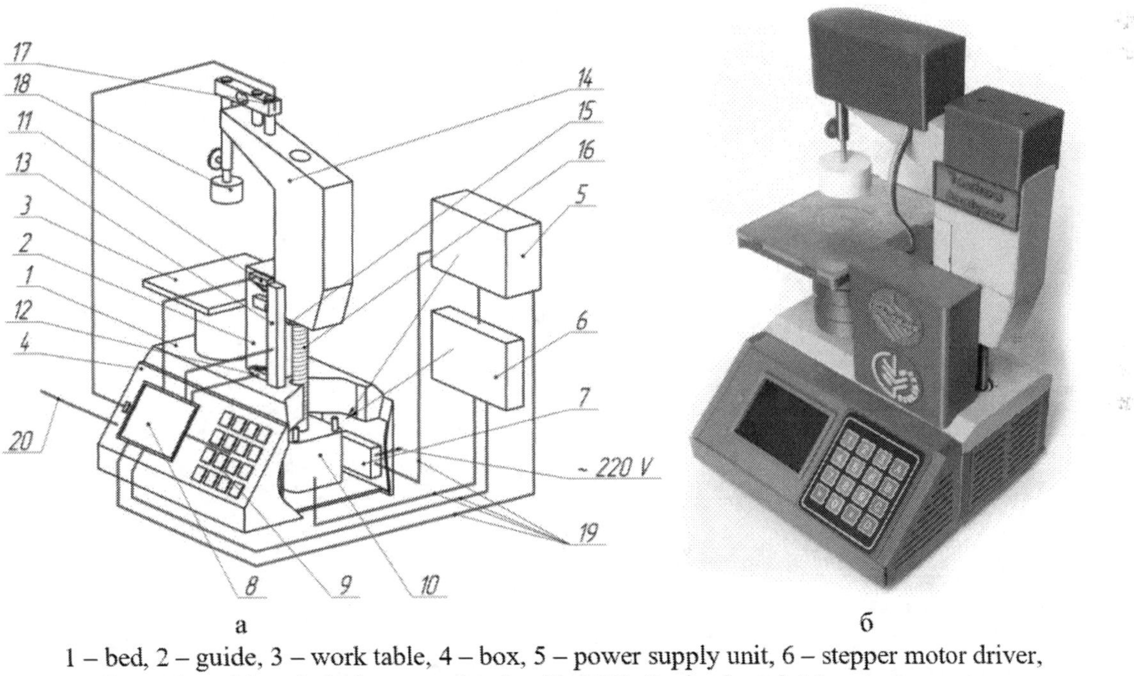

a б

1 – bed, 2 – guide, 3 – work table, 4 – box, 5 – power supply unit, 6 – stepper motor driver, 7 – outlet with switch, 8 – control unit with LCD, 9 – keyboard, 10 – stepper motor, 11 – upper limit switch, 12 – lower limit switch, 13 – linear variable resistor, 14 – tripod, 15 – nut with raceways for balls, 16 – threaded shaft, 17 – strain gauge, 18 – cylindrical indenter, 19 – electrical wires, 20 – USB output

Figure 2. Structural and technological scheme (a) and general view (b) of the device for automatic measurement of food textural properties.

3. Results and discussion

The unique chemical composition of amaranth grain makes it an attractive food source for applying amaranth grain-derived products into bread formulations based on different mixtures of wheat and

IOP Conf. Series: Materials Science and Engineering 1277 (2023) 012021 doi:10.1088/1757-899X/1277/1/012021

amaranth flour. Table 2, Figure 1 show that in this way technological properties, as well as physical and chemical characteristics of bread, were found to show significantly different performance concerning the increase of the amaranth flour in the product. For instance, significantly decreased proofing of the dough highlighted that it can be reduced in 2 or 4 times when the amaranth flour content exceeded 15 and 30% respectively, which is in agreement with previous findings [16]. Despite intensified proofing, the total loss in the weight of bread demonstrated the tendency to increase with 25-30% of amaranth flour content in the bread formulation due to higher drying loss.

Table 2. Quality of wheat-amaranth bread.

Characteristic	Amaranth flour content, %						
	0	5	10	15	20	25	30
Technological properties							
Duration of proofing T_e, min	140±3	70±2	70±2	70±2	35±1	35±1	40±1
Baking loss ω, %	14.4±0.8	13.3±1.4	13.5±1.3	14.4±1.1	15.1±0.6	16.1±2.1	16.6±1.4
Drying loss φ, %	0.73±0.16	2.42±0.25	2.25±0.13	1.14±0.24	0.63±0.05	0.59±0.15	0.36±0.18
Total loss, %	15.1±0.7	15.7±0.1	15.8±1.1	15.6±1.4	15.7±0.6	16.7±1.9	17.0±1.6
Physical and chemical characteristics							
Specific volume ρ, g/cm^3	2.39±0.06	2.43±0.11	2.75±0.12	3.09±0.22	3.47±0.13	3.33±0.20	3.28±0.18
Shape stability H:B	0.33±0.09	0.45±0.02	0.43±0.02	0.46±0.02	0.37±0.08	0.34±0.04	0.14±0.08
Crumb moisture content W, %	42.3±1.3	43.0±0.8	42.5±1.3	42.8±0.8	42.8±0.8	42.8±0.8	42.0±1.5
Titratable acidity TA, °	1.1±0.1	1.4±0.2	1.5±0.2	1.8±0.2	2.0±0.3	2.0±0.3	2.0±0.3
Crumb porosity δ, %	70.6±1.7	74.5±2.1	75.6±1.6	76.4±3.6	76.7±1.1	80.4±2.6	79.8±3.1
Complex quality Q, units	71.0±1.7	88.4±2.4	94.3±2.8	97.8±2.9	93.3±2.5	82.2±1.9	69.0±1.5
Texture characteristics							
Stress value at relative strain σ (25 %), kPa	53.4±2.8	36.4±1.6	34.5±1.5	23.7±2.4	21.6±1.6	15.6±0.7	9.2±1.6
Elastic modulus E, kPa	212.8±11.0	145.4±6.4	138.2±5.9	94.9±9.5	86.3±6.5	62.3±2.7	36.9±6.4
Static elastic hysteresis index S, kPa·%	1412±81	989±86	965±70	668±70	634±48	454±28	286±48

Sensory properties play a crucial role in the acceptance of food by customers, and thus sensory radar (Figure 4) reveals the most important properties of bread on wheat-amaranth composite flour such as crust colour and appearance; crumb colour, structure, and chewiness; aroma and taste of the product. Therefore, the application of the amaranth flour could substantially affect the quality of the bread evaluated by the customer when the amaranth flour content in the composite mixtures reached 20% and more: crust appearance and aroma were the most sensitive attributes of the product deteriorated with the increase of the amaranth flour content in the formulations.

Figure 3. Wheat-amaranth bread with 0 (control), 5, 10, 15, 20, 25, 30 % of amaranth flour substitution.

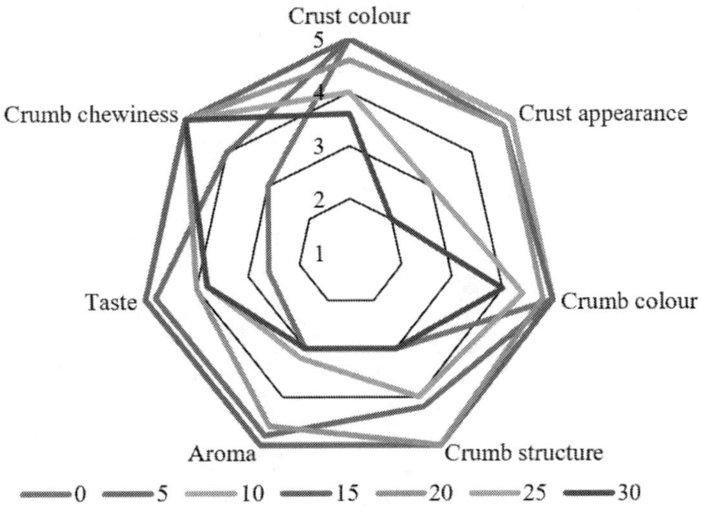

Figure 4. Sensory radar chart of wheat-amaranth bread with 0 (control), 5, 10, 15, 20, 25, and 30 % of amaranth flour.

The graphical interpretation of the obtained relations (Table 2) is shown in Figures 5-7. Thus, the increase in the amaranth flour content α led to a change in the quality characteristics of the bread. Hence, the complex organoleptic index was the highest ($Q = 97.8$ points) for the bread with 15 % of the amaranth flour. In turn, the baking loss ω, crumb porosity δ, and titratable acidity increased along with higher amaranth flour content. The opposite pattern was observed for proofing duration T_e, drying loss φ, and shape stability H:B of the loaf.

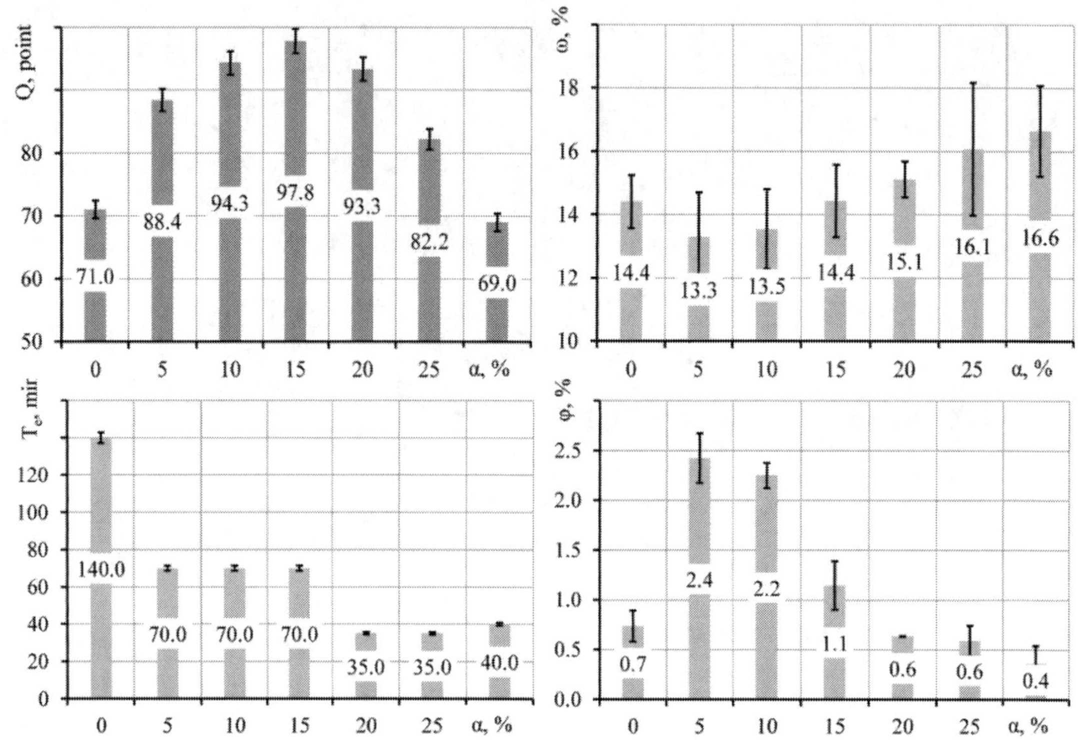

Figure 5. Effect of amaranth flour content α on bread complex organoleptic score Q, baking loss ω, duration of proofing T_e, drying loss φ.

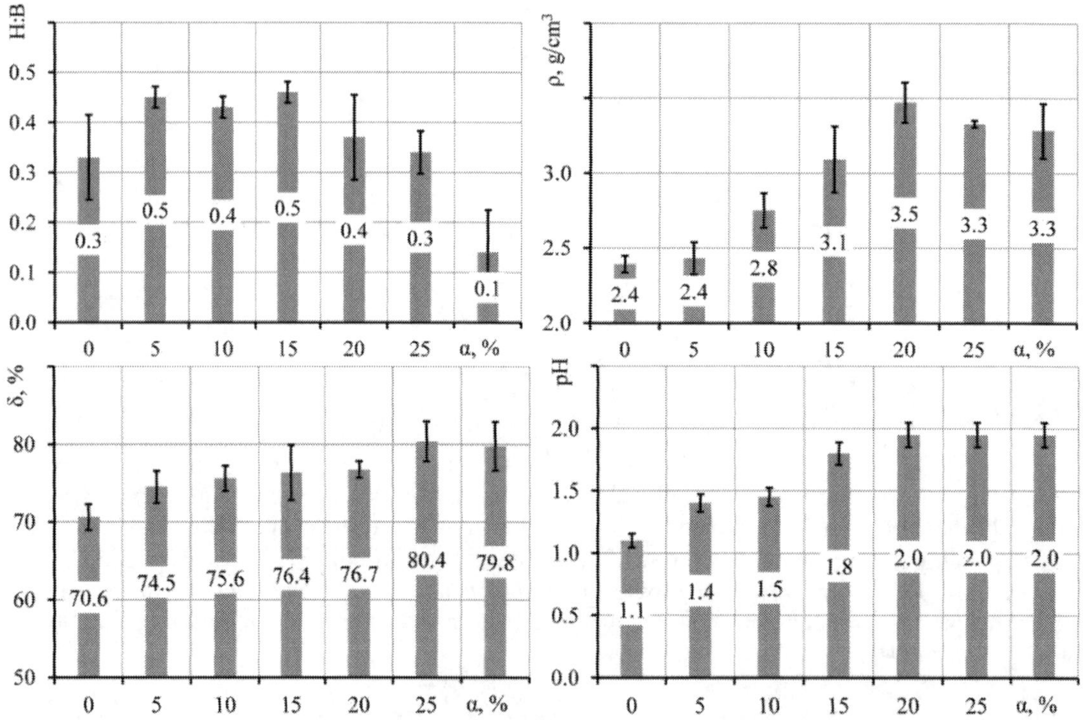

Figure 6. Effect of amaranth flour content α on bread shape stability H: B, specific volume ρ, crumb porosity δ, titratable acidity TA.

Figure 7 presents the dependences of changes in the elastic strength of fresh bread crumb F(t) on time t and the tension of fresh bread crumb σ (ε) on its relative deformation ε with different content of amaranth flour α in the bread formulations. The dependence of the fresh wheat-amaranth bread tension on its relative deformation σ(ε) was presented in the form of elastic hysteresis. When the amaranth flour content α increased, the hysteresis area decreased, which is confirmed by Figure 8, e.g. by the static elastic hysteresis index. A similar pattern was observed for the elastic modulus E and the stress at relative strain σ (25%). So, a decrease in their values was demonstrated with an increase in amaranth flour content α. Obviously, this was related to higher crumb porosity δ and the specific volume ρ of the products.

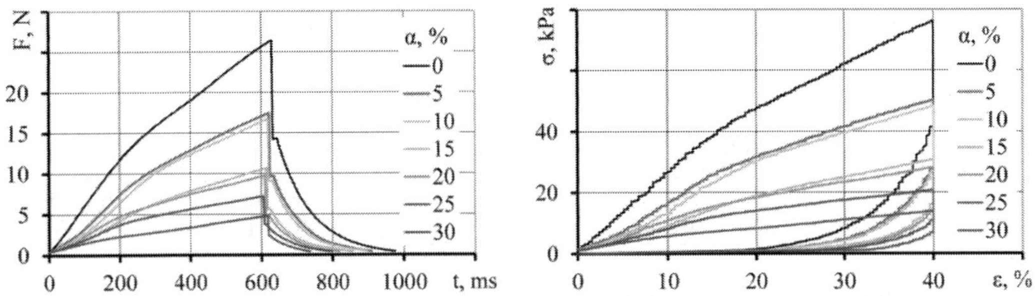

Figure 7. Dynamics of changes in elastic strength of fresh bread crumb F(t) and tension of fresh bread σ (ε) depending on amaranth flour content α.

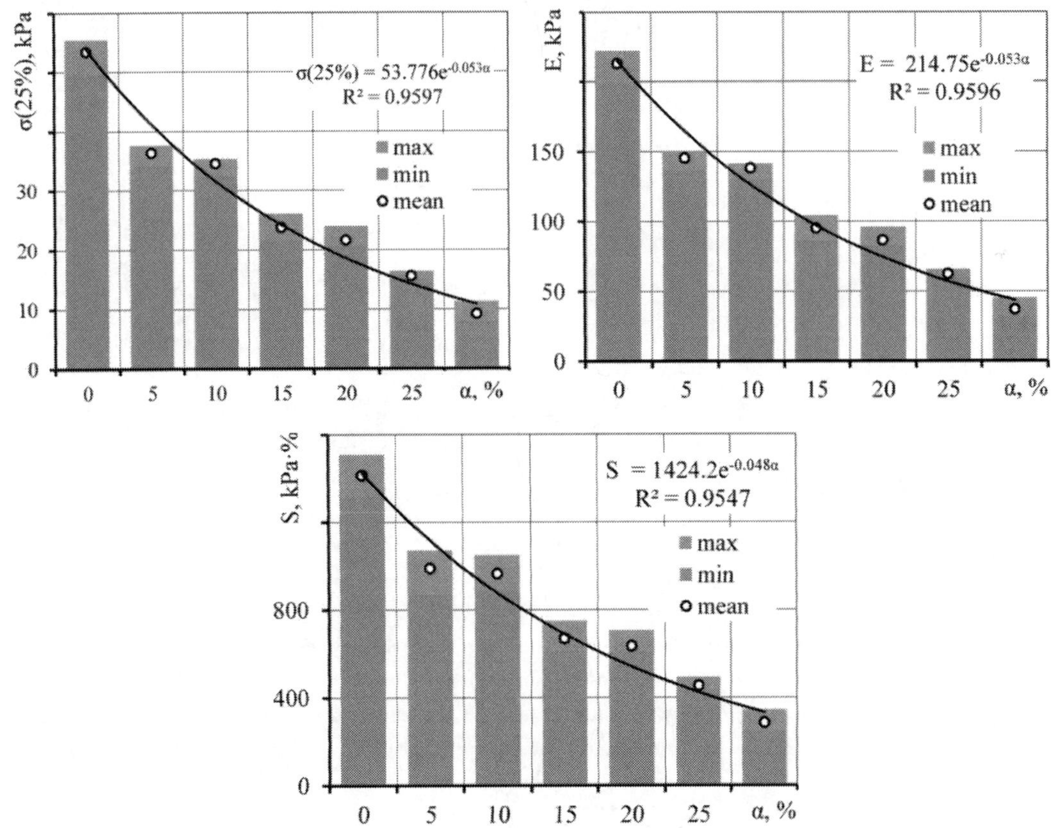

Figure 8. Stress value at relative strain σ (25%), elastic modulus E and static elastic hysteresis index S depending on amaranth flour content α.

Moreover, the results of changes in the crumb textural properties of the bread produced on different wheat-amaranth composite flour were obtained after 2 storage days of the products without access to oxygen under ambient conditions. A comparative assessment of the textural changes of the bread crumb is given in Table 3, Figures 9–11.

Table 3. Textural characteristics of bread with different amaranth flour content during storage.

Storage, hours	Amaranth flour content α, %						
	0	5	10	15	20	25	30
Stress value at relative strain σ (25%), kPa							
24	53.4±2.8	36.4±1.6	34.5±1.5	23.7±2.4	21.6±1.6	15.6±0.7	9.2±1.6
48	95.8±3.2	65.3±2.8	63.7±2.8	43.0±2.6	35.0±1.8	27.5±1.7	17.4±0.9
Elastic modulus E, kPa							
24	212.8±11.0	145.4±6.4	138.2±5.9	94.9±9.5	86.3±6.5	62.3±2.7	36.9±6.4
48	382.6±12.1	261.4±10.2	254.9±7.8	172.0±10.1	139.9±8.2	110.1±4.3	69.8±4.9
Static elastic hysteresis index S, kPa·%							
24	1412±81	989±86	965±70	668±70	634±48	454±28	286±48
48	2733±101	1807±99	1810±80	1239±82	1067±66	827±35	576±34

Figures 9–11 demonstrate that during the storage of the wheat-amaranth bread, the textural properties changed drastically. Thus, in 2 days, the stress value at relative strain σ (25%) and the elastic modulus E increased by 1.8–1.9 times, when the static elastic hysteresis index S accelerated by 1.9–2.0 times. Such a change in the textural properties of the wheat-amaranth bread indicated its styling and related changes in the crumb microstructure. It should be noted that the presence of amaranth flour led to a decrease in the stress value at relative strain (25%) and the elastic modulus E, which was probably associated with lower starch retrogradation and softer structure of the crumb when gluten content decreased with higher amaranth flour substitution.

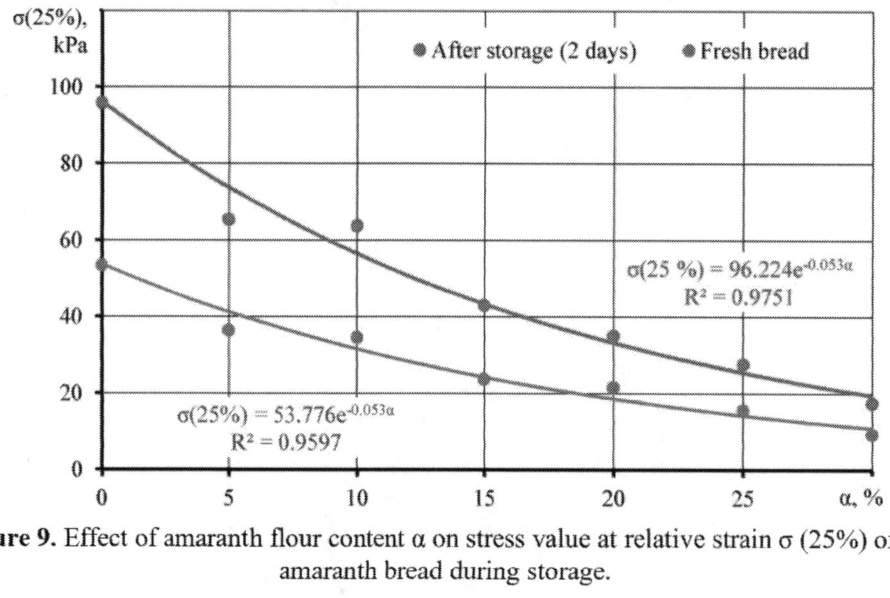

Figure 9. Effect of amaranth flour content α on stress value at relative strain σ (25%) of wheat-amaranth bread during storage.

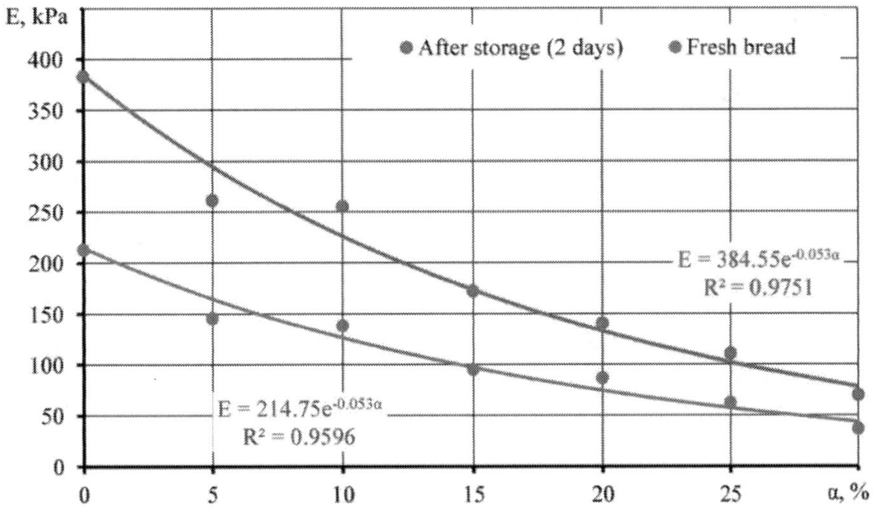

Figure 10. Effect of amaranth flour content α on elastic modulus E of wheat-amaranth bread during storage.

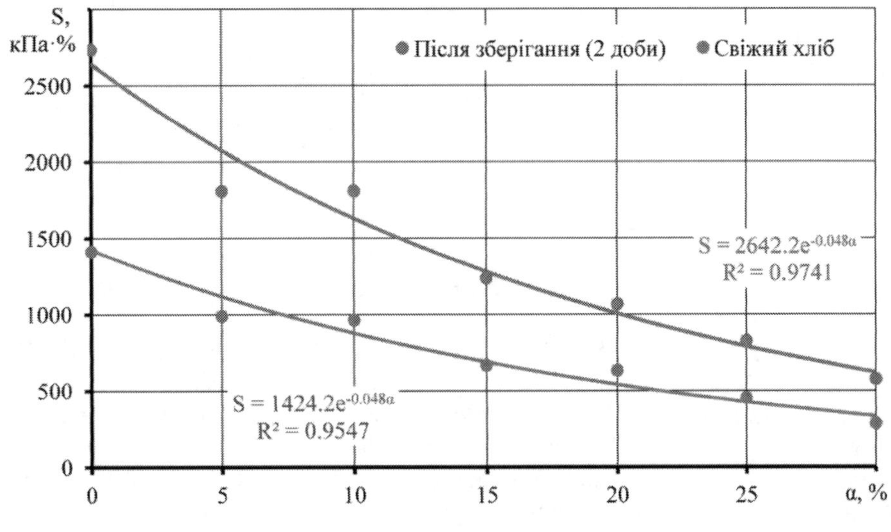

Figure 11. Effect of amaranth flour content α on static elastic hysteresis index S of wheat-amaranth bread duration storage.

The obtained results of the study are in agreement with the research [14] and the reported properties of another amaranth grain-derived products [11–13]. Therefore, amaranth flour application in the bread formulation could make a considerable change to the textural properties of the wheat-amaranth bread, possibly prolonging its shelf life without the deterioration of the bread quality when the wheat flour was 15 % substituted by the amaranth groat flour.

4. Conclusion

The benefits of applying amaranth flour to wheat bread were proved in terms of improving processing efficiency and bread quality, thereby prolonging product freshness without including artificial improvers in the formulations. Taking into account the textural properties of wheat amaranth bread during storage, amaranth groat flour as a product of the grain processing showed a positive impact on changes in the textural properties during storage of the wheat-amaranth bread.

Amaranth flour produced from debranned amaranth grain as an ingredient of wheat-amaranth composite flour improved the organoleptic, and physicochemical properties of the bread at 15–20% content in the mixture. This way higher complex organoleptic scores and decreasing the duration of proofing in bread making were reached without the deterioration of the quality. Despite intensified proofing, the total loss in the weight of bread, and titratable acidity demonstrated an increase when the amaranth flour content exceeded 25–30%.

Changes in the elastic strength of the fresh bread F(t) as a function of time t and the tension of fresh bread σ (ε) as a function of its relative deformation ε for different content of amaranth flour α revealed the formation of the softer structure of the product enriched with the amaranth groat flour. The area of static elastic hysteresis decreased with a higher concentration of the amaranth flour in the wheat-amaranth bread. Application of the amaranth groat flour in the bread formulations could contribute to the freshness of wheat-amaranth during storage enhancing the quality and nutritional value of the product.

References

[1] Gallagher E, Keehan D and Butler F 2005 Development of organic breads and confectionery p 32.

[2] David C, Abecassis J, Carcea M, Celette F, Friedel J K, Hellou G, Hiltbrunner, Messmer, Narducci V, Peigné J, Samson M F, Schweinzer A, Thomsen I K and Thommen A 2012 Organic Bread Wheat Production and Market in Europe *Sustainable Agriculture Reviews* **11** pp 43–62 DOI: 10.1007/978-94-007-5449-23

[3] Aliiev E, Aliieva O and Maliehin R 2020 Technical and technological provision of complex waste processing of plant raw oil cultures in food for organic animals *Scientific Horizons* **07 (92)** pp 112–119 DOI: 10.33249/2663-2144-2020-92-7-112-119

[4] Aliiev E, Maliehin R, Ivliev V and Aliieva O 2021 Simulation of the process of cavitation treatment of liquid feed *Scientific Horizons* **24(2)** pp 16–26 DOI: 10.48077/scihor.24(2).2021.16-26

[5] Aipova R, Abdykadyrova A B and Kurmanbayev A A 2020 Biological products in organic agriculture *Plant biotechnology and breeding* **2(4)** pp 36–41 DOI: 10.30901/2658-6266-2019-4-o4

[6] Mazzoncin M, Belloni P, Risaliti R and Antich D 2007 Organic Vs Conventional Winter Wheat Quality and Organoleptic Bread Test *3rd QLIF Congress Hohenheim Germany* March 20-23 https://orgprints.org/id/eprint/9753/1/Mazzoncini-etal-2007-WheatBreadQuality.pdf

[7] Bussmann R W, Paniagua-Zambrana N Y and Njoroge G N 2021 Amaranthus hybridus L Amaranthus spinosus L Amaranthaceae *Ethnobotany of the Mountain Regions of Africa Ethnobotany of Mountain Regions Springer Cham* DOI: 10.1007/978-3-030-38386-215

[8] Nyonje W A, Schafleitner R, Abukutsa-Onyango M, Yang R-Y, Makokha A and Owino W 2021 Precision phenotyping and association between morphological traits and nutritional content in Vegetable Amaranth (Amaranthus spp.) *Journal of Agriculture and Food Research* **5** pp 1–8 DOI: 10.1016/j.jafr.2021.100165

[9] Moreno C R, Cuevas-Rodriguez E-O and Reyes Fernandez P C 2019 Amaranth: Processing Product Development, and Nutritional Aspects *Whole Grains* **23** DOI: 10.1201/9781351104760-1

[10] Singh M and Liu S X 2021 Evaluation of amaranth flour processing for noodle making 2021 *Journal of Food Processing and Preservation* **45 (4)** DOI: 10.1111/jfpp.15270

[11] Mykolenko S, Zhygunov D and Rudenko T 2020 Baking properties of different amaranth flours as wheat bread ingredients *Food Science and Technology*, **14 (4)** pp 62–71 DOI: 10.15673/fst.v14i4.1896

[12] Nasir S, Allai F M, Gani M, Ganaie S, Gul K, Jabeen A and Majeed D 2020 Physical, Textural, Rheological, and Sensory Characteristics of Amaranth-Based Wheat Flour Bread *International Journal of Food Science* pp 1–9 DOI: 10.1155/2020/8874872

[13] Stankevich G, Valentiuk N, Ovsiannykova L and Zhygunov D 2021 Changes in quality of amaranth grain in the process of post-harvesting processing and storage *Food Science and Technology* **15(1)** pp 80–90 DOI: 10.15673/fst.v15i1.1959
[14] Stakhurlova A A, Derkanosova N M, Aristov A V, Ponomareva I N and Sutolkin A A 2020 Experimental evaluation of amaranth food products in terms of functional food ingredients *BIO Web of Conferences 17 00147* DOI: 10.1051/bioconf/20201700147
[15] Drobot V I 2002 Bakery production technology *Publishing House «Logos»* p 365
[16] Mykolenko S, Hez and Pivovarov O 2021 Effect of bioactivated amaranth grain on the quality and amino acid composition of bread *Ukrainian Food Journal* **10 (3)** pp 576–591 DOI: 10.24263/2304-974X-2021-10-3-11

Acknowledgments
This work is based upon the work from COST Action 18101 SOURDOMICS—Sourdough biotechnology network towards novel, healthier and sustainable food and bioprocesses (https://sourdomics.com/;https://www.cost.eu/actions/CA18101/), where the author S.M. is a member of Working Group 6. COST is a funding agency for research and innovation networks. Also, this work was supported by the Ukrainian State Project for Young Scientists of the Ministry of Education and Science of Ukraine (No. 0120U100322). The authors are also grateful to Olexandr Duda, President of the Association of Amaranth and Amaranth Products, for providing the amaranth grain.

IOP Conf. Series: Materials Science and Engineering 1277 (2023) 012022 doi:10.1088/1757-899X/1277/1/012022

The influence of the design of the upper cover of the cyclone on its characteristics

V V Maistruk

Department of Designing and Operation of Machines, Lviv Polytechnic National University, Lviv, Ukraine

Abstract. An analysis of the works in which the designs of widespread dust-capturing devices are studied. It has been established that the prediction of the operation of dust-collecting devices in certain conditions is most effectively carried out by the methods of numerical modeling and simulation of the separation process, which are widely used for the research of devices of this type. With the help of numerical modeling methods, the structure of the gas flow in the cyclone for different designs of the top cover, the influence of the gas flow velocity in the inlet nozzle on the hydraulic resistance of the device and compared it with experimental data were determined.

1. Introduction

Cyclone dust collectors are relatively simple centrifugal devices used to remove solid particles from gas streams. They are used in various industries and production technologies as independent devices in technological schemes for carrying out various technological processes, and in the form of dust cleaning systems. Growing requirements for environmental indicators of modern productions, quality of finished products, energy and resource conservation and creation of zero-waste technologies require further improvement of the efficiency of dust cleaning devices. One of the options for increasing the efficiency of cyclones is to optimize their design and geometric dimensions.

Among the various types of cyclone separators, countercurrent cyclones are the most studied today. Devices of this type are characterized by relatively low hydraulic resistance and sufficiently high efficiency of capturing dust particles.

The parameters that determine the efficiency of the cyclone can be classified according to operating conditions (inlet flow velocity, temperature, viscosity and gas density, dust particle sizes and their physical properties) and cyclone design. The design dimensions of standard cyclones are described by geometric parameters, which are expressed in fractions of the diameter of the device. This is the height of the cylindrical part, the conical part, the inlet and outlet nozzles, the dimensions of the bunker, etc. In many works, the influence of the main geometric parameters on the value of the hydraulic resistance and the efficiency of the cyclone was studied [1-7].

The analysis of these studies shows that the most determining cyclone geometric parameters are the height of the cyclone (cylindrical and conical parts), the geometry of inlet pipe and exhaust pipes. In connection with rather complex and time-consuming laboratory methods for optimizing the design of cyclones, today the methods of numerical modeling and simulation of the process are widely used, rejecting incorrect ratios of the geometry of the apparatus at the initial stage of design. Such an innovative approach makes it possible not only to approach the urgent task of optimizing the design of dust cleaning equipment in a new way, but also to significantly reduce the time needed to predict their main operating characteristics.

Currently, hydrodynamic modeling of gas flows in devices of various designs based on CFD modeling has been widely studied [8,9]. This method is used, for example, for the optimization of the central exhaust pipe in large cyclone separators for cement production [10], studies of the characteristics of the large industrial Stairmand cyclone [11], studies of the gas flow structure depending on the design of the inlet and outlet nozzle [12], optimization geometries of cyclones of the first stage, that are used in the technological line of clinker firing of cement production [13], study of hydrodynamic modes of cyclone operation for capturing low-pressure polyethylene dust [14], study of the effect of high dustiness of the gas stream at ambient temperatures and elevated temperatures on the operation of cyclone separators [15]. In [16], numerical modeling methods were used to analyze the influence of operational and design parameters on the hydraulic resistance of a straight-flow cyclone. Based on the obtained results, for the considered research conditions, the optimal geometry of the industrial-type apparatus is proposed. The results of computer simulations are in good agreement with the experimental data.

In order to optimize the design of the CN-15 cyclone to reduce the hydraulic resistance, we propose to study the influence of the design of the upper cover of the cyclone on the hydrodynamic structure of the flow in the working zone of the cyclone. It is known that the quality of separation of dusty gas and hydraulic resistance in a cyclone of any design is primarily determined by the velocity, swirling of the flow and the design of the device.

For all the variety of design schemes and types of gas flow intake, there is no clear answer as concerns the best way to ensure high-quality gas swirling. In the general case, swirling of the flow in cyclones of various designs is carried out in three main ways: tangential supply, use of axial swirling inserts and rotation of the separator itself or its working elements.

The most common among cyclones is the CN-15 cyclone (figure 1a), in which the inlet pipe is connected to a spiral cover. Making such a cover is time-consuming. Therefore, it is proposed to replace the spiral cover with a flat cover, which is installed at an tilt angle of the inlet nozzle relative to a plane perpendicular to the axis of the device (figure 1c).

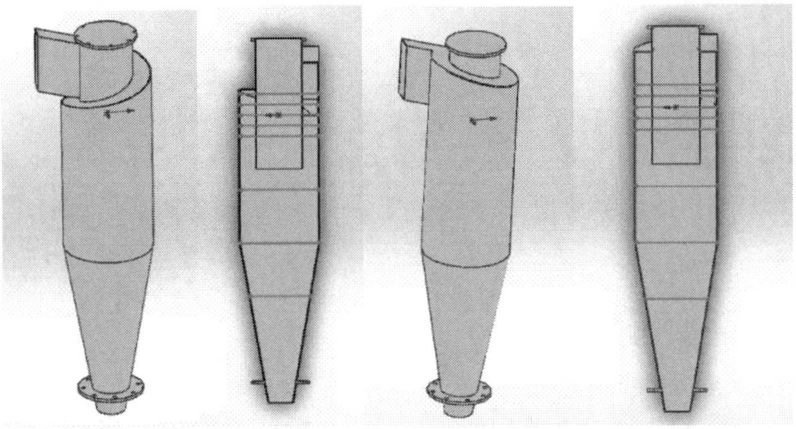

a) *b)* *c)* *d)*

Figure 1. 3D - models of cyclones: a – 3D model of the CN-15 cyclone; b - planes of research in the CN-15 cyclone; c – 3D model of a cyclone with a flat cover; d - planes of research in a cyclone with a flat cover.

Within the framework of theoretical studies, the following tasks were solved by means of numerical modeling:

- a comparison of the studied structure of the gas flow of the CN-15 cyclone with a cyclone with a flat cover was carried out;
- based on the obtained experimental studies and theoretical calculations of gas flow parameters, the energy efficiency of using the design of a cyclone with a flat cover was estimated.

2. Methodology

For CFD modeling of cyclones were built the geometric 3D models of cyclones with a diameter of the cylindrical part of 150 mm (figures 1a, 1c). A computational mesh was created for these models using the mesh generator. All geometric transformations were performed in a CAD program, and grids were created in a CFD program. Aerodynamic studies of these devices were carried out according to the known methodology [17]. The standard k-ε model was used for research, which is one of the most widely used turbulence models due to its simplicity in modeling the structure of turbulent flows. It does not require large computing power compared to other k-ε turbulence models.

The initial parameters for the study of these cyclones were: gas flow through the device - air; atmospheric pressure under normal conditions P_0=101325 Pa; ambient temperature under normal conditions T_0=293 K; density of gas (air) ρ_{air} = 1.293 kg/m^3.

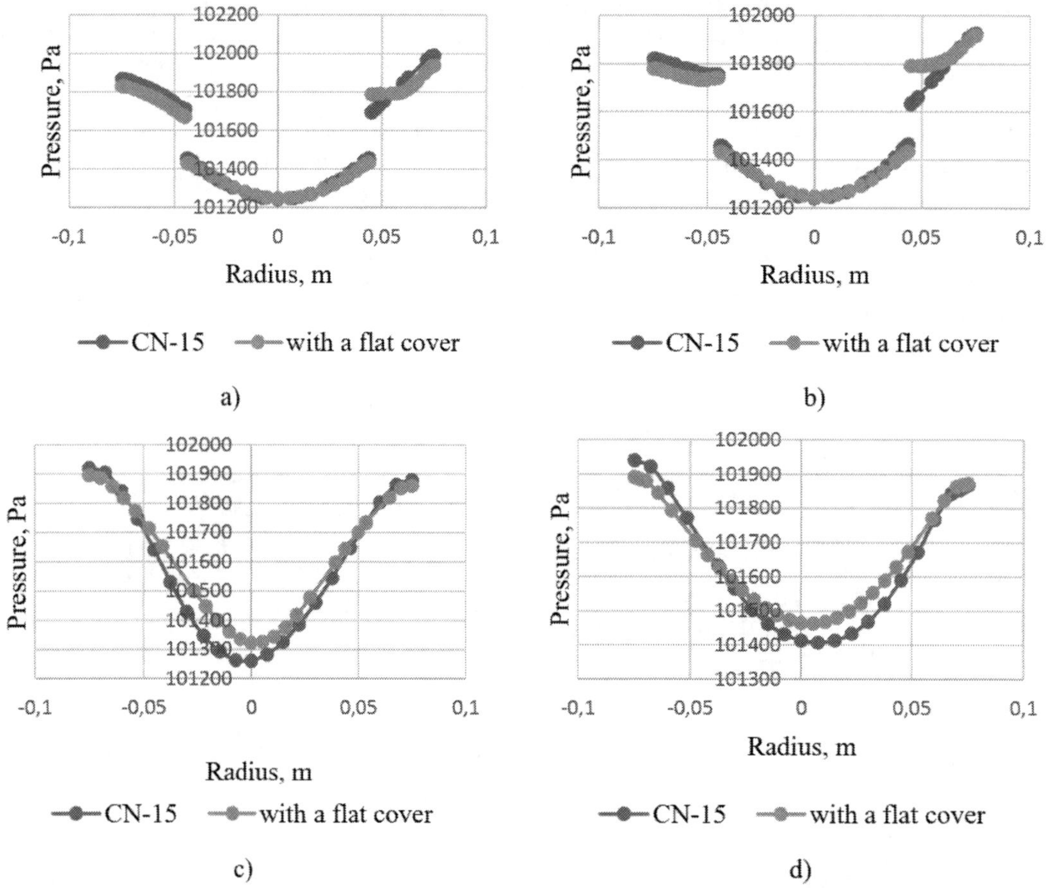

Figure 2. Change of pressure in the devices along the radius and along the height: a) height 40 mm; b) height 0 mm; c) height -140 mm; d) height -240 mm.

For the calculation the following boundary conditions were set: open flow - volume flow at the outlet (flows corresponded to a fictitious air velocity of 1 - 4 m/s); open pressure - ambient pressure; the wall - real wall. It should be noted that studies using CFD programs were conducted for the case when cyclones work on the injection line.

The main parameters analyzed in this work were the change in the radius of the pressure device, as well as the tangential and axial components of the velocity in the studied cyclones. A comparison of the hydraulic resistance of the investigated devices was also carried out. The hydraulic resistance of the devices was determined by the difference in static pressures, since the inlet and outlet pipelines have the same diameter.

The pressure and velocity components of the studied cyclones were examined in horizontal planes relative to the axis of the apparatus every 20 mm in height in the upper part of the apparatus, as well as every 100 mm in the cylindrical and conical parts of the apparatus (figures 1b, 1d). The zero point in height is the plane of the entry of the lower face of the rectangular inlet pipe into the cylindrical part of the apparatus body (figure 1). Positive height values correspond to the positive direction of the vertical Z axis (figure 1). Based on the results of the research, graphs of the dependence of the pressure, the tangential and axial components of the flow velocity on the radius of the device were constructed.

3. Results and discussion

The results of the comparison of the pressure change in the devices along the radius and along the height are presented in figure 2. As can be seen from the figure, in a cyclone with a flat cover, the pressure drop along the radius of the device in the upper part of the separation zone of the device in the downward gas flow is greater compared to the CN-15 cyclone. In the lower part of the separation zone of the apparatus, as well as in the central pipe, the change in pressure along the radius of the apparatus is approximately the same.

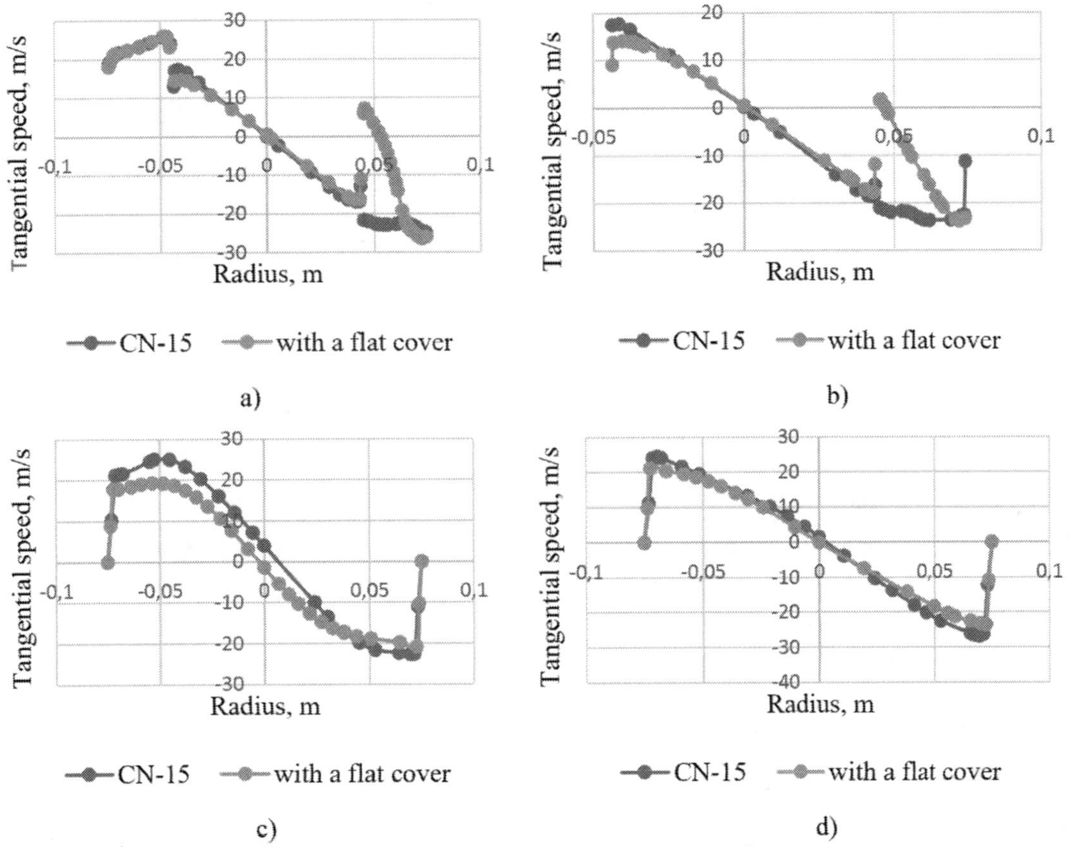

Figure 3. Changes in the tangential velocity in the devices along the radius and along the height: a) height 40mm; b) height 0 mm; c) height -140 mm; d) height -240 mm.

When considering the tangential velocity distribution, it is usually used the hypothesis that the tangential velocity field formed in the middle of the cyclone is the same as in a free vortex. With this approach, it is assumed that due to the low component of the axial velocity, the average tangential velocity is almost equal to the velocity of the gas flow entering the cyclone, i.e. equal to the velocity in the inlet nozzle, and the flow does not stretch along the vertical axis and has the form of a cylindrical spiral with a constant pitch.

The presented results of theoretical studies of tangential velocity indicate different natures of changes in this component of absolute velocity in upstream cyclones. The tangential velocity profile (Fig. 3) in

IOP Conf. Series: Materials Science and Engineering 1277 (2023) 012022 doi:10.1088/1757-899X/1277/1/012022

the working zone of the cyclone is heterogeneous in height and radius. After passing through the forming zone, the structure of the tangential velocity distribution has a maximum along the radius located closer to the cylindrical wall, the result of which is the result of the action of centrifugal forces.

The shape of the upper cover of the apparatus has a great influence on the change of the tangential velocity along the radius of the apparatus in the upper part of the separation zone. As can be seen from figure 3, in the CN-15 cyclone with a spiral cover, the gas flow has tangential velocity values that are close to the velocity in the inlet nozzle. And in a cyclone with a flat cover, the tangential velocity along the radius changes from a positive value near the wall of the central pipe to a negative value near the wall of the cylindrical body. In the lower part of the separation zone of the apparatus, as well as in the central pipe, the nature of the change in the tangential velocity along the radius of the apparatus is approximately the same in both apparatuses.

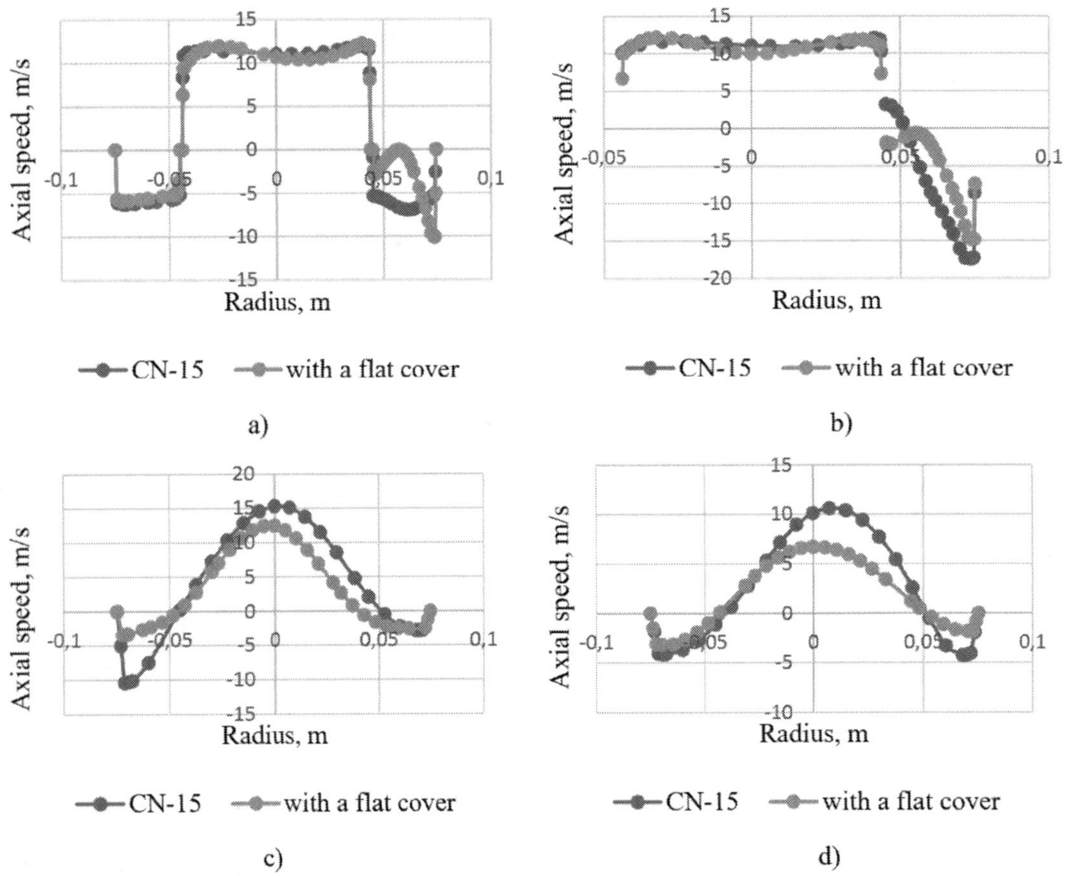

Figure 4. Change in axial velocity in the devices on radius and height: a) height 40 mm; b)height 0 mm; c)height -140 mm; d)height -240 mm.

As can be seen from figure 3, the nature of the change in the axial velocity along the radius of the device in the examined cyclones in the upper part of the separation zone of the device is different. Axial velocity values in a cyclone with a flat cover are lower than in a CN-15 cyclone. In the lower part of the separation zone of the apparatus, as well as in the central tube, the nature of the axial velocity change along the radius of the apparatus is approximately the same in both apparatuses.

It follows from the conducted analysis that the experimental results in terms of tangential and axial velocities for the compared cyclones depend on the shape of the upper cover. As the flow moves along the axis of the cyclone due to the friction of the gas phase against the solid wall, the tangential component decreases, accordingly, the centrifugal force decreases, but at the same time the axial component of the velocity increases.

Therefore, compared to the CN-15 cyclone, in a cyclone with a flat cover, the tangential velocity has more stable values along the radius, and the axial values are relatively low. It is assumed that such a ratio of velocities should have a positive effect on the overall hydraulic resistance of the cyclone, i.e. the resistance of the cyclone with a flat cover will be lower than that of the CN-15 cyclone. The results of research on the dependence of hydraulic resistance on fictitious velocity, i.e. gas consumption, confirm this fact (figure 5).

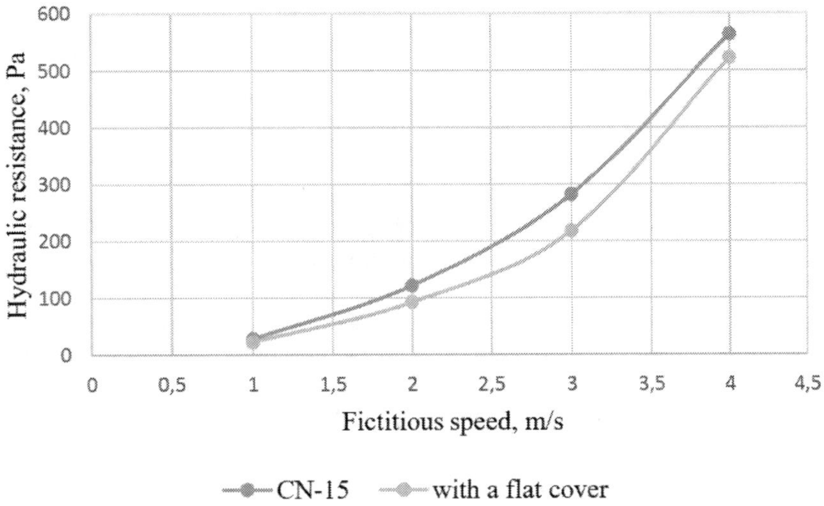

Figure 5. Comparison of the hydraulic resistance of dust collectors with different top covers.

The results presented above allow to make assumptions about the distribution of the flow in the cyclone in the form of certain zones. In the entrance zone, after the introduction of gas, the components of the absolute velocity are reformed and an axially symmetric swirling flow is formed. In the working zone, where is already a formed field of velocities, occurs deposition of solid particles under working conditions. In the outlet zone occur the compression of the gas phase and the transition of purified air into the upward flow.

Thus, the geometry of the upper cover leads to the reorganization of the swirling flow inside the working zones of the cyclone. This, in turn, significantly changes the overall separation capacity of the device and the energy consumption of the cleaning process. The effect of these parameters on the efficiency of the cyclone requires further experimental studies.

4. Conclusions

The influence of the design of the inlet nozzle on the distribution of pressure, tangential and axial velocity in the working zone of the cyclone was studied using the numerical modeling.

The presented comparative results for the CN-15 cyclone and a cyclone with a flat cover show that the features of the flow geometry at the entrance to the cyclone affect not only the general gas flow in the separation zone of the device, but also its hydraulic resistance.

Based on the conducted theoretical studies of countercurrent cyclones, in order to reduce energy costs during the purification of highly concentrated dusty gases from large particles, it is advisable to use a dust collector with a flat top cover.

References

[1] Yao Y, Shang M, Huang Z, Zhou T, Zhang M, Yang H, Lyu J 2022 Effects of the inlet duct length on the performance of a dense medium cyclone: An experimental and numerical study, *Chemical Engineering Research and Design* (In press)

[2] Wang Z, Sun G, Song Z, Yuan S, Qian Z 2022 Effect of inlet volute wrap angle on the flow field and performance of double inlet gas cyclones *Particuology* (In press)

[3] Babaoğlu N U, Hosseini S H, Ahmadi G, Elsayed K 2022 The effect of axial cyclone inlet velocity and geometrical dimensions on the flow pattern, performance, and acoustic noise *Powder Technology* **407** A117692

[4] Yaxin S, Zhenga A, Zhao B 2011 Numerical simulation of effect of inlet configuration on square cyclone separator performance *Powder Technology* **210(3)** 293-303

[5] Qian F, Wu Y 2009 Effects of the inlet section angle on the separation performance of a cyclone *Chemical Engineering Research and Design* **87(12)** 1567-72

[6] Elsayed K, Lacor C 2012 Modeling and Pareto optimization of gas cyclone separator performance using RBF type artificial neural networks and genetic algorithms *Powder Tech.* **217** 84-99

[7] Elsayed K, Lacor C 2011 The effect of cyclone inlet dimensions on the flow pattern and performance *Applied Mathematical Modelling* **35(4)** 1952-68

[8] Artyukhov A, Sklabinsriy V 2015 Theoretical Analysis of Granules Movement Hydrodynamics in the Vortex Granulators of Ammonium Nitrate and Carbamide Production *Chemistry & Chemical Technology* **9** 175

[9] Sklabinsriy V, Liaposhchenko O, Logvyn A, Al-Rammahi M 2014 Modeling of Gas Separator Inertial and Filter Elements for Natural Gas Fine Cleaning *Chemistry & Chemical Technology* **8** 479

[10] Cristea E, Conti P 2013 CFD simulation of large dust collection cyclones positioned vertically in staggered downward cascade arrangement *Fluids Engineering Division* FEDSM(2013)001

[11] Erol H I, Turgut O, Unal R 2019 Experimental and numerical study of Stairmand cyclone separators: a comparison of the results of small-scale and large-scale cyclones. *Heat Mass Transfer* **55** 2341–54

[12] Misiulia D, Antonyuk S, Andersson A G, Lundström T S 2020 High-efficiency industrial cyclone separator: A CFD study *Powder Technology* **364** 943-953

[13] Wasilewski M, Brar L S 2017 Optimization of the geometry of cyclone separators used in clinker burning process: A case study *Powder Technology* **313** 293-302

[14] Alavi S A, Lay N E, Makhmali Z S 2018 A CFD study of industrial double-cyclone in HDPE drying process. *Emerging Science Journal* **2(1)** 31-38

[15] Nakhaei M, Lu B, Tian Y, Wang W, Dam-Johansen K, Wu H 2020 CFD modeling of gas-solid cyclone separators at ambient and elevated temperatures *Processes*. **8(2)** A228

[16] Maistruk V V, Gavryliv R I, Popil A S, Basisty A M 2012 Estimation of energy consumption during the operation of a direct-flow cyclone using the FLOW SIMULATION software package *Eastern European journal of advanced technologies* **6/8(60)** 28-30.

[17] Alyamovsky A A, Sobachkin A A, Odintsov E V, Kharitonovych A I, Ponomarev N B 2005 *SolidWorks. Computer modeling in engineering practice* St. Petersburg: BHV-Petersburg

Rational choice of torques distribution between the front and back electric motors of automobile wheels drive

M Podryhalo[1], R Kaidalov[2] and V Omelchenko[3]

[1] Department of Mechanical Engineering and Machine Repair Technology, Kharkiv National Automobile and Highway University, Ukraine
[2] National Academy of the National Guard of Ukraine, Ukraine
[3] Department of International Relations, Information and Communication, National Academy of the National Guard of Ukraine, Ukraine
[1] E-mail: pmikhab@gmail.com

Abstract. Two-axle all-wheel drive automobiles and the procedure for determining the torques between the front and back driving wheels are considered. The influence of the distribution of torques between the front and back driving wheels of all-wheel drive two-axle automobiles on their controllability, stability and energy efficiency is analyzed. A study was conducted to improve the energy efficiency of two-axle automobiles by rationally choosing the torque distribution coefficient between the front and back wheels. A mathematical model of the efficiency ratio of the wheel drive of a two-axle automobile was obtained, which makes it possible to determine the ratio of the rational distribution of torques between the front and back wheels through its optimization. It has been proven that the use of electric automobiles and automobiles with a combined power unit will allow to control the distribution of torques between their front and back wheels controlled by electric motors. The proposed rational choice of the distribution of torques between the front and rear electric motors of the wheel drive will allow to increase the efficiency of the wheel drive and improve the energy efficiency not only of single electric automobiles and hybrid automobiles, but also of road trains that use a combined electromechanical drive of the driving wheels.
Key words: all-wheel drive automobile, torque, energy efficiency, hybrid automobile

1. Intoduction

The distribution of road contact reactions between the front and back wheels due to the distribution of torques creates an impact not only on controllability and stability but also on the energy efficiency of automobiles. All-wheel drive two-axle automobiles usually have one of the two axles that can be disengaged (usually the front). This is due to the difficulty of adjusting the torque distribution between the front and back axles with a mechanical transmission. If there is a front drive axle that turns off, it can cause an automobile skidding on a slippery road [1, 2].

The appearance of electric and hybrid automobiles with a drive on separate axles or wheels opens up the possibility of distributing torques between the front and back wheels. Stepless smooth control of electric motors allows not to disconnect the drive of one of axles even on roads with a high coefficient of rolling resistance of wheels. At the same time, there is a rational distribution of torques between the front and back wheels, which ensures the maximum efficiency of automobile chassis. [3,4,5].

In this article, a mathematical model of the wheel chassis is proposed and optimized, which made it possible to determine the rational distribution of torques between the axles of a two-axle automobile.

2. Analysis of previous achievements and publications

The energy efficiency of the wheel propulsion unit is determined by its efficiency. In work [1, 6], a dependence for determining the efficiency of a single drive wheel was obtained, which has the following form:

$$\eta_k^{mom} = \eta_{fk}^{mom} \cdot \eta_{elast}^{mom} = \left(1 - \frac{M_{fk}}{M_k}\right)\left[1 - \frac{M_k}{C_{cir}}\left(1 - \frac{M_{fk}}{M_k}\right)\right], \tag{1}$$

де η_{fk}^{mom} – momentary power efficiency of a wheel,

$$\eta_{fk}^{mom} = 1 - \frac{M_{fk}}{M_k}; \tag{2}$$

M_{fk} – wheel rolling resistance moment,

$$M_{fk} = R_{zk} \cdot \alpha = R_{zk} \cdot f \cdot r_d; \tag{3}$$

R_{zk} – normal reaction of the road to the wheel;

f – rolling resistance coefficient;

α – rolling friction coefficient;

r_d – dynamic radius of a wheel;

M_k – torque on a wheel;

η_{elestk}^{mom} – momentary elastic efficiency of the wheel,

$$\eta_{elastk}^{mom} = \frac{r_k}{r_d} = 1 - \frac{M_k}{C_{cir}}\left(1 - \frac{M_{fk}}{M_k}\right); \tag{4}$$

r_k – kinematic radius of the wheel;

C_{cir} – circular stiffness of the tire.

In work [7], an analytical expression for determining the efficiency of a wheel drive was obtained, which contains n_1 rolling wheels and n_2 drive wheels.

$$\eta_{dr}^{mom} = \left[1 - \frac{M_{k2}}{C_{cir}}\left(1 - \frac{R_{z2} f r_d}{M_{k2}}\right)\right]\left[n_2\left(1 - \frac{R_{z2} f r_d}{M_{k2}}\right)\right] - n_1 \frac{R_{z1} f r_d \omega_{k2}}{\eta_{tr}^{mom} N_e}\left(1 + \frac{1}{1 + \frac{C_{cir}}{f R_{z1} r_d}}\right), \tag{5}$$

де R_{z1} – normal reaction of road on rolling wheels;

ω_{k2} – angular speed of drive wheels;

R_{z2}; M_{k2} – normal reaction of road and torque on drive wheel;

η_{tr}^{mom} – momentary transmission efficiency.

However, the work [7] does not consider the influence of the choice of torque distribution between the wheels of the front and back axles on the efficiency of the wheel drive.

2.1. The aim and formulation of research objectives

The aim of the research is to improve the energy efficiency of two-axle automobiles by rational choise of torque distribution ratio between the front and back wheels.

To achieve the aim, it is necessary to determine the dependence for calculating the efficiency of the wheel drive of a four-wheel drive two-axle automobile.

3. Presentation of the main material

For a two-axle all-wheel drive automobile, equation (5) will have the following form:

$$\eta_{dr}^{mom}=\beta_m\left(1-\frac{M_{fp}}{\beta_m M_e U_{tr}\eta_{tr}^{mom}}\right)\left[1-\beta_m\frac{M_e U_{tr}\eta_{tr}^{mom}}{C_{cir}}\left(1-\frac{M_{fp}}{\beta_m M_e U_{tr}\eta_{tr}^{mom}}\right)\right]+ \\ +\left(1-\beta_m\right)\left(\frac{M_{f3}}{\left(1-\beta_m\right)M_e U_{tr}\eta_{tr}^{mom}}\right)\left[1-\frac{\left(1-\beta_m\right)M_e U_{tr}\eta_{tr}^{mom}}{C_{cir}}\left(1-\frac{M_{f3}}{\left(1-\beta_m\right)M_e U_{tr}\eta_{tr}^{mom}}\right)\right], \quad (6)$$

where, M_{fp}; M_{f3} – total moments of resistance to rolling of the front and back wheels;

M_e – effective drive torque;

U_{tr} – transmission ratio;

β_M – coefficient of distribution of the total torque on the front axle of automobile,

$$\beta_M=\frac{M_{\text{kp}}}{M_{\text{kp}}+M_{k3}}, \quad (7)$$

M_{kp}; M_{k3} – total torques on the wheels of the front and rear axles of the car, respectively.

Expression (6) is transformed into the form:

$$\eta_{dr}^{mom}=b_1b_2-a_1a_2-\beta_M^2\left(a_3+b_3\right)+\beta_m\left(a_2+a_1a_3-b_2+b_1b_3\right), \quad (8)$$

where a_1; a_2; a_3;b_1; b_2; b_3;– coefficients,

$$a_1=\frac{M_{fp}}{M_e n_{tr}\eta_{tr}^{mom}}; \quad (9)$$

$$a_2=1+\frac{M_{fp}}{C_{cir}}; \quad (10)$$

$$a_3=\frac{M_e U_{tr}\eta_{tr}^{mom}}{C_{cir}}; \quad (11)$$

$$b_1=1-\frac{M_{f3}}{M_e U_{tr}\eta_{tr}^{mom}}; \quad (12)$$

$$b_2=1+\frac{M_{f3}}{C_{cir}}-\frac{M_e U_{tr}\eta_{tr}^{mom}}{C_{cir}}; \quad (13)$$

$$b_3=\frac{M_e U_{tr}\eta_{tr}^{mom}}{C_{cir}}=a_3. \quad (14)$$

The condition for obtaining the extremum of the function $\eta_{dr}^{mom}\left(\beta_m\right)$ is the equality of its derivative to zero:

$$\frac{\partial\eta_{dr}^{mom}}{\partial\beta_M}=\left(a_2+a_1a_3-b_2+b_1b_3\right)-2\left(a_3+b_3\right)\beta_M=0 \quad (15)$$

Where we find:

$$(\beta_M)_{opt} = \frac{a_2 + a_1 a_3 - b_2 + b_1 b_3}{2 + (a_3 + b_3)}. \tag{16}$$

After substituting ratios (9)-(14) into equation (16), we get:

$$(\beta_M)_{opt} = 0,5\left(1 + \frac{M_{fp} - M_{f3}}{M_e U_{tr} \eta_{tr}^{mom}}\right). \tag{17}$$

The second derivative of a function $\eta_{tr}^{mom}(\beta_m)$:

$$\frac{\partial^2 \eta_{dr}^{mom}}{\partial \beta_M} = -2(a_3 + b_3) = -4\frac{M_e U_{dr} \eta_{tr}^{mom}}{C_{cir}} < 0. \tag{18}$$

The latter means that it is optimal issue $(\beta_M)_{opt}$ corresponds to the maximum point of the function η_{dr}^{mom} i.e $(\eta_{dr}^{mom})_{\max}$.

After substituting expression (16) into equation (8), we get:

$$(\eta_{dr}^{mom})_{\max} = b_1 b_2 - a_1 a_2. \tag{19}$$

Equation (19) taking into account ratios (9), (10), (12), (13) have the form:

$$(\eta_{dr}^{mom})_{\max} = 1 - \left(\frac{\frac{M_{fp}^2 + M_{f3}^2}{C_{cir}} + M_{fp}^2 + M_{f3}}{M_e U_{tr} \eta_{tr}^{mom}} + \frac{M_e U_{tr} \eta_{tr}^{mom} - 2M_{f3}}{C_{cir}}\right) = 1 - (\zeta_{dr}^{mom})_{\min}, \tag{20}$$

where $(\zeta_{dr}^{mom})_{\min}$ – the minimum momentary coefficient of the wheel drive,

$$(\zeta_{dr}^{mom})_{\min} = \frac{\frac{M_{fp}^2 + M_{f3}^2}{C_{cir}} + M_{fp} + M_{f3}}{M_e U_{tr} \eta_{tr}^{mom}} + \frac{M_e U_{tr} \eta_{tr}^{mom} - 2M_{f3}}{C_{cir}}. \tag{21}$$

After substituting expressions (9)-(14) into equation (8), we get:

$$\eta_{dr}^{mom} = 1 - \left[\begin{array}{l} \frac{M_{fp}^2 + M_{f3}^2}{C_{cir} M_e U_{tr} \eta_{tr}^{mom}} + \frac{M_{fp} + M_{f3}}{M_e U_{tr} \eta_{tr}^{mom}} - \frac{2M_{f3}}{C_{cir}} + \frac{M_e U_{tr} \eta_{tr}^{mom}}{C_{cir}} + \\ +2\beta_M^2 \frac{M_e U_{tr} \eta_{tr}^{mom}}{C_{cir}} - 2\beta_M\left(\frac{M_{fp} - M_{f3}}{C_{cir}} + \frac{M_e U_{tr} \eta_{tr}^{mom}}{C_{cir}}\right) \end{array}\right]. \tag{22}$$

The rolling resistance moments of the front and back wheels can be defined as:

$$M_{fp} = mfr_{\partial} g \frac{b}{L}; \tag{23}$$

$$M_{f3} = mfr_d g \frac{a}{L}, \tag{24}$$

where m – mass of automobile;

g – free fall acceleration, g=9.81 m/s^2;

L – longitudinal wheelbase of automobile;

a, b – the distance from the front and back axles of the car to the projection of its center of mass on a horizontal plane passing through the indicated axles..

After substituting ratios (23) and (24) into equation (22), we will obtain:

$$\eta_{dr}^{mom}=1-\left[\begin{array}{l}\dfrac{m^2g^2f^2r_\partial^2(a^2+b^2)}{L^2C_{cir}M_eU_{tr}\eta_{tr}^{mom}}+\dfrac{mgfr_\partial}{M_eU_{tr}\eta_{tr}^{mom}}-\dfrac{2mgfr_d}{LC_{cir}}+\dfrac{M_eU_{tr}\eta_{tr}^{mom}}{C_{cir}}+\\+2\beta_M^2\dfrac{M_eU_{tr}\eta_{tr}^{mom}}{C_{cir}}-2\beta_M\left(\dfrac{mgfr_d(b-a)}{LC_{cir}}+\dfrac{M_eU_{tr}\eta_{tr}^{mom}}{C_{cir}}\right)\end{array}\right]. \quad (25)$$

Comparing equations (20) and (22) with each other, we get:

$$\eta_{dr}^{mom}=\left(\eta_{dr}^{mom}\right)_{\max}-\Delta\eta_{dr}^{mom}, \quad (26)$$

where $\Delta\eta_{dr}^{mom}$ – an absolute decrease of efficiency of wheel drive due to a suboptimal choice of the distribution of torques between the front and back drive axles,

$$\Delta\eta_{dr}^{mom}=2\beta_M^2\frac{M_eU_{tr}\eta_{tr}^{mom}}{C_{cir}}-2\beta_M\left(\frac{M_{fp}-M_{f3}}{C_{cir}}+\frac{M_eU_{tr}\eta_{tr}^{mom}}{C_{cir}}\right). \quad (27)$$

When $\beta_M=\left(\beta_M\right)_{opt}$ issue $\Delta\eta_{dr}^{mom}=0$.

After substituting equations (23) and (24) into expression (17), we get:

$$\left(\beta_M\right)_{opt}=0,5\left[1+\frac{mgfr_d\left(b-a\right)}{LM_en_{tr}\eta_{tr}^{mom}}\right]. \quad (28)$$

The analysis of equation (28) allows us to draw the following conclusions:

- parameters *m, f; b-a; M_e;n_{tr}* change during operation; it means that to obtain the maximum efficiency of the wheel drive, $\left(\beta_M\right)_{opt}$ should change;
- for automobiles with a mechanical transmission, change of $\left(\beta_M\right)_{opt}$ is a difficult technical task that requires the creation of a transfer box with a measurable distribution of changing torques between the front and back axles.
- the appearance of electric automobiles and automobiles with a combined electromechanical drive of the front and back wheels opens up great opportunities for increasing the momentary efficiency of the wheel drive due to stepless control of the current in the circuits of the front and back electric motors.

Considering the driving mode of automobile with separate control of the front and back electric motors and also the acceleration mode.

In the established movement mode:

$$M_eU_{tr}\eta_{tr}^{mom}/r_d=P_{thr\Sigma}=\frac{N_e\eta_{tr}^{mom}}{V_a}, \quad (29)$$

where V_a – linear speed of automobile, m/s.

After substituting (29) into (28), we determine:

$$\left(\beta_M\right)_{opt}=0,5\left[1+\frac{mgf\left(b-a\right)V_a}{LN_e\eta_{tr}^{mom}}\right]. \quad (30)$$

From expression (30) it is clear that during operation it is necessary to ensure the operation of the device that changes the distribution of torques between the axles, which uses sensors of normal loads on the front and back axles, vehicle speed and effective engine power. The number of sensors can be reduced by linking the effective power developed by the engine and the vehicle's resistance. Assuming that the power losses in the tires are included in the total losses of the transmission, we determine[8, 9]:

$$N_e\eta_{tr}^{mom}=P_w\cdot V_a. \quad (31)$$

where P_w – force of aerodynamic resistance to the movement of automobile [3],

$$P_w = \frac{C_x}{2} \cdot \rho F V_a^2; \tag{32}$$

C_x – coefficient of frontal aerodynamic resistance;
ρ – air density;
F – automobile model.
After substituting (32) into (31) and then into (30), we get:

$$\left(\beta_M\right)_{opt} = 0,5\left[1 + \frac{2mgf\left(b-a\right)}{C_x \rho F V_a L}\right]. \tag{33}$$

4. Results

Figure 1 shows diagrams of the dependence $\left(\beta_M\right)_{opt} = F_1\left(V_a\right)$ obtained at different measurements of the coefficient f of the rolling resistance of automobile wheels with the following parameters: C_x= 0.375; F= 1.777 m^2; m=190 kg; L=2.36 m; a= 1.036 m; b=1.324 m. As it can be seen from the indicated diagrams, at low automobile speeds, the coefficient $\left(\beta_M\right)_{opt}$ significantly exceeds to 0.5, approaching close to one. It means that the back axle should be disconnected for the considered automobile when moving up to speed $V_a = 10$ to m/c (36 km/h). In the range of speeds of 10-30 m/s (36-108 km/h), the front wheels should generate large torques in relation to the back wheels. At moving speeds more than 30 m/s (108 km/h), different torques can be created on the front and back wheels.

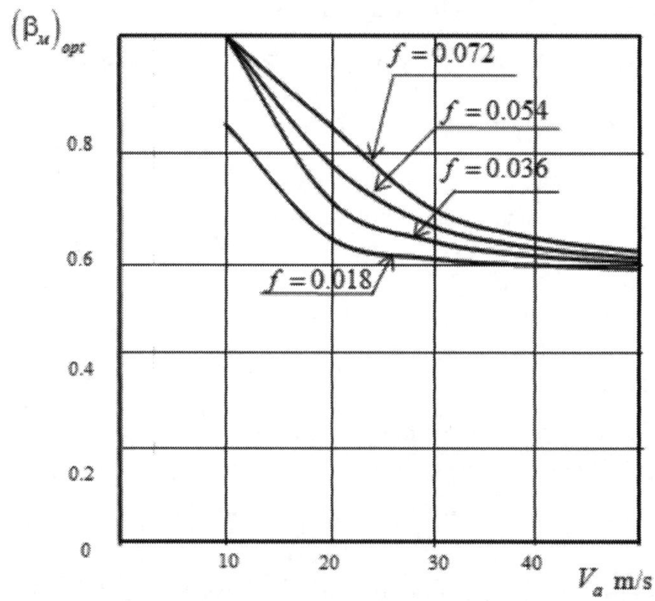

Figure 1. Dependence $\left(\beta_M\right)_{opt} = F_1\left(V_a\right)$ for automobiles with parameters: C_x= 0.375; F= 1.777 m^2; m=190 kg; L=2.36 m; a= 1.036; b=1.324 m at different measures of coefficient f wheel rolling resistance

Fulfillment of the specified conditions, which is easily implemented on automobiles with a separate electric drive of the front and back wheels, allows to get the maximum efficiency of the wheel drive, improve the dynamics of the automobile and reduce energy consumption for movement at a constant speed.

It can be seen from expression (33) that the parameters that change during the operation of automobile are *m, f, (b-a)* and speed V_a. With increasing vehicle speed V_a there is a decrease of the torque on the front axle and an increase in the specified share on back axle. When *a=b*, $(\beta_M)_{opt} = 0,5$ at any mass *m*, speed V_a and rolling resistance coefficients *f*.

When accelerating an automobile:

$$N_e \cdot \eta_{tr}^{mom} = V_a(P_w + \dot{V}_a), \tag{34}$$

where $\dot{V}_a$ – linear acceleration of an automobile.

$$M_{fp} - M_{f3} = \frac{1}{L}\left[mgfr_\partial (b-a) + 2m\dot{V}_a \frac{h - r_\partial}{L} fr_\partial \right]. \tag{35}$$

When an automobile accelerates, expression (33) takes the form:

$$(\beta_M)_{opt} = 0,5\left[1 + \frac{fr_\partial}{L} \frac{g(b-a) + 2\dot{V}_a(h - r_\partial)}{\frac{C_x}{2m}\rho F V_a^2 + \dot{V}_a} \right]. \tag{36}$$

During accelerating, it is also necessary to take into account the linear acceleration of an automobile. The resulting equation (36) will allow further development of a control system for the distribution of torques between the front and back wheels controlled by electric motors. It will increase the efficiency of the wheel drive and improve the energy efficiency of electric automobiles and hybrid automobiles.

5. Conclusions

1. The obtained mathematical model of the efficiency of wheel drive of a two-axle automobile made it possible to determine the ratio of the rational distribution of torques between the front and back wheels through its optimization.
2. For automobiles in which the center of mass is located in the middle between the front and back wheels, the maximum efficiency of the wheel drive can be realized with an equal distribution of torque (traction) moments between the front and back axles during the set movement.
3. During the operation of automobile at $a \neq b$ parameters *m, f, b-a;* M_e; U_{mp} can change, to get $(\eta_{dr}^{mom})_{max}$, β_M needs regulation. For automobiles with mechanical transmission, it is a difficult mechanical task, but for electricautomobiles and hybrid automobiles with a split drive from electric motors to the front and back wheels, it opens up a lot of possibilities. During adjusting the current, it is possible to adjust the torques.

References

[1] Podryhalo M, Bazhynov O, Kaidalov R, Hushchenko V and Kudimov S 2018 Synthesis of rational laws for controlling the distribution of torques between axles during acceleration of automobile *Bulletin of machine building and transport*, issue 2 (Kharkiv) p 85-91

[2] Podryhalo M, Kaidalov R, Kudimov S and Yefimchuk V 2018 Rational management of acceleration of all-wheel drive electric vehicle *Automobile and Electronics. Modern*

technologies, issue 15(Kharkiv) p 58-63

[3] Podrigalo M, Kaidalov R,. Klets D, N Podrigalo,. Makovetskyi A,. Hatsko V,. Abramov D,. Tarasov Yu, Lytovchenko D and Lytvynov O 2018 Synthesis of energyefficient acceleration control law of automobile *Eastern-european gournal of enterprise technologies*, issue 91 (Kharkiv) p 62-70

[4] Podrigalo M, Klets D, Boroshko O, Rudzinskyi V, Kaidalov R,. Nikorchuk A, Adamchuk M., Luhovskyi I, Bilenko O, Kriukov O and Besedin Y 2019 Breaking Performance of the One Side Wheels for the Realization of the Crab Motion for the Front-Wheel Vehicle: SAE International *SAE International AE Technical Paper*, (Kharkiv)

[5] Podrigalo M, Bogomolov V, Kholodov V, Koryk O, Turenko A, Kaidalov R, Verbitskiy V, Nikorchuk A, Volodarets M, Kudimov S and Khodyriev S. 2020Energy efficiency of vehicles with combined electromechanical drive of driving wheels *SAE international* (Kharkiv) p 31-39

[6] Abdulgazis A, Abramov D and Artiomov M 2019 *Dynamics of automobile wheel* (Kharkiv: KhNARU) p 199

[7] Podryhalo M, Kaidalov R and Omelchenko V 2022 Efficiency estimation of wheel automobile propulsion unit *Automobile and Electronics. Modern technologies*, issue 21 (Kharkiv) p 31-39

[8] Podryhalo M, Kaidalov R, Litvinov O, Kudimov S, Korobko A and Tarasov YU 2019 Improvement of the method of experimental and theoretical determination of the parameters of the aerodynamic resistance of an automobile *Collection of scientific papers,* issue 1 (Kharkiv: KhNARU) p 21-30

[9] Podryhalo M , Kaidalov R and Kudimov S. The influence of aerodynamic indicators on the rational dynamic characteristics of an automobile *Collection of scientific papers*, issue 2 (Kharkiv: NANGU) p 22-28

Specificity of the stress-strain and thermodynamic state simulation of the titanium alloy workpiece in the cutting zone

V Stupnytskyy[1,3], Xianning She[1] and N Stupnytska[2]

[1] Department of Robotics and Integrated Mechanical Engineering Technologies, Lviv Polytechnic National University, Lviv, 12 Bandery street, 79013, Ukraine
[2] Department of Civil Safety, Lviv Polytechnic National University, Lviv, 12 Bandery street, 79013, Ukraine

[3] vadym.v.stupnytskyi@lpnu.ua

Abstract. Titanium alloys have specific mechanical and physical-chemical properties that significantly reduce their machinability. Providing high productivity and accuracy has a great sense at the stage of technological preparation of machine-building production, taking into account the high cost of products from titanium alloys, as well as the need to provide increased operational properties. Simulation of machining processes of titanium alloys has many differences from the simulation of cutting of medium-carbon steels. This article focuses on recommendations for selecting an effective cutting model, describing the specific rheological properties of titanium alloys, optimizing solver selection, and cutting fracture criteria. The stages of adiabatic shear are described. The use of these recommendations contributes to the development of a more adequate model of machining titanium alloys and the choice of optimal cutting parameters to ensure a given quality of the machined surface layer of the workpiece.

1. Introduction

Four main factors influence the quality of precision machined surfaces. These are the technological characteristics of the machining process, the functionality of the essential technological equipment, the workpiece material's characteristics, and the cutting tool's geometry. A scientific approach is necessary to provide a theoretical basis for establishing an adequate relationship between the obtained physical-mechanical and microgeometric parameters of the machined surface with the defining parameters of these four main factors to further investigate surface engineering to ensure its priority operational properties. Therefore, it is crucial to conduct a scientifically based analysis of the machining of the most loaded surfaces of the product to ensure efficient production with given repeatability, predictability, reproducibility, and productivity. The ultimate goal of scientifically implementing such an approach is to investigate, simulate, optimize, and control the accuracy of machined surfaces, including their texture, topography, and functional properties.

The main goal of modeling functionally-oriented processes of machining workpieces of titanium alloys is to evaluate the mutual influence of the main technological parameters (machining parameters) on the formation of the quality state of the product surface layer by the cutting tool. It is realized based on a systematic analysis of the influence of technological factors and tool geometry parameters on the stress-strain and thermodynamic state of the machined surface, chip, and cutting edge, formed in machining a product from titanium alloy. Of course, the adequate research method is systematic experimental research. However, this method has some significant disadvantages, namely:

- complexity of implementation of experimental installations, their high cost, and duration of research;
- the impossibility of simultaneous analysis of stress, strain, and temperature conditions in all parts of the chip-forming and shaping zone;
- Difficulty in taking into account a large number of technological and design factors and cutting conditions;
- The problems of determining the dominant factors in forming the stress-strain and thermodynamic state of elements of the technological system "Machine-Fixture-Tool-Workpiece" (MFTW);
- insufficient adequacy of the obtained regression relations in the case of changes in the material of the workpiece, tool geometry, lubricants-coolants, etc.

On the other hand, the methodology of simulation modeling of the cutting process has some weaknesses, among which the most important is the lack of adequacy of the research results to the actual indicators of machining. It is especially true for machining such materials as titanium alloys, where the number of influencing factors on the physical-mechanical and stress-strain parameters of cutting is very large and diverse [1]. In addition, it is difficult to choose adequate fracture criteria, deformation conditions, rheological properties of the processed materials in the power process and thermodynamic loading. Therefore, the most effective is the combination of simulation modeling with subsequent adjustment of the models based on the results of the experimental studies.

2. Methodology of simulation modeling of titanium alloy cutting process based on finite element theory

To adequately numerically simulate the machining of titanium alloys, it is necessary to know the mechanical and thermal behavior of the workpiece material under extreme deformation conditions (2-5), strain rates (10^4-10^6 s^{-1}), temperatures in the cutting zone (up to 1200 °C), etc. That is, material models for machining must consider the deformation response to high dynamic power loading (high strain rates), deformation stress (strain hardening), and large temperature fluctuations (temperature softening). For moderate force loading, in scientific practice, several models are used that adequately interpret the machined material's rheological properties quite adequately with varying accuracy. Examples of such models are the phenomenological models of Johnson-Cook [2], Zerilli-Armstrong [3], Steinberg [4], and Follansbee-Kocks [5].

Analysis of the use of these models allows us to conclude that the most effective method for formalizing the rheological properties of titanium alloys is the Johnson-Cook phenomenological model [2] because this model takes into account both kinematic hardening and the adiabatic thermodynamic shear of titanium alloy as a dependence of the cutting temperature and effective stress. And it is these indicators that are most significant in forming the rheological machining pattern of titanium alloys [6]. In addition, this model assumes that the strength is isotropic and dominantly depends on the average stress, in contrast to machining, for example, chromium-nickel alloys, where hardness and brittleness parameters play a significant role in the machined material fracture. In this model, the equivalent plastic stress is described by the equation:

$$\sigma = \left(A + B\varepsilon_2^n\right)\left[1 + Ch\left(\frac{\dot{\varepsilon}_2}{\left(\dot{\varepsilon}_2\right)_0}\right)\right]\cdot\left(\frac{\dot{\varepsilon}_2}{\left(\dot{\varepsilon}_2\right)_0}\right)^{\alpha}\left(D - E\left(\frac{T - T_{room}}{T_{melt} - T_{room}}\right)^m\right), \quad (1)$$

where A is the yield strength at slow load (A = 862 MPa for Ti-6Al-4V [7]); B is the static hardening describing the deformation property of the material (B=331 MPa); ε_2 is the equivalent plastic strain; n is the coefficient characterizing the hardening property of titanium alloy (n=0.34); C is the strain rate factor (C=0.012) [7]; $\dot{\varepsilon}_2$ is the rate of plastic deformation; $\left(\dot{\varepsilon}_2\right)_0$ is the strain rate in the static mode of deformation [7]; T_{room}, T_{melt} the initial environmental temperature and the melting temperature of the material, respectively; m is the degree index, taking into account the phenomenon of thermal softening of the material (m = 0.97).

In equation (1), the first part describes the static hardening effect, the second part characterizes the dynamic strengthening, and the third part formalizes the thermal softening effect.

The primary approach is to analyze, step by step, the influence of the material's effective strain, strain rate, temperature, and strength characteristics on the current chip shear stress. In fact, the effect of titanium alloy hardness parameters on the shear stress can be divided into two different effects: the force factor, which increases the strength of the material in the forming zone, and the temperature factor, which affects the deformation softening of the machined titanium alloy.

After the condition is reached, when the value of the resulting plastic deformation ($\Delta\bar{\varepsilon}$) exceeds the limiting strain of material fracture ($\overline{\varepsilon_f}$), the controlled cutting process begins (material fracture in the forming zone), characterized by the energy level $[U_f]$. This energy is determined by the ratio of the fracture resistance index K_c to Young's modulus E for a specific material of the workpiece:

$$[U_f] = \frac{K_c^2 \cdot (1-\mu^2)}{E} \tag{2}$$

where μ is the Poisson's ratio, which determines the elastic properties of the material

Using the finite element method assumes that when the energy criterion index at the top of the lattice reaches a critical value [8], this element disappears from the lattice, and a cavity crack appears in its place, which develops in the feed direction of the cutting edge. As a result, some of the material separates, forming a chip. Exceeding the energy level across the entire cross-section of the chip contributes to its fracture and separation from the workpiece in the form of segments [9].

In a simplified form, the deformational model of local fracture can be described by the Cowper-Symonds material deformation law, which considers the Yield Strength's direct dependence on the strain rate. To consider the strain rate at, we introduce a scaling factor β:

$$\sigma_{YS}^D = \sigma_{YS} \cdot \beta; \tag{3}$$

$$\beta = 1 + \left(\frac{\dot{\varepsilon}}{C}\right)^{1/P}, \tag{4}$$

where σ_{YS}^D is the current (dynamic) yield strength; P, C are the constants depending on the material of the workpiece [10].

The fracture criterion is the strain value at which the plastic strain ε^P reaches the limit value $[\varepsilon_B]$ [11]:

$$\varepsilon^P > [\varepsilon_B]; \tag{5}$$

$$\varepsilon^P = \int_o^t \left(\frac{2}{3}\dot{\varepsilon}_{ij}^P\right)^{0,5} dt, \tag{6}$$

where $\dot{\varepsilon}_{ij}$ are the components of the plastic strain rate tensors.

The dependence of Thermal conductivity and Young's modulus of titanium alloy on temperature is created in DEFORM-2D using a particular procedure and based on statistical data [11]. An example of data input when forming the model of rheological (stress-strain) properties of titanium alloy Ti6Al4V is shown in Fig. 1.

IOP Conf. Series: Materials Science and Engineering 1277 (2023) 012024 doi:10.1088/1757-899X/1277/1/012024

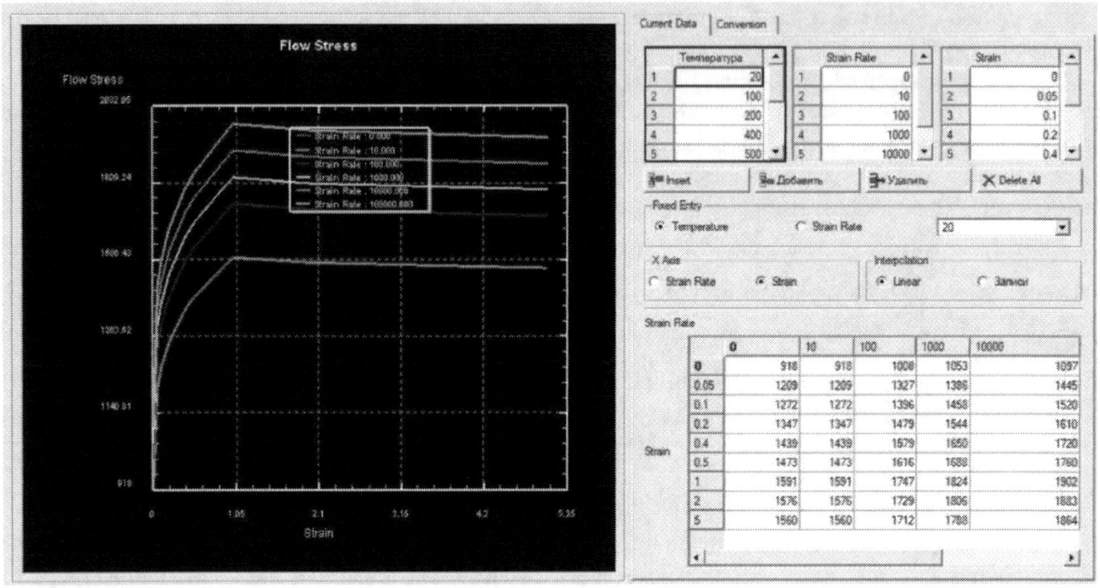

Figure 1. Example of data input when forming a model of rheological properties of titanium alloy Ti6Al4V.

3. Specific features of titanium alloys machining simulation in DEFORM 2D

Modern software DEFORM-2D has been used for predictive studies of the machining process [12]. This system is a multiple-purpose finite element analysis program used to solve two- and three-dimensional dynamic nonlinear problems of strained solid body mechanics, as well as problems associated with this process. They implement an explicit and implicit finite element method with the possibility of constructing Lagrangian, Eulerian, and hybrid meshes, a multicomponent fluid dynamics method, and a lattice smoothing method based on the Galerkin method [13]. The software uses procedures for automatic adjustment and smoothing of meshes from ordinary elements when they have degenerated, highly effective algorithms for solving contact problems, a wide range of problem-oriented material specifications, and possibilities for user programming of special procedures related to mechanical machining of materials, etc.

Input data for DEFORM 2D preprocessor when simulating machining (cutting) operation are: 2D model of the workpiece (surface) to be machined; cutting parameters (feed rate, cutting speed, depth of cut); geometry of the cutting edge, mechanical properties of the tool material and type and depth of coating; mechanical and thermophysical characteristics of the machined material (for example, Ti6Al4V); tool wear model (the Archard and Ushui recommended functions); criteria and limitations of the general transformation of finite-element lattices (calculated as the sum of modeling errors on the power vector, velocity vector, and allowable geometric error); type of deformation modeling (Lagrange Incremental or Steady-State Machining); time increment of the study (constant, or function of processing time); iterative method of analysis (direct iterative or Newton-Rafsen method), type of strain and temperature resolver (Skyline method or sparse matrix method), solution convergence errors (by force vector and by strain rate vector), failure criterion (Rice-Tracy, McCintock, Cockroft-Laytham etc. п.), selection method and specifics of generating a Lagrangian or Eulerian mesh for FEM analysis.

An essential step in studying the stress-strain state of a titanium alloy workpiece during cutting is the correct choice of the most effective format of the FEA solver (including irregular matrices). The author's experience with similar problems shows that the Skyline conjugate gradient works faster and requires less memory than the decoupler operating on the sparse matrix algorithm. However, when combined with the iterative solver, it reveals difficulties in converging solutions due to the problem of the insufficient number of contact points. It was found that, for example, when modeling a process with a low cutting depth, once a small number of nodes are detected in the contact, then when the Skyline solver is used, there are problems with the convergence of the study results. Therefore, for this class of problems, it is appropriate to use the algorithm of the Sparse Method []. However, for issues with a large number of tetraelements (four-node lattice Lagrangian grid), the Sparse matrix solver requires more memory than the computer can allocate. Therefore the calculation procedure can be terminated without notice. It has been experimentally established that the maximum limit for the decoupler using the sparse matrix method is 140 thousand elements. Therefore, the Skyline solver is more effective for simulating cutting processes with a cutting depth of more than 0.5 mm.

The dissipation rate component that characterizes the plastic deformation of the workpiece is a homogeneous first-order function of the plastic strain rate corresponding to the elastoplastic state case, regardless of the time scale of the cutting process. Obviously, the plastic strain increases with dynamic loading. The cutting resistance is physically characterized by the elastic modulus (usually the shear modulus G and the yield strength σ_{YS}), which, in addition to temperature, also depends on an additional parameter of the state, characterized by the fracture criterion D.

Based on the high plastic properties of titanium alloys, it can be concluded that, according to the existing cutting models, the basic strain and energy fracture criteria (McClintock, Oyada, Ayada, Osakada and other methods [My paper]) generally adequately describe the rheological pattern of cutting. However, it may be noted that models using energy criteria for fracture, such as the normalized Cockcroft-Latham criterion (D1) or the exponential Rice-Tracy criterion (D2) are more adequate [9]. These fracture criteria are based on the calculation of the potential energy of plastic deformation, that is, the area of the figure bounded by the stress-strain curve:

$$D_1 = \int^{\bar{\varepsilon}} \frac{\sigma_{\max}}{\bar{\sigma}} d\bar{\varepsilon} \tag{7}$$

$$D_2 = \int^{\bar{\varepsilon}} e^{\frac{\alpha \sigma_m}{\bar{\sigma}}} d\bar{\varepsilon} \tag{8}$$

where $\bar{\varepsilon}$ is the accumulated equivalent strain; $\sigma_{\max}$ is the maximum principal stress; $\bar{\sigma}$ is the equivalent Mises stress; α is a factor that depends on material properties; σ_m is the hydrostatic pressure.

The DEFORM 2D system uses an intelligent lattice structure generator to build a finite element mesh, capable of reconstructing the mesh during computation and separating the cutting zone into areas with different mesh sizes. Reducing the size of the grid edges is used for the chip-forming zone and close to it areas of the workpiece and tool, and increasing the mesh size is used for the outer parts of the workpiece and tool. This mesh gradient allows obtaining a more accurate picture of the cutting process without excessive consumption of processor memory resources to calculate the elastic-deformed and thermodynamic state in the whole workpiece volume. Analyzing the simulation results will make it possible to establish both the current and the residual stress-strain and thermodynamic parameters of the cutting process.

When cutting titanium alloys, it is necessary to consider the occurrence of a number of the negative factors that will significantly affect the machinability of this material [15, 16]:

- High temperature in the cutting zone leads to softening of the hard alloys of the tool. It contributes to a decrease in tool wear resistance and cutting speed, which, in turn, negatively affects overall machinability;
- high rigidity and strength of titanium alloys which create high contact stresses in the cutting zone, increasing the probability of the cutting tool brittle fracture;
- the low thermal conductivity and high strength of titanium materials lead to an increase in the intensity of heat flow in the tool direction. It leads to a decrease in tool life as a result of plastic destruction of the cutting edge;
- the high chemical activity of titanium materials with complex machinability, especially at high cutting temperatures, leads to activation of physical and chemical processes on the tool edges, and this is the main reason for increased adhesion fatigue and diffusion tool wear;

- The tendency of titanium alloys to intense abrasive tool wear in case of dry cutting of the workpiece.

A comparison of the mechanical characteristics of titanium alloys and typical carbon steels shows that the value of the real tensile strength and HB hardness at average temperature and in the absence of deformation (hardening) are approximately equal. Therefore, the worse machinability of titanium alloys is caused by other physical-mechanical and chemical properties and, first of all, by the structure, mechanical characteristics that determine their properties not only in the initial state but also in the hardened state and when heated. The reasons for low machinability are also the thermodynamic features of the material's strength formed at higher machining temperatures [16].

The machining process for titanium alloys has several special features compared to other alloys, such as low-carbon or chromium-nickel steels:

1. The actual reason for the complexity of machining the above materials is the significant hardening (compaction) of the material as a result of the process of plastic deformation of the metal during cutting [16]. The increased hardening is explained by the specific features of the crystal lattice structure of these materials. The characteristic that determines the plasticity or hardening ability of the material is the value of the conditional yield strength, corresponding to 0.2 percent of the residual stress to the ultimate strength. The lower the ratio, the more ductile the material is and the more work and power it requires to cut efficiently. The value of this ratio for titanium alloys is 0.4...0.45, while for the conventional low and middle carbon steels, this value is 0.6...0.65 and more [17]. Due to the increased ability to hardening during plastic deformation of such alloys, values can increase by two times (from 600 to 1200 MPa). Still, at the same time, the relative elongation decreases from 40-65% to 5-10%.

2. Another important reason for the low machinability of titanium alloys is the relatively low thermal conductivity, which contributes to an increased temperature in the contact zone "tool-chip-machined surface" and, consequently, to the activation of diffusion processes, intensive adhesion of the contact surfaces and more intense wear of the tool [18].

3. An important reason for the low machinability of most titanium alloys is the titanium alloy's ability to retain its original strength and hardness at high temperatures. This property leads to high specific stresses on tool contact surfaces during the cutting process. The low thermal conductivity of these materials amplifies this factor. Therefore, high temperatures on the contact surfaces do not increase plasticity and reduce the hardness of the layer to be cut.

4. Reduced vibration resistance of the MFTW system is due to the high hardening factor of titanium materials with non-uniformity during their plastic deformation process [19]. The occurrence of vibrations leads to high-frequency oscillations of power and thermal parameters of cutting and additional vibration loading on the tool. It contributes to microcracks with subsequent destruction of the cutting edge. The phenomena of chip build-up with the tool's rake face are intensified in the presence of vibrations, which, in turn, causes a particularly negative effect on tool life.

Thus, physically, the process of cutting titanium alloys can be described by stages [6]. First, the working surfaces of the tool collide with relatively soft, not hardened metal, and under their powerful impact, there is a plastic deformation of the cut layer. It is accompanied by significant energy absorption by the tool. At this stage, the sheared layer is considerably hardened due to the compaction of the crystal lattice. It acquires the properties of hardened metal and becomes more brittle. After some time, the plasticity reserve is exhausted, and shear occurs - material destruction in the primary shear zone with subsequent separation of the chip segment. The low thermal conductivity of titanium alloys leads to a sharp decrease in heat dissipation into the chip and the workpiece to be machined, and hence to an increase in temperature in the contact zone of the cutting part of the tool and the workpiece. A kind of bubble with strongly compacted walls is created, with high temperature and low plasticity, and increased material pressure inside this zone. Then its fracture occurs, and the process of energy accumulation is repeated. Activation of adhesion and diffusion processes is characteristic of this process. Tool wear increases significantly, and the build-up phenomena of the machined material are strongly activated.

Considering such a shaping process allows us to decide on the practicality of using cutting parameters in the range of resonant vibrational frequencies, contributing to an increase in tool vibrations and the appearance of local zones of residual stresses. It leads, on the one hand, to an increase in surface roughness and, on the other hand, to a decrease in the fatigue strength of titanium alloy products.

4. Conclusions

1. Analysis of advantages and disadvantages of methods of scientific research of function-oriented technological processes of products machining made of titanium alloys allows drawing the main conclusion that the most practical combination of simulation and analytical modeling methodology with subsequent adjustment of models based on the results of experimental research

2. The usage of DEFORM-2D software for prediction studies of force, thermodynamic and stress-strain parameters of mechanical machining of titanium-intensive alloys is proposed. This system is a multipurpose finite element analysis program for solving nonlinear dynamic problems of deformed solid mechanics. The effectiveness of solving the set problems exactly with the use of this software is due to the presence of the procedure of automatic regulation and smoothing of grids at degeneration of finite elements, application of highly effective algorithms of solving contact problems, and the presence of a large number of specifications of machined and tool materials with formalized thermo-deformation and rheological properties and with the possibility to correct them.

3. A comprehensive analysis of the application of software solvers that implement the finite element method for simulating the machining process of titanium alloys has proved that the Skyline conjugate gradient is faster and requires less memory than the solver working according to the iterative sparse matrix algorithm. However, the application of this solution is problematic in the case related to the insufficient number of points in the contact zone between the tool and the workpiece (for example, when modeling the machining process with a small cutting depth). It is reasonable for this class of tasks to use an algorithm based on the sparse matrix method. Therefore, Skyline is more effective for simulating machining processes with a cutting depth of more than 0.5 mm.

4. The most effective methodology for describing the rheological state of the material during the process of chip formation and the appearance of the power and thermal state in the forming zone is the Johnson-Cook phenomenological model since it is this model that most comprehensively formalizes such processes as kinematic hardening and the adiabatic shear of titanium alloy, arising due to the specific dependence of the efficient stress on strain rate and temperature

5. Based on the specific mechanical properties of titanium alloys (first of all, because of the high ratio of yield strength to temporary resistance to fracture), we can conclude that, following the existing strain cutting models, the deformation and energy fracture criteria are more adequate models that use energy criteria - such as the normative Cockcroft-Latham criterion or the exponential Rice-Tracy criterion. These fracture criteria are based on the calculation of the potential energy of plastic strain, that is, the area of the area bounded by the stress-strain curve.

References

[1] Davim J P 2010 *Surface Integrity in Machining* (London: Springer)

[2] Johnson G R and Cook W H 1983 A Constitutive Model and Data for Metals Subjected to Large Strains, High Strain Rates, and High Temperatures *Proc. 7 Int. Symp. on Ballistics (Hague)* **1** 541-547

[3] Zerilli F G and Armstrong R W 1987 Dislocation-mechanics-based constitutive relations for material dynamics calculations *J. of Appl. Phys.* **61**(5) 1816

[4] Steinberg D J and Lund C M 1989 A constitutive model for strain rates from 10^{-4} to 106 s^{-1} *J. of Appl. Phys.* **65** 1528

[5] Follansbee P S and Kocks U F 1988 A constitutive description of the deformation of copper based on the use of the mechanical threshold stress as an internal state variable *Acta Metallurgica* **36**(1) 81-93

[6] Stupnytskyy V, Hrytsay I and Xianning She 2021 Finite Element Analysis of Thermal and Stress-Strain State During Titanium Alloys Machining *Lecture Notes in Mechanical Engineering* **1** 629-639
[7] Yancheng Zhang, Outeiro J C and Tarek Mabrouki 2015 On the Selection of Johnson-cook Constitutive Model Parameters for Ti-6Al-4V Using Three Types of Numerical Models of Orthogonal Cutting *Procedia CIRP* **31** 112-117
[8] Tekkaya A E and Soyarslan C 2014 *Finite Element Method*(Berlin, Heidelberg: CIRP Encyclopedia of Production Engineering. Springer)
[9] Coppola T, Iob F and Campanelli F 2014 Critical Review of Ductile Fracture Criteria for Steels *Procedia Materials Science* **3** 1548-1553
[10] Gyliene V and Ostasevicius V 2011 Cowper-Symonds material deformation law application in material cutting process using LS-DYNA FE code: turning and milling *8 European LS-DYNA Users' Conference* (Strassburg) **1** 1-12
[11] DEFORM V11.0 (PC) *Documentation Table of Contents 1 Introduction to DEFORM*
[12] Stupnytskyy V and Hrytsay I 2020 Computer-Aided Conception for Planning and Researching of the Functional-Oriented Manufacturing Process *Advanced Manufacturing Processes: Lecture Notes in Mechanical Engineering* **1** 309-320
[13] Ern A and Guermond J L 2004 *Theory and practice of finite elements* (NY: Springer)
[14] Davim J P 2014 *Machining of Titanium Alloys* (Berlin, Heidelberg: Springer)
[15] Moaz H A A, Basim A K, Bashir M, Azim A O 2011 Machining of Titanium Alloys: A Review. *Proceedings of The Institution of Mechanical Engineers Part B-journal of Engineering Manufacture* 1-60
[16] Niknam S A, Khettabi R and Songmene V 2014 *Machinability and Machining of Titanium Alloys* (Berlin, Heidelberg: Springer)
[17] Ivasishin O M, Markovsky P E, Matviychuk Yu V, Semiatin S L, Ward C H and Fox S 2008 A comparative study of the mechanical properties of high-strength β-titanium alloys *Journal of Alloys and Compounds* **457**(1–2) 296-309
[18] Stupnytskyy V and Xianning She 2020 Comparative Analysis of Simulation Results of Hard-to-Cut Materials Machining by Coated Cutting Tools *Journal of Mechanical Engineering - Strojnícky časopis* **70**(2) 153 – 166
[19] Cheng Kai 2009 *Machining Dynamics Fundamentals, Applications and Practices* (London: Springer)

Special features of the dynamic cutting process analysis in the machining of titanium alloys

Xianning She[1,3] and N Yevtushenko[2]

[1] Department of Robotics and Integrated Mechanical Engineering Technologies, Lviv Polytechnic National University, Lviv, Bandery str. 12, 79013, Ukraine
[2] Department of Occupational and Environmental Safety, National Technical University «Kharkiv Polytechnic Institute», Kharkiv, Kyrpychova str. 2, 61002, Ukraine

[3] lvivshe@gmail.com

Abstract. The achievement of high accuracy parameters and the surface layer quality of the machined surface greatly depends on the dynamic stability of the system "Machine-Fixture-Tool-Workpiece". It is especially important for machining titanium alloys because intensive tool vibrations are generated due to the adiabatic shear phenomenon. The causes and consequences of such effects are described in the article. The paper also describes a specific method for studying the dynamic state of the cutting system of titanium alloys, based on the system integration of the classical method of the analytical description of the vibration processes arising when machining and the method of rheological simulation of the cutting process. Two important problems are proposed to be solved for the effective analysis of the system modeling results: first, it is the filtering of the noise signal with preservation of the limiting values of modeling results and maximum sensitivity using the Savitsky-Golay mathematical apparatus; second, the problem of approximation and interpolation of the filtered function of cutting parameters by Fourier transform in time.

1. Introduction

An important aspect of the formation of microgeometry, strength, tribomechanical, and stress-strain properties of the surface layer of products is to ensure the vibration stability of the technological system "Machine-Fixture-Tool-Workpiece" (MFTW) [1,2]. Moreover, the dynamics of the machine tool system components and the dynamics of the machining process are two inherently integrated parts that should be considered simultaneously when optimizing the MFTW system. It is especially difficult to ensure the vibration stability of the MFTW system when titanium alloys are being machined. In addition to the rigidity parameters and damping properties of the MFTW system elements, the particular forming pattern due to the high plasticity of the machined material layer also influences the occurrence of vibrations. This is expressed in the high-frequency change of longitudinal and transverse cutting forces and, as a consequence, the formation of a toothed (serrated) chip shape [3,4,5]. The disturbing forces, depending on the physical nature of the vibration excitation mechanism acting on the elastic MFTW system, can create both forced and self-oscillations.

The operating quality of the machined surface depends equally on the machining method and the cutting conditions. That is, the roughness and residual stress-strain state of the machined layer is a consequence of both the geometrical and kinematic component of cutting (i.e., the reflection of the cutting tool trace) and the plastic deformations of the machined metal, as well as vibrations of the technological system [6]. In addition, elastoplastic deformation when cutting metals (especially titanium alloys) is very complex. This is due to the relationship with other factors and phenomena (temperature in the cutting zone, tribomechanical, structural-phase state of the machined material, etc.) that accompany the cutting process. Therefore, a more complete characterization of the physical foundations of titanium alloys cutting can be obtained only through a comprehensive study and isolation of the dominant processes such as surface layer deformation, cutting forces, frictional processes, and the condition of the cutting tool. Analytical modeling of such processes is not adequate in this case [7].

2. Literature review

Advanced cutting simulation systems, also called computer-aided forming systems (Deform 3D, LS-Dyna, AdvantEdge, Abaqus), allow one to predict with a certain accuracy the interrelated dynamics of force and stress-strain phenomena that accompany the machining process. However, firstly, the results of modeling such systems, unfortunately, are not always adequate to the real cutting process parameters, and secondly, the initial state of the workpiece and tool is a priori considered "rigid", and the damping properties of the MFTW system are not taken into consideration at all.

When machining titanium alloys, the contact area between chip and tool is much smaller than when machining carbon steels. Thus, the contact pressure increases, and the thermal energy generated by the friction of the chips sliding on the rake face of the tool is concentrated in a smaller area. This leads to extremely high temperatures in the contact area of the tool with the chips since titanium has a very low thermal conductivity [6]. For example, for alloy Ti6Al4V the thermal conductivity coefficient is 11 W/(mK), while for steel AISI 1040 it is 45 - 40 W/(mK). Therefore when cutting medium-carbon steel the temperature in the contact zone " tool-workpiece" reaches 300-350°C, and when cutting titanium alloys under the same conditions the temperature exceeds 900-1200°C. This is one of the strong reasons for the recommendation of relatively low cutting speeds for titanium (from 20 to 60 m/min) to reduce the thermodynamic consequences of intensive tool wear, as well as to provide high machining accuracy [6].

Forced vibration is usually caused by an unbalanced force caused by external (external to the MFTW) factors, while self-excited vibration is created and short-circuited within the processing system. It is spontaneous and grows rapidly from a low vibration amplitude to a large one at the time when the forced vibration in most cases leads to oscillations of constant amplitude. The work [2] summarizes all types of machining instability and their features. The loss of vibration stability by the technological system MFTW is classified as random (or parametric) vibration and forced vibration. Random or free vibration usually also includes any impact or impulse loads on the machine tool. A typical example of random vibration is tool oscillation, such as when that tool collides with a cavity or compaction in the structure of the material being machined.

A characteristic feature in the machining of titanium alloys is the high tendency of the elastic technological system MFTW to intense oscillations. This is explained primarily by the large values of high-frequency oscillations of the radial, longitudinal and transverse components of the cutting force [5]. This is a consequence of the fact that the high plasticity of titanium alloys contributes to the emergence of a specific effect of forming the serrated shape of the chips (Fig. 1).

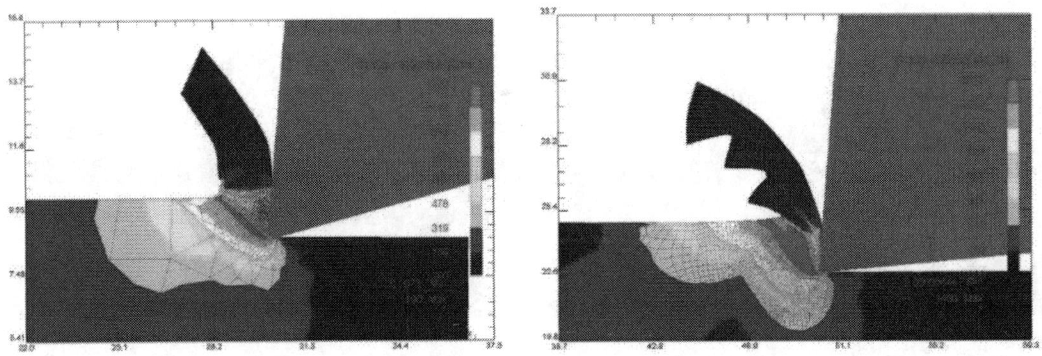

a b

Figure 1. Comparison of chip formation during machining of medium-carbon steel (a) and titanium alloy (b)

The authors [4] explain that phenomenon as follows. When machining materials with normal ductility (e.g., medium-carbon steels such as AISI 1020), the cut material layer is first compressed (compression process) and then displaced (shear process). The compression process and the shear process take place simultaneously during further cutting. That is, there is always a layer of compressed material in front of the shear layer. This is not the case when cutting titanium alloys. First, the sheared layer of metal is compressed. When the compression process reaches a critical limit, the shear process begins. However, unlike the traditional pattern of cutting medium ductility material, the compressed material is almost completely removed along with the chips and a new layer of uncompressed metal appears in front of the tool, then the process repeats.

This complex pattern is accompanied by the appearance of an adiabatic shear zone [8]. In this case, high-frequency oscillations of the longitudinal and transverse components of the cutting force are produced asynchronously, which causes high-frequency oscillations in the tool and workpiece (Fig. 3) [9]. This diagram shows the results of adiabatic shear zone emergence: at a certain moment of the study (simulation) (stage 1), the transverse cutting force has a maximum value. At the same time, the longitudinal cutting force is minimal. This stage of chip formation occurs with a predominant process of compression of the primary deformation zone with the convexity of the chip root over the outer surface. At the moment of cutting (stage 2), the opposite phenomena occur: the transverse cutting force takes on a minimum value and the longitudinal cutting force takes on a maximum value. This step corresponds to the priority predominance of the shear mechanism with the characteristic concavity of the chip over the outer surface. At the moment of cutting (stage 3), the step of dominant compression of the chip formation zone, etc., is repeated.

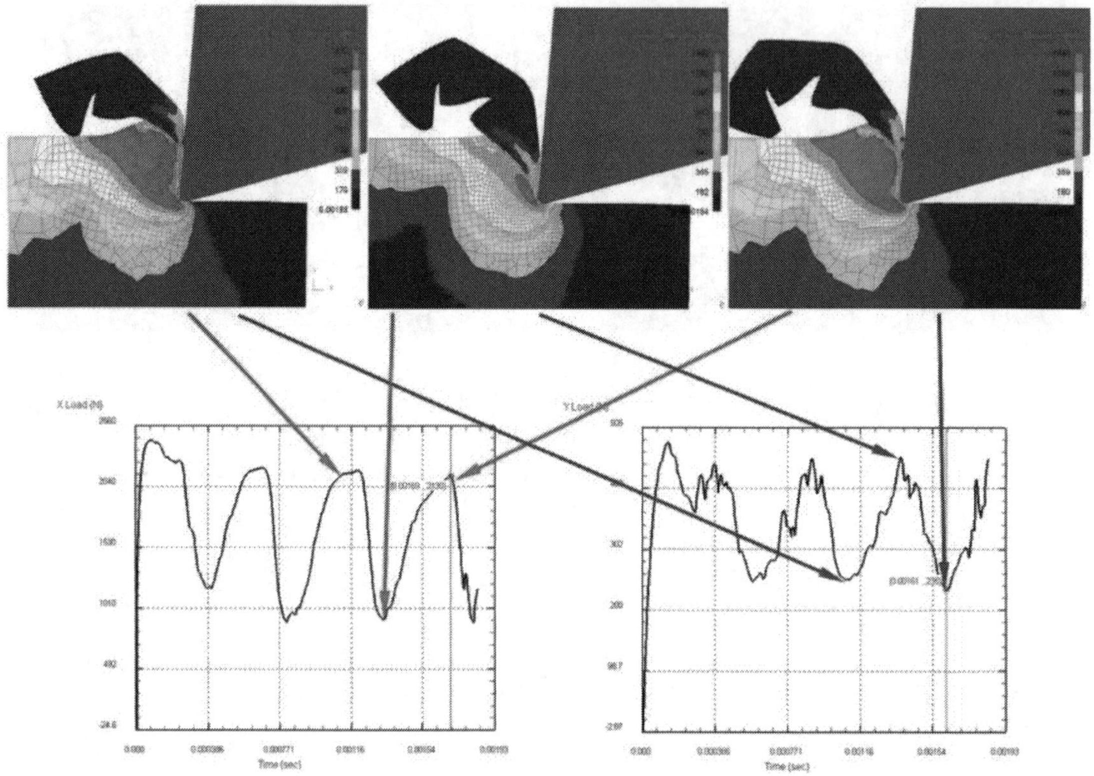

Figure 2. Interpretation of the mechanism of adiabatic shear formation during the processing of titanium alloy (stages 1, 2, and 3 are marked from left to right) [9].

The physical mechanism of perturbation of the MFTW technological system oscillations functions in the following sequence [1, 2, 10]. Any random perturbation (e.g., non-uniformity of the allowance, heterogeneity of the machined material, a machining gap, workpiece or tool run-out, etc.) leads to the appearance of the natural damping oscillations of the technological system. These oscillations are always accompanied by a change in cutting thickness and chip sliding velocity on the rake face of the tool. The change in these values leads to a corresponding dissonant change in the components of the cutting force of the titanium alloy. And if the change in cutting force lags in time or is phase-shifted relative to the change in cutting thickness, then the damped natural oscillations are transformed into un-damped auto-oscillations. The energy required to support such self-oscillations is created by the variable component of the cutting force. A similar cause of excitation of self-oscillations occurs if the radial component of the cutting force decreases as a result of an increase in the cutting speed [2].

The works [11, 12] describe how the excitation of oscillations of the technological system MFTW occurs due to the ratio of vertical and horizontal movements of the tool relative to the workpiece. The authors distinguish two dominant vibrational systems during the cutting process (Fig.3) [11]. Firstly, it is the system "Workpiece - Fixture". This system, as a rule, makes low-frequency vibrations (50 - 300 Hz). However, the frequency spectrum of titanium alloy processing in this subsystem is higher and amounts to 600 -1800 Hz. The second is the subsystem "Tool - Toolholder" which performs high-frequency vibrations (800 - 3000 Hz). The cutting zone is the closing link in these oscillation systems. Causal connection of these systems consists in the fact that because of the vibrations of cutter and workpiece the thickness of cutting continuously changes from the minimal (t_{min}) to the maximal (t_{max}) value (Fig.3). In this case the first step of the tool movement from point A to point B the cutter moves towards the cutting force R and the system additionally consumes part of the energy E_1. In the second step of the reverse motion from point B to point A, the direction of the cutter motion coincides with the direction of the cutting force. The system thus consumes additional energy E_2. Since the thickness of the cut layer on the second section is greater, $E_2 > E_1$. Therefore, the excess energy $\Delta E = E_2 - E_1$ is stored in the overall balance, supporting the oscillation of the technological system MFTW

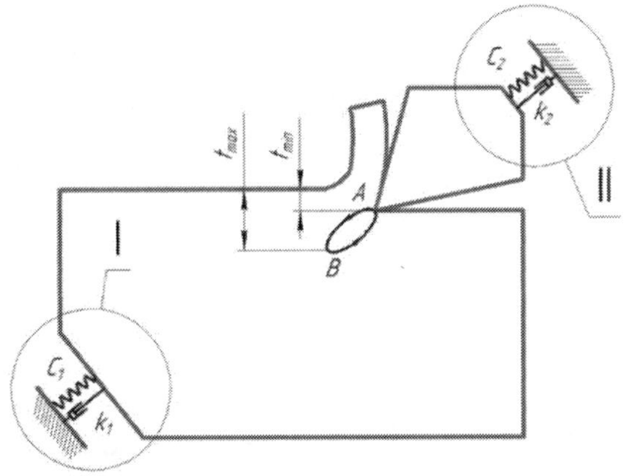

Figure 3. Scheme of formation of self-oscillations in the technological system MFTW: I - subsystem "Workpiece - Spindle"; II - subsystem "Tool - Support"; k_1, k_2 - generalized damping coefficients of I and II elastic subsystems; C_1, C_2 - rigidity of the first and second vibrational subsystems.

The existing methods of reducing the amplitude of self-oscillations can be classified into two classes - technological and design methods. Technological methods include such as the choice of appropriate cutting parameters and tool sharpening angles. The design techniques include increasing the resistance in the oscillating system and the use of dynamic vibration dampers. Thus, the problem-oriented analysis of literary sources allows us to draw the following conclusions:

1. Mechanical machining of titanium alloys has many specific features due to the mechanical and physicochemical properties of these materials. One of the most significant factors is the presence of large vibrations in the tool due to the specific phenomenon of the adiabatic effect in the primary shaping zone and the complex thermodynamic processes during the cutting of these materials.

2. The control of dynamic cutting processes is of particular importance, because, firstly, it has a significant influence on the formation of roughness and waviness of the surface layer just for titanium alloys; secondly, the low cutting speed caused by dynamic instability leads to a decrease in machining productivity; thirdly, the formation of cyclic loads leads to the appearance of local residual stress and strain zones, which have an extremely negative effect on the fatigue strength of the product [9].

3. For effective influence on the occurrence and character of vibrational processes in the machining of titanium alloys, it is necessary to analyze the priority of factors affecting the real pattern of the physical processes of the elastic-dynamic system "Machine-Fixture-Tool-Workpiece". This system, on the one hand, is described by the well-known equations of mechanical dynamics [2, 10, 12]. However, the analytical description of the system of dynamic equations in classical form cannot take into account the peculiarities of physical phenomena in the machining of titanium alloys (adiabatic shear, specific dynamics of thermodynamic cutting processes, etc.). Its influence can be taken into consideration only with the help of simulation modeling of the cutting process. But, on the other hand, the results of such simulation do not take into account the elastic-strain state of the MFTW system.

Therefore, the combination of analytical modeling of vibration processes in the titanium alloy cutting system and simulation will be able to take into account both the main features and the specifics of this process. The results of such research are subject to adequacy analysis as a result of comparison with the results of experimental studies

3. Methodology for researching the dynamic processes of the MFTW system when machining products from titanium alloys

Unsteady dynamic processes of cutting titanium alloys are the cause of generating periodic and aperiodic vibrations of elements of technological system MFTW, which leads to increased intensity of cutting tool wear, reduction of dimensional accuracy, and profile microroughness, as well as deterioration of the product surface layer quality. The appearance of significant (affecting the quality operational indicators of the product) vibration is caused by disturbing forces, inertial and elastic-dissipative properties of the technological system "Machine-Fixture-Tool-Workpiece". Moreover, given the complexity and multifactorial cutting of titanium alloys, it can be stated that at present there is still no consensus in the analytical interpretations of the excitation mechanism of vibrations in the technological system, the determination of the dominant factors of their amplitude values.

The analytical description of the vibration processes determined by the cutting process and combining dynamic subsystems MFTW is based on the well-known principles of the cutting dynamics theory [2]. It is accepted that the dynamic relation formed by the cutting process largely determines the constancy of the trajectories of stationary forming movements of the tool on the workpiece, which, in turn, affects the transformation of the machine tool drives shaping movements as a result of the implementation of the CNC program. To study the dynamics of the cutting process, the generalized equation of the subsystem "Tool-Workpiece" in the system of one z-coordinate is used (Figure 4):

$$\frac{d^2X}{dt^2}\cdot m_X + C_X X + k_X \cdot \frac{dX}{dt} = F_z\left(X,V,S,h\right) \tag{1}$$

where $X = \{x_{iz}, x_{3z}\}^T$ is the transposed vector of elastic strain displacements of the cutting tool tip and the workpiece at the point of application of the cutting force in the tangential direction of machining *z*; $C_X = \left[c_{iz};c_{3z}\right]$, $k_X = \left[k_{iz};k_{3z}\right]$ are the functional matrices of inertial and dissipative factors depending on the vectors of strain displacements of the elastic subsystem "Tool - Workpiece" respectively; $F_z\left(X,V,S,h\right)$ is the vector-functions of the cutting process dynamic characteristic, which Interprets the dependence of the tangential component of cutting force on machining parameters, namely: from the cutting speed *V*, the feed rate *S*, and the cutting depth *h*.

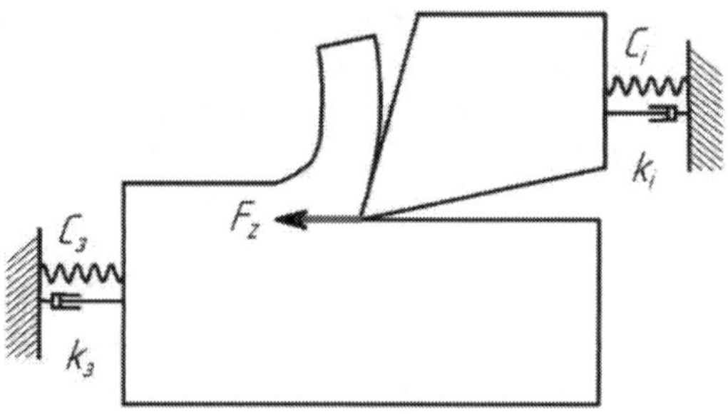

Figure 4. Schematic diagram of a single-coordinate dynamic system "Tool-Workpiece"

The value of the dissipative coefficient of vibration extinction is determined by the equation:

$$k_i = \frac{m_i \delta_i \omega}{\pi}, \tag{2}$$

where m_i is the reduced mass of the *i*-th element of the system (workpiece or tool); δ_i is the logarithmic decrement of vibrations of the *i*-th element of the vibrational scheme, which characterizes the damping rate of the vibration process in the "Tool - Workpiece" system; ω is the angular frequency of vibrations.

The average value of the logarithmic decrement of vibration in the universal lathe (based on its structural and functional properties) can be assigned in the range of 0.32-0.34, and the milling machine 0.36-0.38.

An important step in determining the damping indices of the elements of the MFTW process system is the analysis of the internal friction indices. This index is directly related to such phenomena as plastic intercrystalline strains, dislocation motion, thermoelastic effect, magnetostriction, diffusion processes, etc. In this case, this coefficient of internal friction depends considerably on the stress in the forming zone and the thermodynamic state of the machined material. The recommended internal friction parameters [12] for the contact pair "machined surface - tool's flank face" for the majority of titanium alloys are $\mu_1 = 0.15$, and $\mu_2 = 0.25$ for the contact pair "tool's rake face - chip". In the plastic deformation zone of the machined titanium alloy, the dry friction coefficient acquires the value μ_3=0.55-0.65.

The generalized logarithmic decrement of oscillation damping can be described by the formula:

$$\delta_\Sigma = \sqrt{v_1 \cdot \mu_1^2 + v_2 \cdot \mu_2^2 + v_3 \cdot \mu_3^2}, \quad (3)$$

where v_1 - v_3 are the correlation coefficients

Based on the analysis of the intensity of the action of various factors in the turning of titanium alloys roughly take $v_1 = 0.8$; $v_3 = 0.5$; $v_3 = 0.1$.

Another important component of equation (1) is the vector-function of the dynamic characteristic of the cutting process, which interprets the dependence of the tangential component of the cutting force on the machining parameters. It is well known that the cutting force is a value, the nonlinearity of which depends on the change in the cutting area (thickness a and width b):

$$F_z = k\left(A_0 + \frac{B_0}{a(t)^{y_p}}\right) \cdot ba(t)^{y_p} \quad (4)$$

where k, A_o, B_0, y_p are the constants that depend on the properties of the machined material, the tool material, and its geometry [12].

That is, the dynamics of the cutting force are determined primarily by the change in the volume of the removed material per unit of cutting time. Due to the influence on the technological system of an external variable force, which causes an oscillatory process with a frequency equal to the frequency of the disturbing force, or complex steady-state processes due to the nonlinear properties of the MFTW system, there are forced and parametric vibrations of the elements of this technological system. The intensity of such oscillations is especially great at resonance modes, which, as a rule, are not allowed in metal-cutting machines as working conditions. Reducing the intensity of vibrations in the technological system is a pressing problem, solved based on the dynamic synthesis of non-critical (away from resonance) parameters of cutting. Determination of this value depends on many complexly formalized parameters, so the definition of the disturbing force of the vibrational process of titanium alloy cutting is best realized in the system of simulation modeling DEFORM 2D [13].

The complexity in formalizing the load characteristic of the cutting process consists in the significant noisiness of the results of calculations performed by the finite element method for most high-frequency dynamic problems [14]. For effective processing of the DEFORM 2D system results, it is necessary to solve two most important problems - firstly, it is to filter the noise signal with preservation of the limit values of cutting force parameters and maximum sensitivity; secondly, to solve the problem of approximation and interpolation of the filtered cutting force function by time (based on the Fourier transformation).

As mathematical support for the filtering procedure for smoothing the rheological modeling data in the DEFORM 2D system, that is, to increase the accuracy of the data without distorting the trend of the vibration signal, it is proposed to use the Savitzky-Golay mathematical apparatus [15, 16].

The algorithm for solving such a problem involves fitting successive sets of neighboring data points by a low-degree polynomial using the method of least linear squares. The idea of realization of this method consists of the assumption that for data points that are in the same time range as each other, it is possible to find the analytical solution of the least squares equations in the form of one set of "convolution factors" which can be applied to all data sets to obtain estimates of smoothed signal (or derivatives of smoothed signal) in the central point of each such set. That is, it is necessary to construct an approximating polynomial of the n-th degree by $(2k+1)$ consecutive equidistant points of the rheological study and use it to calculate the smoothed value of the polynomial value at the $(k+1)$-th point. Mathematically, such a value is calculated by determining a moving weighted average with point weights that are positive in the center of the filtering window and negative at the periphery. The results of Savitzky-Golay filtering are the same for polynomials of degrees $2n$ and $(2n+1)$. For example, the moving average value of the force parameter (according to the polynomial of degree zero and the first stage) will give the average value of the signal, at which the approximating curve will always pass through the center of the grouping of research results, equal to the arithmetic mean value of cutting force parameters in a certain time range, which is interpreted as a smoothed value of the approximation curve.

In the general case, the simple sliding average value of cutting force is equal to the arithmetical mean of values of the investigated spectral characteristic of this parameter in the predetermined range and is calculated by the equation:

$$F_{sZ}(k_i)=\frac{1}{n}\sum_{j=1}^{n-1}F_Z(k_{i-j})=\frac{F_Z(k_i)+F_Z(k_{i-1})+F_Z(k_{i-2})+\ldots+F_Z(k_{i-j+1})}{n} \tag{5}$$

where F_{sZ} (k_i) is the calculated (modified as a result of filtering) value of the force parameter in the spectral field range k_i after recalculation; F_Z (k_i) is the simulated in DEFORM 2D value of the force function for the given step k_i; n is the number of values of the parameter set for calculation of the moving average parameter (smoothing interval).

The width of the smoothing interval determines the degree of smoothness of the spectral characteristic of sensitivity after the application of a given type of filtering and is a multiplicity value in the selected incircle of rheological modeling time. That is, for example, if the chosen time incircle for the conducted research is $\tau=3\cdot10^{-5}$ s and the smoothing interval is taken $\tau=3\cdot10^{-3}$ s, then the value $n=100$.

The modified moving average parameter method proposed in [16] is a recurrence relation for which the sensitivity value at a given wave number is related to a pre-calculated sensitivity value:

$$F_{smZ}(k_i)=\frac{F_Z(k_i)+(n-1)F_{smZ}(k_{i-1})}{n} \tag{6}$$

where F_{smZ} (k_i), F_{smZ} (k_{i-1}) are the modified parameters obtained by filtering the values of the force parameter for a pair of spectral field values k_i and k_{i-1}

An example of the Savitzky-Golay filtering for the cutting force parameter calculated in DEFORM 2D for the titanium alloy Ti6Al4V is shown in Figure 5.

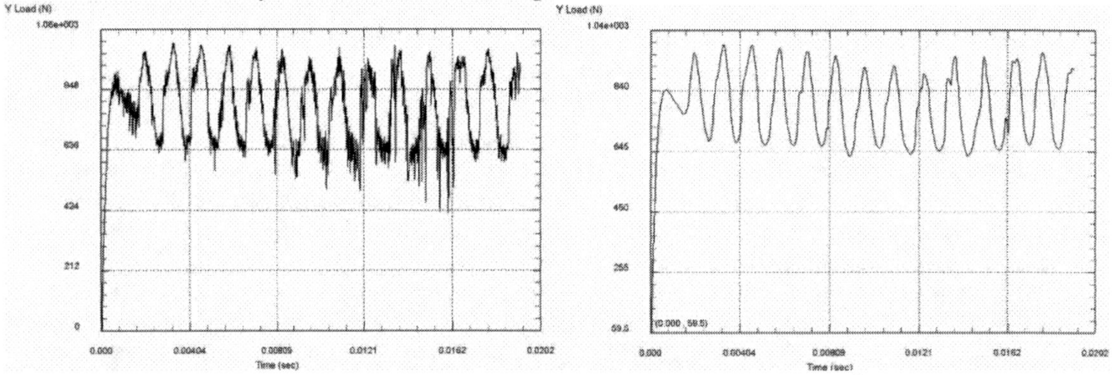

a b

Figure 5. Results of filtering the simulated cutting force parameter by the Savitzky-Golay method (smoothing parameter n=100): a - noisy (original) signal; b - filtered (output) signal

4. Conclusions

The most effective methodology for analyzing the dynamic state of the technological system "Machine-tool-tool-blank" when machining products from titanium alloys is a combination of the methodology of scientific research of dynamic shaping processes, obtained as a result of simulation modeling with classical analytical modeling of dynamic cutting processes. Analyzing the causes of oscillations, this article focuses on the oscillation generating factors arising from the specifics of machining of titanium alloys - for example, the phenomenon of adiabatic shear when cutting specifically titanium alloys.

The complexity in formalizing the force, thermal and stress-strain results of the simulation modeling of the cutting process consists of considerable noise content in the results of calculations performed by the finite element method, which is characteristic of most high-frequency dynamic problems. For effective processing of DEFORM 2D system results, it is necessary to solve two most important problems - firstly, it is filtering of a noise signal with preservation of limiting values of simulation results and maximum sensitivity; secondly, the solution of a problem of approximation and interpolation of the filtered function of cutting parameters by time (by Fourier transform). As mathematical support of the filtering procedure for smoothing the rheological simulation data in the DEFORM 2D system, i.e. to increase the data accuracy without distorting the signal trend, it is proposed to use the Savitzky-Golay mathematical apparatus.

References

[1] Schmitz T and Smith K 2019 *Machining Dynamics: Frequency Response to Improved Productivity* (Cham: Springer)

[2] Cheng Kai 2009 *Machining Dynamics Fundamentals, Applications and Practices* (London: Springer)

[3] Xu W J, Zhang X M, Leopold J and Ding H 2017 Mechanism of Serrated Chip Formation in Cutting Process Using Digital Image Correlation Technique *Procedia CIRP* **58** 146-151

[4] Wang B and Zhanqiang L 2014. Investigations on the chip formation mechanism and shear localization sensitivity of high-speed machining Ti6Al4V. *The International Journal of Advanced Manufacturing Technology*. **75** 1065–1076.

[5] Stupnytskyy V, Hrytsay I and Xianning She 2021 Finite Element Analysis of Thermal and Stress-Strain State During Titanium Alloys Machining *Lecture Notes in Mechanical Engineering* **1** 629-639

[6] Davim J P 2014 *Machining of Titanium Alloys* (Berlin, Heidelberg: Springer)

[7] Hazir O 2015 A review of analytical models, approaches and decision support tools in project monitoring and control *International Journal of Project Management* **33**(4) 808-815

[8] Yang Q, Liu Z, Shi Z et al. 2014 Analytical modeling of adiabatic shear band spacing for serrated chip in high-speed machining *Int J Adv Manuf Technol* **71** 1901–1908

[9] Stupnytskyy V, Xianning S 2021 Comprehensive analysis of tribological factor influence on stress-strain and thermal state of workpiece during titanium alloys machining *Archive of Mechanical Engineering* **68**(2) 227–248

[10] Eliseev S V, Eliseev A V 2020 *Theory of Oscillations: Structural Mathematical Modeling in Problems of Dynamics of Technical Objects* (Cham: Springer Nature)

[11] Mazur N P, Vnukov Yu N, Grabchenko A I et al. 2003 *Basics of metal cutting theory* (Kharkiv: NTU KhPI) (in Russian)

[12] Zharkov I G 1986 *Vibrations during processing with a blade tool* (Leningrad: Mashinostroyenie) (in Russian)

[13] Stupnytskyy V and Hrytsay I 2020 Computer-Aided Conception for Planning and Researching of the Functional-Oriented Manufacturing Process *Advanced Manufacturing Processes: Lecture Notes in Mechanical Engineering* **1** 309-320.

[14] Ern A and Guermond J L 2004 *Theory and practice of finite elements* (NY: Springer)

[15] Nikitas and Pappa-Louisi 2000 Comments on the two-dimensional smoothing of data *Analytica Chimica Acta* **415**(1–2) 117–125

[16] Luo J, Ying K, He P and Bai J 2005 Properties of Savitzky–Golay digital differentiators *Digital Signal Processing* **15**(2) 122–136

The method of experimental determination of tangential stresses on the surface of flow around with a gaseous medium

T V Dmytriv [1], O S Lanets [1], V T Dmytriv [1, 4], I V Dmytriv [2], I M Horodetskyy [3] and N G Horodetska [3]

[1] Department of Machine Design and Automotive Engineering, Lviv Polytechnic National University, Ukraine
[2] Department of Motor Vehicle Transport, Lviv Polytechnic National University, Ukraine
[3] Lviv National Environmental University, Ukraine
[4] E-mail: Dmytriv_V@ukr.net

Abstract. Experimental research of the force load on the flow around surface is important for the design of aircraft and gas transport systems. The main parameter is the tangential stress, which is maximum in the boundary layer on the flow around surface. The velocity in the boundary layer and its thickness affect the tangential stresses that arise during the flow of the medium. They also have an influence on the friction coefficient, which characterizes the pressure losses. In the work, the method of experimental measurement of velocity on the flow around surface is considered, and an information and measurement system is proposed for determining the tangential stresses on the flow around surface based on the flow velocity measurement in the boundary layer. The algorithm of the measurement technique is presented in the form of a functional structural scheme of the laminar boundary layer on the surface of the flow around. The sensor for measuring the velocity in the boundary layer is installed at characteristic points namely at x distance from the front border of the flow around surface, and also at y height from the surface of the boundary layer. The results of the planned factorial experiment of air flow around the surface are given. The factors were the v_0 velocity of the air flow entering the flow around surface, the x distance from the edge of the surface, the y distance from the surface along the thickness of the layer. The response criterion was the calculated value of the tangential stress based on the v_x measured flow velocity at a given point of the boundary layer. The developed method of studying the velocity distribution and determining the tangential stresses on the surface of the flow around under a dynamic boundary layer makes it possible to determine with high discreteness the parameters of the process of the surface around flow of a Newtonian gaseous medium.

1. Introduction

The velocity in the boundary layer and its thickness affect the tangential stresses that arise during the flow of the medium. They also have an influence on the friction coefficient, which characterizes the pressure losses. He and Ariyaratne [1] determined the quasi-stationary shear stress in the boundary layer near the wall by calculating the friction coefficient at the instantaneous Reynolds number. Sundstrom and Cervantes [2] show that tangential shear stresses depend on the thickness of the boundary layer, the nature of the movement, which is not having a single meaning in turbulent and pulsating (forced turbulence) modes of movement. Kong and Kim [3] studied pressure losses and calculated the friction coefficient using empirical equations. This does not always correspond to the physics of the flow process, the characteristics of the boundary layer, and the stresses that affect the pressure losses.

A number of researchers make more exact the empirical equations determined experimentally for specific intervals of Reynolds numbers [4-7]. Ganat and Hrairi [8] use existing dependences of friction

IOP Conf. Series: Materials Science and Engineering 1277 (2023) 012026 doi:10.1088/1757-899X/1277/1/012026

coefficients for liquids, gases, and two-phase flows. The results of such studies more or less distort the accuracy of the calculated indicators. Boundary layer parameters and tangential stresses that arise between layers are absent in the calculation equations. Brkic and Praks [9], based on the results of experiments, derive empirical equations for determining friction coefficients on the interval of the entire Reynolds number series. Ortiz-Vidal et al. [10, 11] approximate the experimental data to get an empirical equation of the friction coefficient, especially for two-phase flows (liquid and gas). Approximated empirical dependences of the coefficient of friction give a significant error in the calculations, which reaches up to 10% when used for different Reynolds numbers from the experiment. Under the condition of changing liquid flows to gas, the physics of the process changes and one cannot used such empirical dependences. Accordingly, the parameters of the boundary layer with a minimal error in their values will not correspond to the real process. A common method of researching of the boundary layer parameters is the numerical research method based on the Navier-Stokes equations [12, 13]. The solutions of such equations also have an approximate result due to the limitation of the initial conditions, which are usually applied for the turbulent supersonic regime of the medium motion.

The examination of the problem showed that the experimental study of the boundary layer on the surfaces of the flow by a Newtonian gaseous medium would allow getting a dynamic analysis of the boundary layer on the surface of the flow.

2. Methodology

2.1. Measurement of the flow rate of the gaseous medium on the flow surface

The measurement of the flow velocity on the surface of the flow around is based on the measurement of the voltage change of the bridge circuit and its transformation into a discrete digital signal. The receiving elements work in the mode of the specified voltage imbalance of the bridge circuit. As a result of heat removal, there is a voltage imbalance, which is measured. The resistance of the circuit elements changes due to the change in the flow rate of the medium on the surface of flow around. The circuit is made in such a way that the ambient temperature and the change in the temperature of the environment do not affect the accuracy of the measurement. The diagram of the bridge connection of the sensor is shown in Figure 1, and the sensor is shown in Figure **2**.

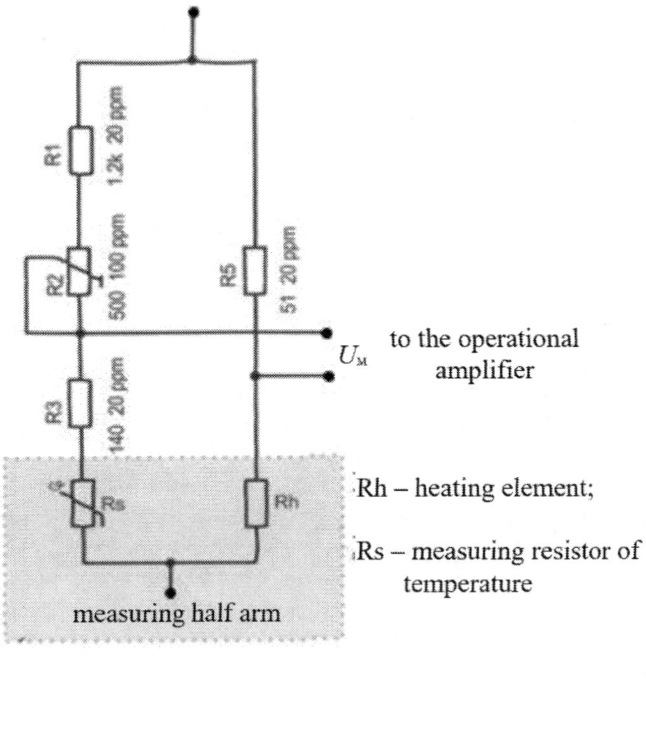

Figure 1. Electrical diagram of a sensor for measuring the flow rate of a gaseous medium on the flow surface

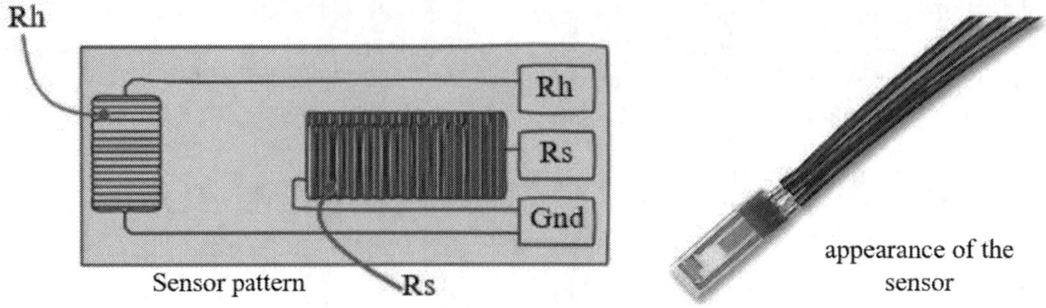

Figure 2. General view of the flow rate sensor

The Rh heating element and the Rs temperature sensor are connected in a bridge circuit that is balanced in the absence of flow and unbalanced for changes in the resistance of the heater. As the flow rate is increased, the heater cools, the bridge becomes unbalanced, and the U_M unbalance signal is sent to the operational amplifier.

Figure 3 shows the calibration graph for the given nominal values of the bridge circuit (**Figure 1**).

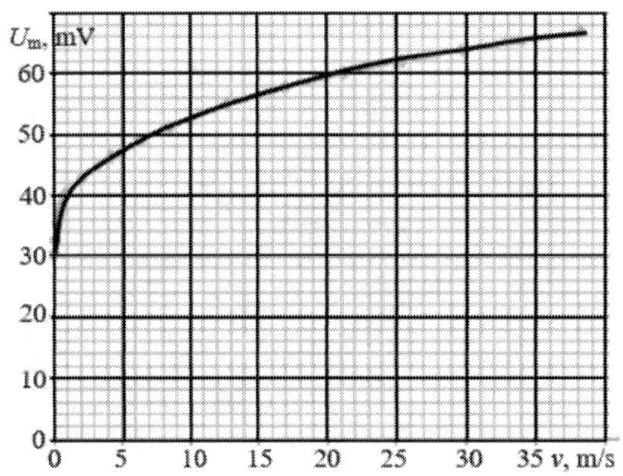

Figure 3. Calibration graph of the air flow velocity sensor
v – the air flow velocity (flow rate); U_m – the output voltage of the bridge circuit (**Figure 1**)

The output signal was set to 30 mV at zero velocity, which corresponds to a temperature difference (Rh heating element temperature minus ambient) of about 30 K. At the temperature difference, the optimal upper part of the heating element is generated and, accordingly, there is a minimum heating power.

The measuring characteristic of the sensor has the highest sensitivity at low flow rates.

The flow velocity was measured at an air temperature of about 22 °C. The sensor was installed in a tube with an inner diameter of 5 mm.

To process the analog signal of the bridge measurement circuit and convert it into a digital code, we will use the electrical processing scheme proposed by researchers Dmytriv et al. [14]. The output signal from the analog signal processing circuit, Figure 4, turns on the power of the bridge circuit with a different amplitude and frequency of alternating current, which accordingly brings the bridge circuit into equilibrium. The temperature difference is determined by the engaging scheme, the parameters of the medium are constant, the heat emission of the heating element depends on the flow rate.

The **Figure 4** scheme of the analog signal processing of the flow velocity measurement sensor works as follows. With the help of an automatic system with closed feedback, which includes a bridge circuit 1 (see **Figure 1**) and other components: the alternating current amplifier 2, the phase detector 3, the null device 4, the control unit 11, the rectangular pulse generator 10. The functioning of the system ensures the average value of the superheat temperatures of Δt_1 and Δt_2 of the thermistor receiving elements of *R*h, *R*s above the level of the ambient temperature. The difference between the maximum and average temperature of the elements (**Figure 5,b**) is given by the transmission coefficient of the amplifying path and the ΔU_0 zone of insensitivity of the null device (**Figure 5,c**).

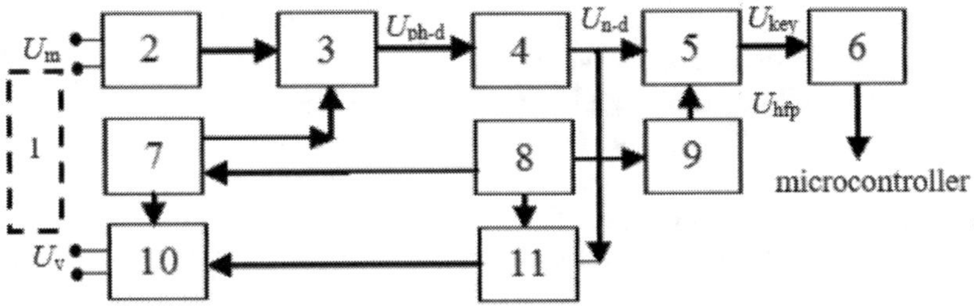

Figure 4. Block scheme of the analog signal processing of the flow velocity measuring sensor on the flow around surface

1 – bridge circuit of the sensor; 2 – alternating current amplifier; 3 – phase detector; 4 – null device; 5 – electronic key; 6 – bus-bar former; 7 – reference generator; 8 – power conditioner; 9 – high frequency pulser/generator; 10 – rectangular oscillation generator, 11 – control unit

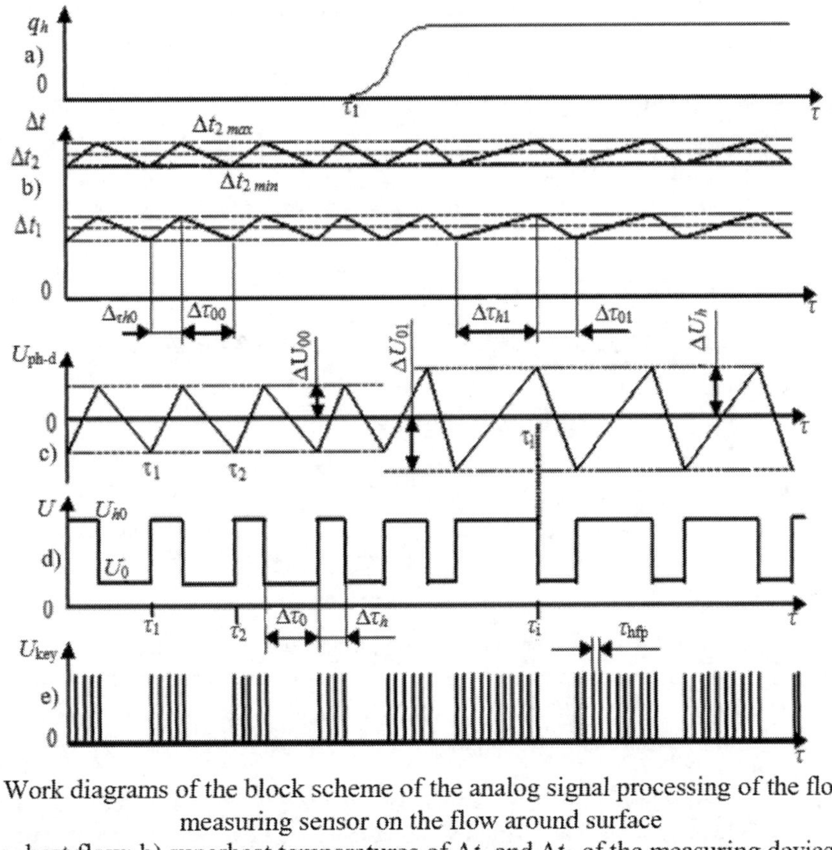

Figure 5. Work diagrams of the block scheme of the analog signal processing of the flow velocity measuring sensor on the flow around surface

a) the q_n heat flow; b) superheat temperatures of Δt_1 and Δt_2 of the measuring device sensors; c) the output voltage of 3 phase detector and input voltage of 4 null device; d) supply voltage of bridge circuit from the 10 rectangular oscillation generator; e) the characteristic of the output voltage of 5 electronic key

In the absence of medium flow ($v = 0$) (**Figure 5,a**, the quantity of heat $q_h = 0, 0 < \tau < \tau_1$) the power that is diverted from the sensors is insignificant.

The superheat temperatures of *R*h and *R*s elements is maintained at level of Δt_1 i Δt_2 (**Figure 5b**).

To heat the elements, a generator of rectangular oscillations is used, the duration of heating is shorter compared to the duration of cooling ($\Delta\tau_{h0} << \tau_{00}$, **Figure 5,b**, for example, $\tau_1<\tau<\tau_2$). At the same time, the rate of voltage increase at the output of the phase detector is much higher than the rate of its decrease. During the short heating time of the $\Delta\tau_{h0}$ elements, the $U_{n\text{-}d}$ output voltage of the 4 null device through the 5 electronic key is filled with high-frequency pulses (**Figure 5, e**) from the 9 generator. Then the signal reaches the microcontroller through the 6 bus-bar former. If there is a medium flow of q_h in the sensor (**Figure 5,a**, $q_h \neq 0$) the power taking from the thermistor elements of *Rh, Rs* increases proportionally to the v velocity to the power of n, where $n < 1$. This leads to an increase in the $\Delta\tau_{h1}$ duration (**Figure 5,b**) of heating the thermistor elements and their power supply from the generator 10 with a voltage of an amplitude of U_h (**Figure 5,d**). Therefore, the number of pulses coming from the 9 high-frequency pulse generator through the 5 electronic key (**Figure 5,e**) will increase during the time of $\Delta\tau_h$.

A synchronous phase pulse detector converts the output alternating voltage of the bridge circuit into a digital signal. In a synchronous phase the two-stroke detector, the voltage of the U_{inp} input signal and the U_{rv} reference voltage are sequentially added and the sum of the voltages is fed to the input of the operational amplifier 4, which is a null device.

2.2. The method of tangential stresses research on the surface of the flow around

We will show the algorithm of the measurement technique in the form of a functional structural scheme of the laminar boundary layer on the surface of the flow around (**Figure 6**). The sensor for measuring the velocity in the boundary layer is installed at characteristic points, at ***x*** distance from the front border of the flow around surface, and also at ***y*** height from the surface of the boundary layer.

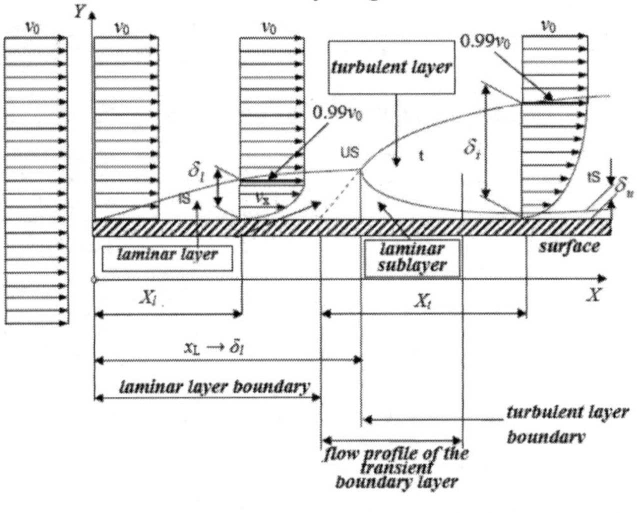

Figure 6. Functional structural scheme of the boundary layer on the flow around surface
δ_l – the thickness of the laminar boundary layer; δ_t – the thickness of the turbulent boundary layer; δ_u – the thickness of the boundary sublayer; v_0 – the velocity of the incoming flow at the front edge of the flow around surface; v_x – velocity in the boundary layer

Previously, we determine the thickness of the boundary layer according to the dependence (1), assuming that the $\boldsymbol{\delta}$ thickness of the boundary layer and the ***y*** coordinate are commensurate, respectively, the velocities of $\boldsymbol{v_0}$ and $\boldsymbol{v_{x|y=\delta}}$ are also commensurate:

$$\frac{v_{x|y=\delta}}{v_0} = f\left(\frac{y}{\delta}\right), \quad (1)$$

which is equivalent of

$$\frac{v_{x|y=\delta}}{v_0} = 1 \approx f\left(\frac{y}{\delta} = 1\right). \quad (2)$$

To determine the thickness of the boundary layer, we use the results of solving the Prandtl equation (3) of the boundary layer [15]:

$$\delta = 5 \cdot \sqrt{\frac{\nu \cdot L}{v_0}}, \tag{3}$$

where ν - kinematic viscosity of air, m²/sec; L - the length of the flow around surface in the flow direction, m.

Knowing the thickness of the boundary layer for specific initial conditions, we measure the flow velocity in the v_x boundary layer at a specific point, which has the following parameters: x – distance from the edge of the surface; y – the distance from the surface along the thickness of the layer. And then we calculate the tangential stresses at a given point according to the formula derived by Dmytriv et al. [16]

$$\tau_{str} = \mu \cdot \frac{1.231}{\delta} \cdot \frac{v_0{}^2}{v_x} \cdot \frac{\frac{\delta}{3} \cdot \left(\frac{y}{3\delta^2 - y^2} + \frac{1}{y}\right) - 1}{\left(\frac{y}{\delta} + \frac{1}{3} \cdot \ln\frac{3\delta^2 - y^2}{y^2}\right)^2}, \tag{4}$$

where τ_{str} – the tangential stresses on the flow around surface in the laminar boundary layer, N/m²; μ – the dynamic viscosity of the flow around medium, N·s/m²; v_o – the velocity of the flow around medium at the outer limit of the boundary layer, m/s; v_x – the velocity of the flow around medium along the height of the boundary layer, m/s; δ – the thickness of the boundary layer, m; y – the coordinate of the thickness of the boundary layer, m.

2.3. Methodology of the experiment

To study of the tangential stresses arising during the flow of a gaseous medium on the flow surface, we applied the method of a planned factorial experiment. Air is used to flow around the surface with a gaseous medium. The main factors were the v_0 velocity of the air flow entering the flow around surface, the x distance from the edge of the surface, the y distance from the surface along the thickness of the layer. The response criterion was the calculated value of the tangential stress based on the v_x measured flow velocity at a given point of the boundary layer. We implemented a three-factor planned experiment at five levels with three-fold repetition of experiments. In order to identify patterns of tangential stresses when the air flow enters the flow around surface and at a distance from the edge of the air entry surface, we adopted two intervals of variation of the factors. The levels of factor variation for two intervals of experimental studies are shown in tables of 1 and 2.

Table1. Factors variation levels and their code values (interval 1)

Factor	Marking	Dimension	Levels of factors				
			upper		null		lower
			Code values				
			-1	-0.5	0	+0.5	+1
the velocity of the air flow entering the surface, v_0	x_1	m/s	1.0	2.0	3.0	4.0	5.0
the distance from the edge of the surface, x	x_2	m	0.01	0.033	0.055	0.077	0.1
the distance from the surface along the thickness of the layer, y	x_3	m	≈0	0.25δ	0.5δ	0.75δ	$y = \delta$

Table 2. Factors variation levels and their code values (interval 2)

Factor	Marking	Dimension	Levels of factors				
			верхній		null		нижній
			Code values				
			-1	-0.5	0	+0.5	+1
the velocity of the air flow entering the surface, v_0	x_1	m/s	1.0	2.0	3,0	4,0	5,0
the distance from the edge of the surface, x	x_2	m	0.1	0.33	0.55	0.77	1.0
the distance from the surface along the thickness of the layer, y	x_3	m	≈0	0.25δ	0.5δ	0.75δ	$y=\delta$

3. Results and discussion

To analyze the interval selection of the velocity measurement and assess the reliability of the experimentally obtained values, we previously assessed the character of the constant coefficient as a multiplier in equation (4). We will determine the tangential stresses based on the measured velocity v_x of the boundary layer at the point with coordinates of x and y. Then we will determine the tangential stress according to the dependence:

$$\tau = \frac{K_V}{v_x}\,, \tag{5}$$

where

$$K_V = \mu \cdot 1.231 \cdot {v_0}^2 \cdot \frac{\frac{1}{3}\cdot\left(\frac{y}{3\delta^2 - y^2} + \frac{1}{y}\right) - 1}{\left(\frac{y}{\delta} + \frac{1}{3}\cdot \ln\frac{3\delta^2 - y^2}{y^2}\right)^2}. \tag{6}$$

Taking into account the relative parameter of the boundary layer thickness measurement, as y/δ, the (6) dependence after the appropriate transformations, will take the form of

$$K_V = \mu \cdot 1.231 \cdot {v_0}^2 \cdot \frac{\frac{1}{\frac{y}{\delta}\cdot 5 \cdot \sqrt{\frac{v\cdot x}{v_0}}} - 3 + \frac{y^2}{\delta^2}}{\left(\frac{y}{\delta} + \frac{1}{3}\cdot \ln\frac{3 - \frac{y^2}{\delta^2}}{\frac{y^2}{\delta^2}}\right)^2}, \tag{7}$$

where $\mathbf{v}$ – kinematic viscosity of air, m^2/s.

The results of the experimental data are presented in the form of a 3-D model (**Figure 7** and **8.**).

The regression correlation equation for a three-factor planned experiment in conventional units is as follows:

$$\tau = 7.0662 \cdot 10^{10} \cdot x_1 + 9.4789 \cdot 10^{12} \cdot x_2 + 2.7544 \cdot 10^{11} \cdot x_3 - \\ -1.1083 \cdot 10^{13} \cdot x_1 \cdot x_2 - 6.8858 \cdot 10^{11} \cdot x_1 \cdot x_3 - 8.0301 \cdot 10^{12} \cdot x_2 \cdot x_3. \tag{8}$$

According to the results of the calculation of equation (8) with the Student's criterion, all coefficients are significant. According to the Cochran criterion, the experiments are reproducible. The model is adequate according to Fisher's test.

At the air flow velocity at the entry of the flow around surface $v_0 = 5$ m/s, the maximum tangential stresses are $\tau = 5.18 \cdot 10^{14}$ N/m² at $x = 0.01$ m and $\tau = 5.18 \cdot 10^{13}$ N/m² at $x = 0.01$ m directly on the surface ($y \approx 0$). The minimum tangential stresses, respectively, are $\tau = 143.496$ N/m² at $x = 0.01$ m and $\tau = 14.348$ N/m² at $x = 0.01$ m at a distance from the surface of $0.005 \cdot \delta$ m.

In a similar manner, at the air flow velocity of $v_0 = 3$ m/s, the maximum tangential stresses are $\tau = 2.408 \cdot 10^{14}$ N/m² at $x = 0{,}01$ m and $\tau = 2.408 \cdot 10^{13}$ N/m² at $x = 0{,}01$ m directly on the surface ($y \approx 0$), and the minimum tangential stresses, respectively, are $\tau = 66.69$ N/м^2 at $x = 0{,}01$ m and $\tau = 6.668$ N/m^2 at $x = 0.01$ m at a distance from the surface of $0.005\cdot\delta$ m.

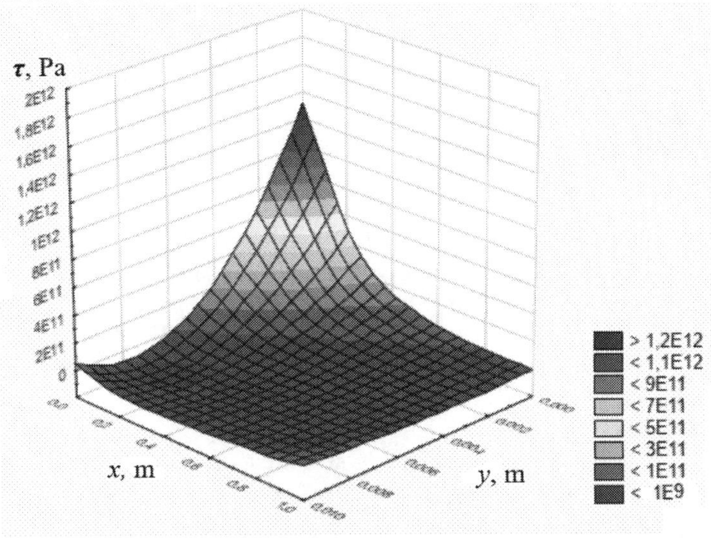

Figure 7. Graphical interpretation of the dependence of tangential stresses τ on the distance x from the beginning of the plate and the relative coordinate y of the thickness of the boundary layer

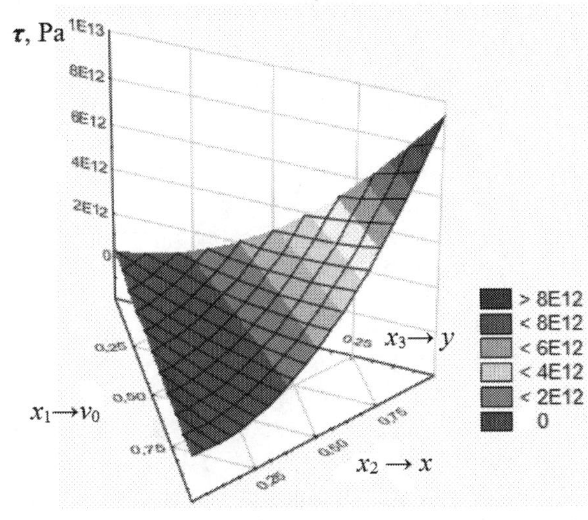

Figure 8. Graphical interpretation of the dependence of tangential stresses τ on the distance x from the beginning of the plate and the relative coordinate y of the thickness of the boundary layer and the v_0 air flow velocities at the entrance to the flow around surface (factors in conventional units)

For the air flow velocity of $v_0 = 1$ m/s, the maximum tangential stresses are $\tau = 4.633\cdot10^{13}$ N/m^2 at $x = 0.01$ m and $\tau = 4.633\cdot10^{12}$ N/m^2 at $x = 0{,}01$ m directly on the surface ($y \approx 0$), and the minimum tangential stresses, respectively, are $\tau = 12.834$ N/m^2 at $x = 0{,}01$ m and $\tau = 1.283$ N/m^2 at $x = 0.01$ m at a distance from the surface of $0.005\cdot\delta$ m.

4. Conclusion

The information and measurement system for determining the tangential stresses on the flow around surface is developed on the basis of the measurement of the flow velocity in the boundary layer. A thermo-anemometric type sensor was used to measure the velocity.

To determine the tangential stress on the flow around surface, the parameters are the flow velocity in the boundary layer, the velocity measurement coordinate along the thickness of the boundary layer, the medium flow velocity at the boundary of the flow surface, and the longitudinal coordinate of the velocity measurement point.

The developed experimental method of studying the velocity distribution and determining the tangential stresses on the surface of the flow around under a dynamic boundary layer makes it possible to determine with high discreteness the parameters of the process of the surface around flow of a Newtonian gaseous medium.

5. References

[1] He S and Ariyaratne C 2011 Wall shear stress in the early stage of unsteady turbulent pipe flow *Journal of Hydraulic Engineering* **137(5)** 606-10

[2] Sundstrom L R J and Cervantes M J 2018 On the similarity of pulsating and accelerating turbulent pipe flows *Flow Turbulence and Combustion* **100(2)** 417-436. https://doi.org/10.1007/s10494-017-9855-5

[3] Kong R and Kim S 2015 Characterization of horizontal air-water two-phase flow *The 16th Int. Topical Meeting on Nuclear Reactor Thermal Hydraulics* (*NURETH-16*) August-September (Chicago, USA) 5559-72

[4] Offor U H and Alabi S B 2016 An accurate and computationally efficient explicit friction factor model *Advances in Chemical Engineering and Science* **6** 237-245 http://dx.doi.org/10.4236/aces.2016.63024

[5] Medina Y C, Fonticiella O M C and Morales O F G 2017 Design and modelation of piping systems by means of use friction factor in the transition turbulent zone *Mathematical Modelling of Engineering Problems* **4(4)** 162-167 doi: 10.18280/mmep.040404

[6] Azizi N, Homayoon R and Hojjati M R 2018 Predicting the colebrook-white friction factor in the pipe flow by new explicit correlations *Journal of Fluids Engineering* **141(5)** doi:10.1115/1.4041232

[7] Pimenta B D, Robaina A D, Peiter M X, Mezzomo W, Kirchner J H and Ben L H B 2018 Performance of explicit approximations of the coefficient of head loss for pressurized conduits *Brazilian Journal of Agricultural and Environmental Engineering* **22(5)** 301-307 Retrieved from http://www.agriambi.com.br

[8] Ganat T A and Hrairi M 2018 Gas–liquid two-phase upward flow through a vertical pipe *Influence of Pressure Drop on the Measurement of Fluid Flow Rate, Energies* (*MDPI*) **11(11)** 1-23 doi:10.3390/en11112937

[9] Brkic D and Praks P 2018 Unified friction formulation from laminar to fully rough turbulent flow *Applied Sciences* **8(11)** 20-36 doi: 10.3390/app8112036

[10] Ortiz-Vidal L E, Mureithi N and Rodriguez O M H 2013 Friction factor in two-phase gas-liquid pipe flow *8-th Int. Conf. on Multiphase Flow (ICMF-2013)* May (Jeju/Korea) Retrieved from https://www.researchgate.net/publication/237079232

[11] Lukman S and Oke I A 2017 Accurate solutions of colebrook-white's friction factor formulae *Nigerian Journal of Technology* (*NIJOTECH*) **36(4)** 1039–48

[12] Kostornoi S D 2014 A flow model that takes into account the peculiarities of the boundary conditions of a real liquid. *Bulletin of NTU «KhPI»* **1(1044)** 214-221

[13] Prykhodko A A and Polevoi O B 2007 Control of Turbulent Supersonic Flow Separation Using Heat and Mass Transfer. *Industrial heat engineering* **29** 15-20.

[14] Dmytriv V, Dmytriv I and Dmytriv T 2018 Recearch in thermo-anemometric measuring device of pulse flow of two-phase medium *17th Int. Scientific Conf.: ENGINEERING FOR RURAL DEVELOPMENT* May 23-25 Proc. **17** (Jelgava Latvia: University of Life Sciences and Technologies Faculty of Engineering) 898-904. DOI: 10.22616/ERDev2018.17.N200

[15] Landau L D and Lifshits E M 1988 *Theoretical Physics* Vol VI *Hydrodynamics* (Moskov: Science)

[16] Dmytriv T V, Mykyychuk M M and Dmytriv V T 2021 Analytical model of dynamic boundary layer on the surface under laminar flow mode *Machinery & Energetics (Journal of Rural Production Research* Vol 12) (Kyiv Ukraine) 97-102.

Modeling of three-link road trains

R Zinko[1], O Lanets[2], Y Skvarok[3]

[1]Department of Equipment Design and Automotive Engineering, Lviv Polytechnic National University, Lviv, 79013, Ukraine
[2]Department of Technical Mechanics and Dynamics of Machines, Lviv Polytechnic National University, Lviv, 79013, Ukraine
[3]Department of Technological and Professional Science, Drohobych Ivan Franko State Pedagogical University, Drohobych, 82100, Ukraine

E-mail: rzinko@polynet.lviv.ua

Abstract. It is proposed to use graphs for developing the mathematical model of multi-link road trains with coupling devices of different designs. The calculation scheme is formed on the basis of the structure graph of the multi-link road train construction scheme. Next, the structure graph of the mathematical model generalized coordinates levels is written. This allows to develop the mathematical model of appropriate complexity depending on the research objectives. The developed mathematical motion model of a multi-link road train with an elastic-damping joints of their main elements was represented by a system of nonlinear inhomogeneous ordinary differential equations of the second order. The acceleration process for three-link road train with pneumatic coupling devices between train links was modeled.

1. Introduction

The share of expenses, associated with the transportation of goods, is 25–40% of the total cost of manufactured products. Transportation efficiency can be improved by using additional trailer wagons for road trains. However, in the process of road trains and technological machines motion there are a number of undesirable influences, in particular, the occurrence alternating shock load at the point of adhesion of tractor and trailer, which leads to a deterioration of the dynamic processes of vehicle motion. In addition, increasing the speed deteriorates indicators of stability, controllability. The maneuverability and reliability characteristics also depend on the design of the coupling devices. This is especially can be seen for three-link road trains.

Therefore, it is important to create such coupling devices (CD) that would provide optimal modes of multi-link road train (MLRT) motion in different operation modes.

2. Analysis of the main achievements and literature

Research of MLRT is conducted towards the improvement of maneuverability [1, 2], course stability [3, 4], steerability [5, 6], safety [7, 8].

Mathematical models of MLRT with CD operation should reflect both the main characteristics of MLRT and the factors related to the efficiency of their operation. Longitudinal dynamic loads on the main elements of MLRT with CD depend not only on changes in speed and motion trajectory, but also on the radial and tangential stiffness of the driving wheels [9], MLRT control systems [10], brake

systems [11], rigidity of the MLRT axle suspension [12], elastic-damping properties of CD [13], road profile [14]. When choosing a calculation scheme, it is necessary to take into account the design features of the machine, its operating conditions, research tasks, etc. [15]. Therefore, it can be argued that there is a minimum level of the calculation scheme simplification, after which it will no longer reflect the machine design. The level of such calculation schemes complexity can be set by various means, for example, by the branched structures [16], graphs [17], kinematic pairs [18].
To facilitate the work of designers and researchers, software environments are being developed that allow solving typical problems in the area of designing new machines and equipment in a faster and more efficient way. Also, in software environments it is possible to explore typical modes or processes. In the case of the development of new equipment models or modernization of existing ones, it is often necessary to solve problems that are not covered by these software environments. This problem is usually solved by creating new programs, requiring the development of new mathematical models capable for solving these exact problems with the necessary and sufficient level of quality. The development of methods for such mathematical models remains an urgent problem.

3. Creating a model of three-link road train using graphs

A kinematic scheme of MLRT Fig. 1 is built using graphs of its structure [17]. The structure graph of the road train construction scheme means a graph that shows the connections of the long combination vehicle (LCV) elements, taking into account its structure and external influences. The scheme element is a rigid inertial body of various constructive content, it is displayed by a circle, a rigid kinematic connection between two elements is shown by a straight line, and an elastic connection is marked by a wavy line.

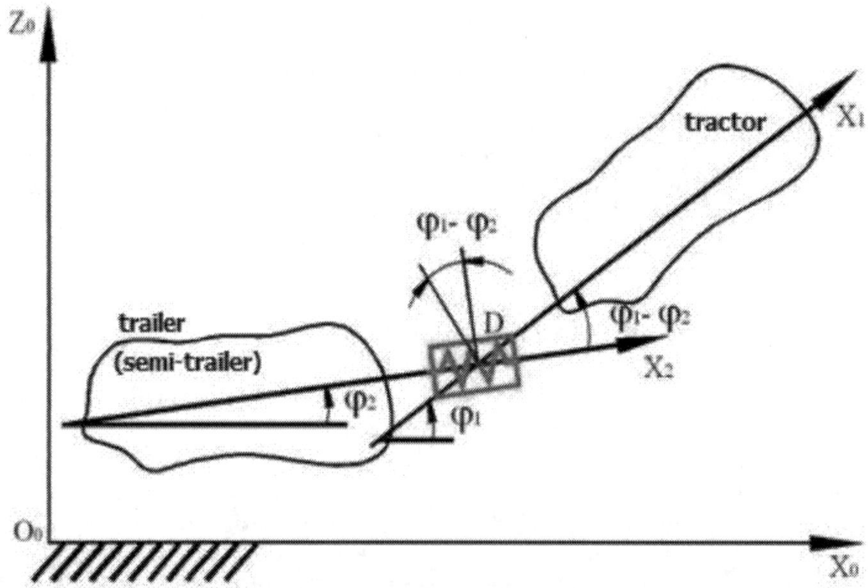

Figure 1. Example of MLRT "Tractor – trailer (semi-trailer)" with CD moving in a fixed vertical plane: $X_0O_0Z_0$ – fixed vertical longitudinal plane of MLRT motion; D – coupling device.

To form the mathematical model of MLRT CD operation, which is maximally adapted to perform the particular research, the structure graph is used for the levels of generalized coordinates of the mathematical model [17]. The structure graph of the levels of generalized coordinates of the mathematical model means a graph, which shows the list of generalized coordinates in the mathematical model and the structure of their levels. To display the structure of the levels of generalized coordinates, they are placed by levels, marked sequentially from bottom to top, starting from zero; one level contains the generalized coordinates of the same physical content.

Structure graphs of the levels of generalized coordinates, shown in Fig. 2, allow to develop mathematical models of MLRT operation, which have the given type of CD, taking into account the external mechanical characteristics of their power units, transmission, pavement quality and road profiles.

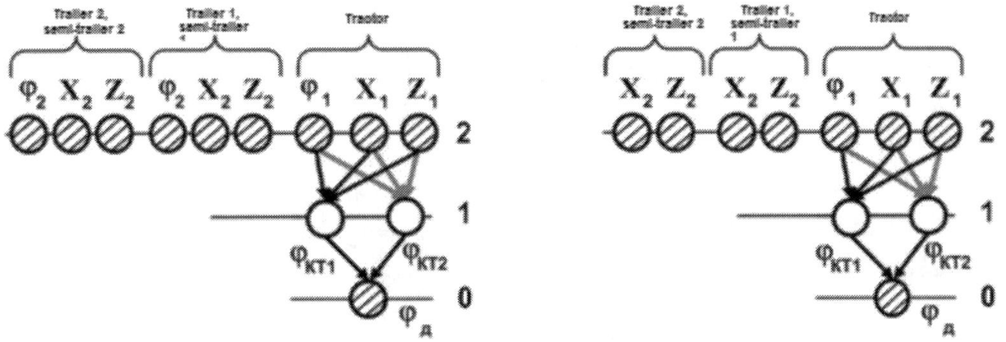

a) b)

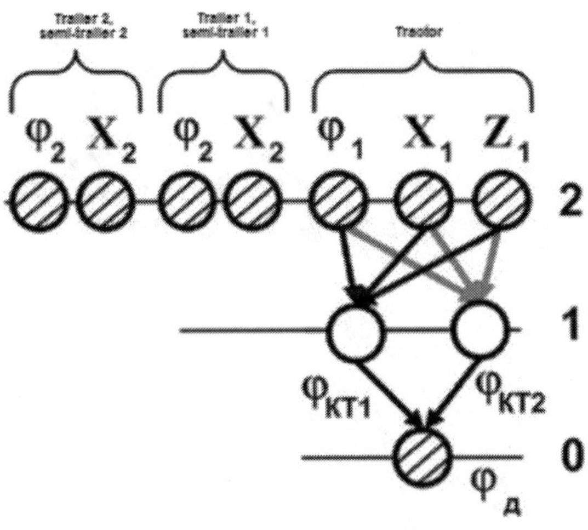

c)

Figure 2. Graph of the levels of generalized coordinates of the mathematical model of MLRT operation with CD, which have constructive schemes of general and partial cases:
a – according to the graphic representation of CD in Fig.1; b – according to the graphic representation of CD in Fig. 3.a; c – according to the graphic representation of CD in Fig. 3.b; $\varphi_д$ – rotation angle of the output shaft of the power unit driven to the semi-axle of the tractor driving wheel; φ_{KT1}, φ_{KT2} – rotation angles of the semi-axles of the tractor driving wheels; X_i, φ_i, Z_i $i=1,2$ – coordinates of the O_i origins of $X_iO_iZ_i$ coordinate systems, which are rigidly connected with the trucking vehicle (i = 1) and semi-trailer frames (i = 2); 0, 1, 2 – numbers of the levels of generalized coordinates of the mathematical model of MLRT.

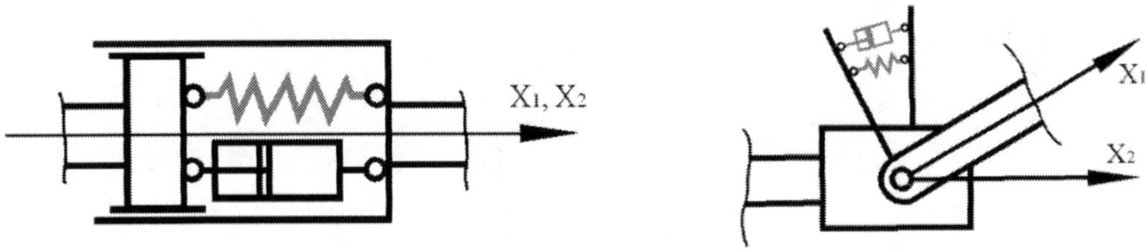

a) b)

Figure 3. Constructive schemes of partial cases of CD at MLRT movement in the vertical longitudinal plane: a – translational pair; b – rotating pair.

Based on the general equation of the mechanical system for inertial rigid bodies dynamics written in generalized coordinates [19] for MLRT with structure graphs of the constructive schemes Fig. 4, in which general case CD Fig. 5 is used and which corresponds to the structure graph of the levels of generalized coordinates of the MLRT operation mathematical model Fig. 2.a, we can build the mathematical model itself, which has the following matrix form:

$$\begin{vmatrix} a_{11} & \cdots & a_{1n} \\ \vdots & \ddots & \vdots \\ a_{k1} & \cdots & a_{kn} \end{vmatrix} * \begin{vmatrix} \ddot{X}_1 \ldots \ddot{X}_p \\ \ddot{Z}_1 \ldots \ddot{Z}_r \\ \ddot{\varphi}_1 \ldots \ddot{\varphi}_w \\ \ddot{\varphi}_{KT1} \ldots \ddot{\varphi}_{KTf} \\ \ddot{\varphi}_d \end{vmatrix} = \left| Q_1 \ldots Q_d \right| \quad (1)$$

where $n, k, p, r, w, f, d = 1 \ldots N$.

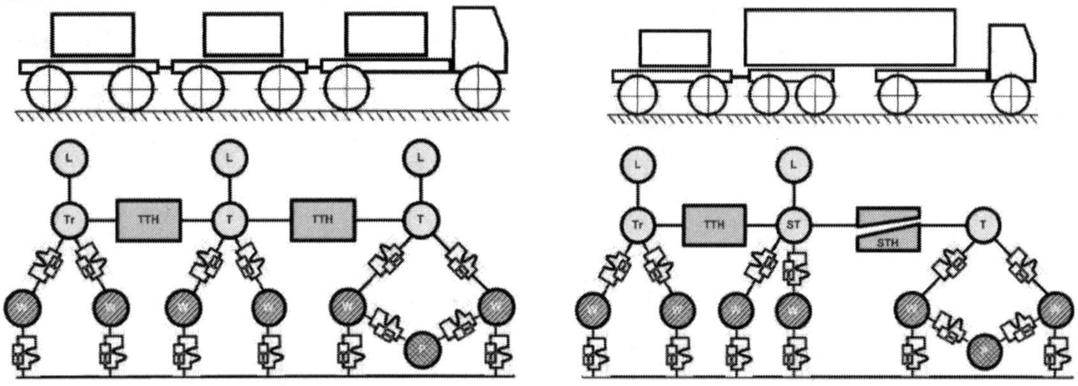

a) б)

Figure 4. Structure graph of the construction schemes of MLRT with a trailer (a) and a semi-trailer (b): T – tractor; Tr – trailer; ST – semi-trailer; P – power unit with transmission; L – load; W – wheel; ▬ – passive rigid kinematic connection between MLRT elements; – passive elastic-dissipative kinematic connection between MLRT elements; ● – element that is affected by external kinematic perturbation (from the road profile) and external power perturbation (torque acting on the engine crankshaft).

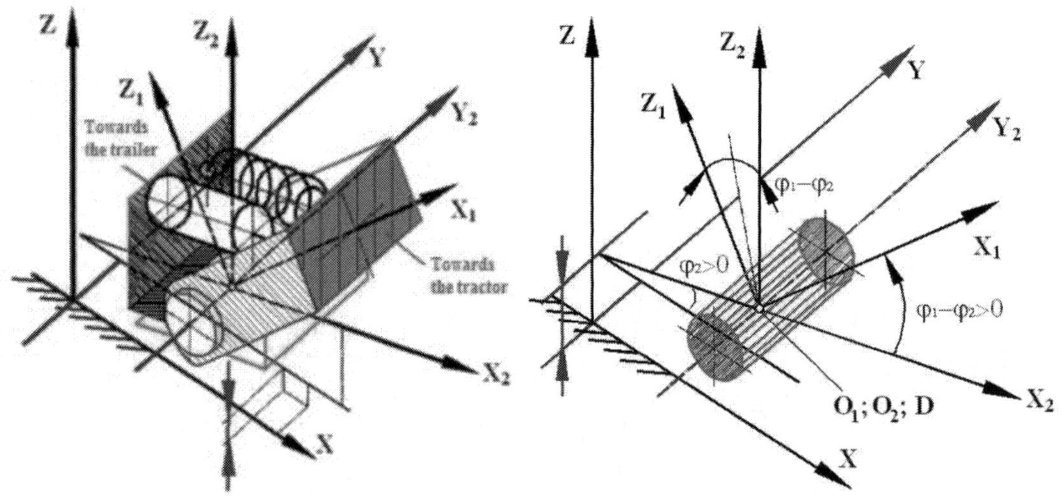

Figure 5. Graphic representation of the general case of CD during MLRT movement in the vertical longitudinal plane.

The mathematical model of MLRT operation, built using the graphs in Fig. 2.a, has nine ordinary differential equations of the second order. This mathematical model can be used to investigate the impact on the efficiency of MLRT operation:

1 – external speed characteristics of the engine, transmission parameters;

2 – characteristics of the transmission components, pavement quality and road profile;
3 – the influence of weight and geometric characteristics of the cargo;
4 – elastic-damping characteristics of the suspension of the trucking vehicle and trailer (semi-trailer);
5 – MLRT characteristics;
6 – factors affecting fuel economy.

As it can be seen from the mathematical model, which is based on the structure graphs of the levels of generalized coordinates in Fig. 2, it gives a lot of opportunities to simulate real processes of MLRT operation. The disadvantages of such mathematical model are its multi-parametericity and cumbersomeness.

In the case when it is necessary to investigate individual aspects of MLRT operation or the operation of its individual components, it is advisable to build mathematical models based on partial cases of structure graphs of MLRT constructive schemes Fig. 4 and structure graphs of the levels of generalized coordinates of the mathematical model of MLRT operation Fig. 2. Thus, when the influence of MLRT transmission and engine on its operation is not researched, it is expedient to develop mathematical models of MLRT operation with different CD on the basis of partial cases of structure graphs of the levels of generalized coordinates, shown in Fig. 2. These mathematical models are not able to simulate the movement of MLRT and gear shifts during their movement, but these models are best suited to perform the tasks of a particular research, i. e. they are easier to use when researching the MLRT in motion on a certain constant transmission gear.

It is expedient to research the operation of MLRT with different CD when the trucking vehicle is moving at a predetermined speed using mathematical models based on partial cases of structure graphs of the levels of generalized coordinates, shown in Fig. 6. They are not able to simulate the impact of trucking vehicle engine and transmission on the movement of MLRT, but the tasks of a particular research are performed.

The mathematical model of MLRT operation, when the trucking vehicle is moving at a predetermined speed with the CD of the structural scheme Fig. 3.b, based on the structure graph of the levels of generalized coordinates Fig. 6 will have the following form:

$$\begin{vmatrix} a_{22} & a_{23} & a_{25} & a_{26} \\ a_{32} & a_{33} & a_{35} & a_{36} \\ a_{52} & a_{53} & a_{55} & a_{56} \\ a_{62} & a_{63} & a_{63} & a_{66} \end{vmatrix} * \begin{vmatrix} \ddot{Z}_1 \\ \ddot{\varphi}_1 \\ \ddot{Z}_2 \\ \ddot{\varphi}_2 \end{vmatrix} = \begin{vmatrix} Q_2 \\ Q_3 \\ Q_5 \\ Q_6 \end{vmatrix} \quad (2)$$

The research of the autonomous operation of MLRT trucking vehicle should be carried out using the mathematical model (3), which is based on the partial case of the structure graph of the levels of generalized coordinates, shown in Fig. 6.

As an illustration of graph usage for building the mathematical model of MLRT, which is best suited to perform the tasks of a particular research, can be an example of a research for a vehicle with a semi-trailer [17], which has a TCD, shown in Fig.5.

4. Model of three-link road train motion

We will form the mathematical model of MLRT with a trailer operation on the basis of the calculation scheme Fig. 7. The calculation scheme identifies the main elements of MLRT with CD and elastic-damping connections between them, driving and steered wheels.

Hereinafter the commonly accepted presumptions for the research of MLRT operation [14,17,19] were used:

1. Limiting the research of unsteady motion of MLRT with flat design, the calculating scheme is considered as flat.
2. The MLRT trucking vehicle and semi-trailer are rigid inertial bodies.
3. The inertia of the drive elements of the MLRT chassis is reduced to the driving wheels and distributed in proportion to the distribution of the torque from power unit between them.

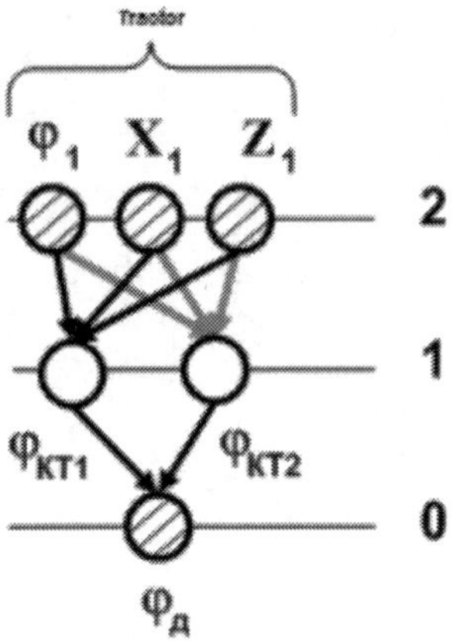

$$\begin{vmatrix} a_{11} & a_{12} & a_{13} & a_{17} & a_{18} & a_{19} \\ a_{21} & a_{22} & a_{23} & a_{27} & a_{28} & a_{29} \\ a_{31} & a_{32} & a_{33} & a_{37} & a_{38} & a_{39} \\ a_{71} & a_{72} & a_{73} & a_{77} & a_{78} & a_{79} \\ a_{81} & a_{82} & a_{83} & a_{87} & a_{88} & a_{89} \\ a_{91} & a_{92} & a_{93} & a_{97} & a_{98} & a_{99} \end{vmatrix} * \begin{vmatrix} \ddot{X} \\ \ddot{Z}_1 \\ \ddot{\varphi}_1 \\ \ddot{\varphi}_{KT1} \\ \ddot{\varphi}_{KT2} \\ \ddot{\varphi}_{\partial} \end{vmatrix} = \begin{vmatrix} Q_1 \\ Q_2 \\ Q_3 \\ Q_7 \\ Q_8 \\ Q_9 \end{vmatrix} \tag{3}$$

Figure 6. Structure graph of the levels of generalized coordinates and mathematical model of autonomous operation of MLRT trucking vehicle.

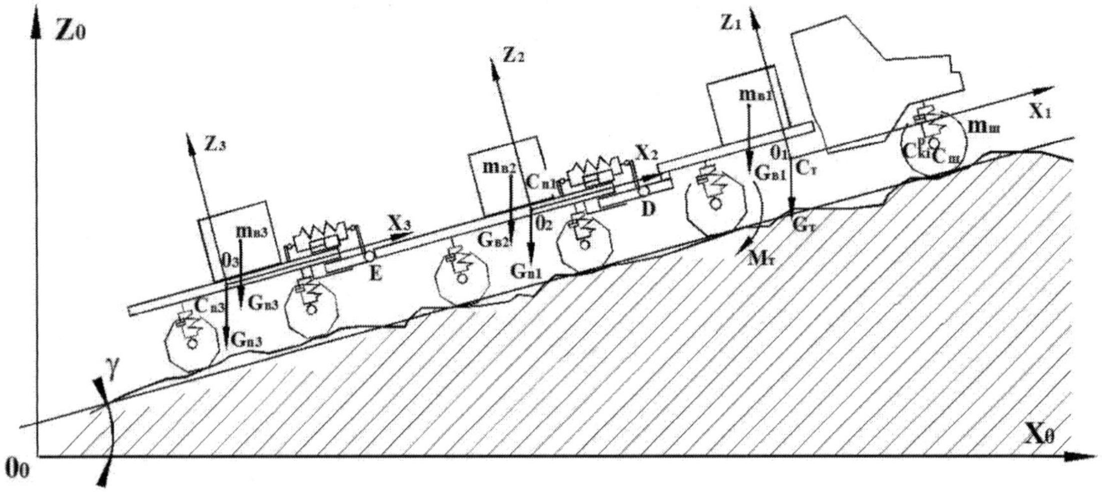

Figure 7. Calculating scheme of the truck jointed with trailers.

4. The mechanical characteristics of the MLRT power unit are based on the traction characteristics of MLRT engines.
5. MLRT wheels have radial pliability. The wheels of its driving axles also have tangential pliability, which depends on the angle of their twist.
6. The total moment of brakes of MLRT wheels functionally depends on time.
7. The magnitude of the resistance forces of the MLRT wheels functionally depend on the forces pressing them to the road.
8. The non-horizontal and non-linear macro-profile of the road is a function of the longitudinal coordinate X_d of the fixed system XOZ, and the micro-profile of the road has the features of the profiles of typical roads or off-road.

Based on the first assumption, the following flat coordinate systems were accepted (Fig. 7):

XOZ – a fixed coordinate system, the Z axis is parallel to the line of gravity and opposite to it, the X axis is perpendicular to the Z axis and directed in such a way that a right-hand coordinate system is formed; we assume that the movement of the MLRT is directed in the direction of X increase.

$X_1O_1Z_1$ – a moving coordinate system that is rigidly connected to the body of MLRT tractor and lies in the XOZ plane; the origin O_1 of this system can change its position in the fixed coordinate system XOZ, i. e. can change its X_{01}, Z_{01} coordinates, and O_1X_1, O_1Z_1 axes can rotate by the φ angle relative to the axes of the fixed coordinate system; when rotating the axes counterclockwise, the angle φ is assumed to be positive; the angle φ origin is assumed to be the case, when O_1X_1, O_1Z_1 axes are respectively parallel to OX, OZ axes; the direction of O_1X_1, O_1Z_1 axes coincides with the direction of the corresponding OX, OZ axes.

$X_2O_2Z_2$ – a moving coordinate system that is rigidly connected to the body of the first semi-trailer and lies in the XOZ plane; the origin O_2 of this system can change its position in the fixed coordinate system XOZ, i. e. can change its X_{02}, Z_{02} coordinates, and O_2X_2, O_2Z_2 axes can rotate by the ψ angle relative to the axes of the moving coordinate system $X_1O_1Z_1$; when rotating the axes counterclockwise, the angle ψ is assumed to be positive; the angle ψ origin is assumed to be the case, when O_2X_2, O_2Z_2 axes are respectively parallel to OX, OZ axes; the direction of O_2X_2, O_2Z_2 axes coincides with the direction of the corresponding OX, OZ axes. The diagrams show: the coordinates of origins O_1, O_2 of moving coordinate systems; C_T, C_P – centers of mass of the tractor and trailer, m_T, m_P – masses of the tractor and trailer, respectively; $\{X_d, Z_d(X_d)\}$ – coordinates of the road profile.

$X_3O_3Z_3$ – a moving coordinate system that is rigidly connected to the body of the second semi-trailer and lies in the XOZ plane; all its characteristics are similar to the $X_2O_2Z_2$ coordinate system.

The trucking vehicle and the MLRT semi-trailer together with the rigidly connected $X_1O_1Z_1$, $X_2O_2Z_2$ coordinate systems move in plane-parallel in a fixed XOZ coordinate system.

An arbitrary road profile on which the MLRT moves or stands is analytically represented by the function $Z_d = Z_d(X_d)$, the argument of which is the longitudinal X_d coordinate. In order to save only the positive Z_d coordinates of the road profile, the OX axis is selected so that it passes below the lowest point of the profile.

To write the equations of motion of the inertial elements of the calculation scheme of MLRT with CD, which has certain elastic-damping characteristics, the general equation of dynamics of the mechanical system in the form of the Lagrange equation of the second kind [19] was used.

On the basis of the accepted calculation scheme and chosen coordinate systems for the generalized coordinates of mechanical system – MLRT we will accept:

$q_1 = x_{01}, q_2 = z_{01}$ – coordinates of the O_1 origin of the $X_1O_1Z_1$ coordinate system in the fixed XOZ coordinate system;

$q_3 = \varphi$ – rotation angle of the moving $X_1O_1Z_1$ coordinate system relative to the fixed coordinate system;

$q_4 = \psi$ – rotation angle of the moving $X_2O_2Z_2$ relative to the moving $X_1O_1Z_1$ coordinate system;

$q_5 = {}^2x_d$ – coordinate of the center of the axis of the hinge CD in the moving $X_2O_2Z_2$ coordinate system;

$q_{5+j} = \varphi_j$ – rotation angle of the axle shaft of the j driving tractor axle, where $j = 1, \ldots m$. Thus, the mechanical system – MLRT has n = 5 + m degrees of freedom.

Then the mathematical model of MLRT will have the following form:

$$
\begin{cases}
\ddot{x}_{01}\cdot M-\ddot{\varphi}\cdot M\cdot L-(\dot{\varphi})^2\cdot M\cdot L_1'=+\sum\limits_{j=1}^{m}\dfrac{C_{kj}^T}{R_k}\cdot \\
\cdot\left[\left(\varphi_j-\varphi_j^{strt}\right)-\int\limits_{t_{strt}}^{t}\dfrac{\sqrt{\left(x_{OP.j}\right)^2+\left(z_{OP.j}\right)^2}\,dt}{R_k}\right]- \\
-\sum\limits_{j=1}^{m_1}\dfrac{f_{kj}\cdot C_{OP.j}}{z_{OP.j}-z_d\left({}^0x_{OP.j}\right)}\cdot\left\{L_{OP.j}^V-\left[z_{OP.j}-z_d\left(x_{OP.j}\right)\right]\right\}; \\
0 \\
\ddot{z}_{01}\cdot M+\ddot{\varphi}\cdot M\cdot L_1'-(\dot{\varphi})^2\cdot M\cdot L_2'=-M_g+ \\
+\sum\limits_{j=1}^{m_1}C_{OP.j}\cdot\left\{L_{OP.j}^V-\left[z_{OP.j}-z_d\left(x_{OP.j}\right)\right]\right\}; \\
\ddot{x}_{01}\cdot M\cdot L_2'+\ddot{z}_{01}\cdot ML_1'+\ddot{\varphi}\cdot\left\{M\cdot\left[\left(L_1'+L_2'\right)^2\right]+L_c\right\}=-g\cdot M\cdot L_1'+ \\
\sum\limits_{j=1}^{m_1}C_{OP.j}\cdot\left\{L_{OP.j}^V-\left[z_{OP.j}-z_d\left(x_{OP.j}\right)\right]\right\}\cdot L_{5j}; \\
I_{Tkj}\ddot{\varphi}_j=M_{kj}-C_{kj}^T\cdot \\
\cdot\left[\left(\varphi_j-\varphi_j^{strt}\right)-\int\limits_{t_{strt}}^{t}\dfrac{\sqrt{\left(x_{OP.j}\right)^2+\left(z_{OP.j}\right)^2}\,dt}{R_k}\right];\ j=1,\ldots,m.
\end{cases}
\qquad (4)
$$

The developed system of differential equations (4) is a mathematical model of MLRT motion with CD of the selected design. In addition to the given 2 (1 + j) functions it includes $14+2m+5m_1$ known parameters: m_T, m_P – masses of tractor, semi-trailers and their moments of inertia relative to their centers of mass; I_{Tkj} – the moment of inertia of the semi-axles of the j-th driving axle together with part of the reduced moment of inertia of the engine and transmission to their geometric axes; C_{Kj}^T, C_{Kj}^P – coefficients of rigidity of the wheel tire of the j-th driving axle of the tractor in the tangential direction and in the radial direction of the wheel of the j-th axle of MLRT; C_{Pj} – coefficient of transverse stiffness of the spring or longitudinal stiffness of the damping spring of the j-th MLRT bridge; f_d – coefficient of resistance to movement of MLRT wheels along the road; α_j – damper damping factor of the j-th MLRT support per mass unit; ${}^1x_{CT}$, ${}^1z_{CT}$, ${}^1x_{CP}$, ${}^1z_{CP}$ – respectively horizontal and vertical coordinates of the centers of mass of the tractor and semi-trailer; ${}^1x_{ch1}$, ${}^2x_{k1}$ – coordinates of fastening of the elastic element to the chassis in the $X_1O_1Z_1$ coordinate system and to the body in the $X_2O_2Z_2$ coordinate system; R_k – wheel radius; $L_{ch,k}$, $L_{i-1,i}$ – lengths of elastic elements in the undeformed state; L_{OPj}^V – free length of the j-th chassis support.

Since the dynamic loads on the main elements of MLRT during their elastic-damping connection with the help of CD will be researched, i. e. in mathematical terms the Cauchy problem for the system of differential equations will be solved, it is necessary to form a set of 2 ($5+m$) initial conditions for

this system of equations.
Depending on the initial motion conditions of MLRT with elastic-damping connection of the main elements, it is possible to model different modes of motion, in particular rise, descent, acceleration, braking, reversal of the transmission moment, skidding, sliding wheels, etc.
A common initial motion condition for MLRT with an elastic-damping connection of its main elements is motion from a state of balanced rest – a state in which all generalized velocities are zero, and the values of generalized coordinates are such that external forces are balanced by the reactions of elastic chassis supports. Other cases of initial motion conditions for MLRT with elastic-damping connection of the main elements give more opportunities to vary the initial values of the generalized coordinates, so setting them is not difficult.
The mathematical motion model of MLRT with elastic-damping connection of their main elements is represented by a system of $5+m$ nonlinear inhomogeneous ordinary differential equations of the second order, which is not solved with respect to the older derivatives of generalized time coordinates. Subsequently, numerical methods were used to integrate systems of ordinary differential equations (for example, in PTC's Mathcad 14).
When compiling a mathematical motion model of MLRT with elastic-damping connection of the main elements, no conditions and restrictions were put forward on the values of parameters and the type of functions. Therefore, there is a real possibility that if the form of the equations themselves does not change, the mathematical model will become "rigid", i. e. the one that matches the features of a "rigid system of differential equations". Among the large number of methods for the numerical solution of the Cauchy problem for a system of first-order differential equations, we paid attention to methods that allow integrating "rigid" systems of differential equations.
The mechanical characteristics of the MLRT power unit, the characteristics of its brakes, as a rule, are presented in the relevant literature in graphical form. For the system of equations it is necessary to have their analytical representation, i. e. it is necessary to perform interpolations of these functions. Based on the dynamic characteristics of the real power unit of the vehicle, we built the mechanical characteristics of the power unit using the method of interpolation of functions for equations in the form of a third-order spline function that characterizes the power unit and brakes of MLRT with elastic-damping connection of basic elements.
Off-roads are poorly described by probabilistic functions, because the law of normal distribution in relation to the off-road profile can be used very conditionally. In addition, our task is related to the research of dynamic processes of MLRT with elastic-damping connection of their main elements at overcoming specific obstacles, specific sections of the road or off-road. Therefore, in the future the profile of the road was represented by specific analytical or transcendent functions.

5. Simulation of three-link road train acceleration process

The acceleration process for three-link road train (as a MLRT option) was modeled. The specificity is the usage of a pneumatic traction coupling device which uses a one-section pneumatic cylinder as an elastic element Fig. 8.
When starting a road train, due to the inertia of the trailer, the engine only loses power on moving the trucking vehicle mass. The trailer at this time remains stationary, because the slider 2 of the drawbar 1 slides alone the guide 3 until the support plate 5 does not bump into the buffer 6 of the guide 3. This also stretches the air cylinder 9 and the shock absorber. After deformation of the buffer 6 through the support plate 5, the traction force from the engine is transmitted to the drawbar 1 of the trailer and the trailer moves. The shear rate of the trailer is determined by the rigidity of the air cylinder 9. Possible jerks of the trailer are smoothed by the shock absorber 10. The rigidity of the air cylinder 9 is determined by the amount of air supplied from the receiver 10 by the control valve 11 based on the signal from the control unit 12. The control unit 12 receives information from the sensors of load 13 and acceleration 14, placed on the trailer and on the tractor. Accordingly, the device works during braking. Only in this case the support plate 7 bumps into the buffer 8.

IOP Conf. Series: Materials Science and Engineering 1277 (2023) 012027 doi:10.1088/1757-899X/1277/1/012027

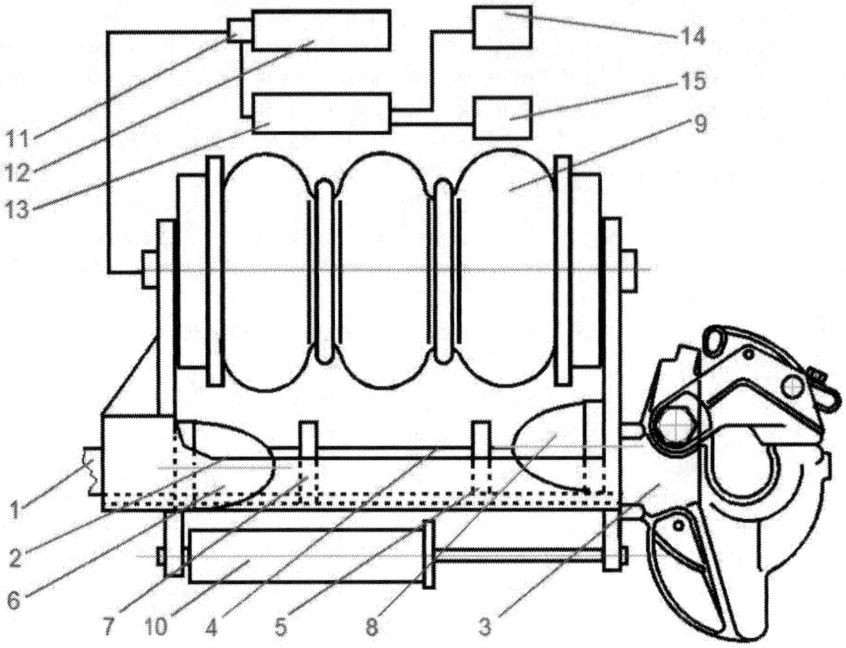

Figure 8. Coupling device.

The description of the operation of a single-section pneumatic cylinder is described by a polynomial of quadratic degree:

$$y = 0{,}0006x^4 - 0{,}0123x^3 + 0{,}0755x^2 - 0{,}0978x + 0{,}1026$$

Modeling results are shown in Fig. 9 and Fig. 10.

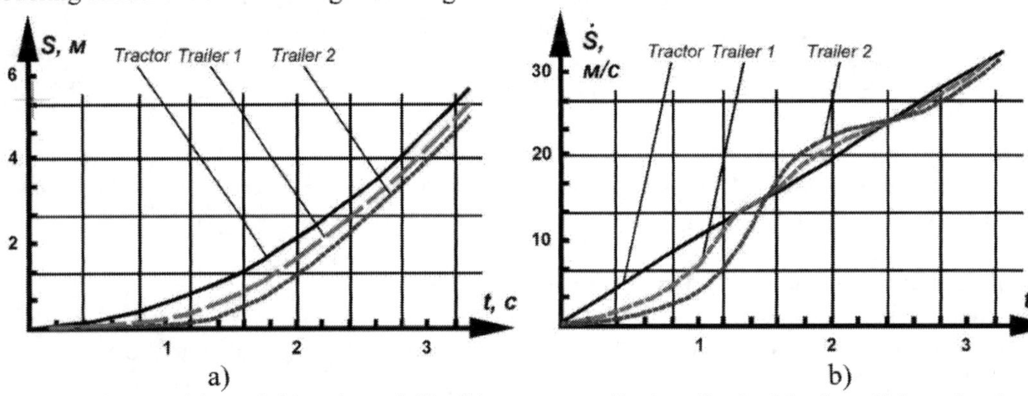

Figure 9. Change of the path (a) and speed (b) of the tractor, trailer 1, trailer 2 of the three-link road train when starting.

6. Analysis of the obtained results

The maximum oscillations were observed in the last trailer, so the stability of its movement determined the stability of the entire road train Fig. 9. A sharp increase in the amplitude of oscillations was observed with increasing number of trailers and speed. The speed of a road train with one trailer is about 50 km/h, a road train with two trailers – about 43 km/h. As it can be seen from the graph, with increasing the speed up to 20 km/h, the amplitude of oscillations increases by 24%, up to 30 km/h – by 27%, up to 43 km/h – by 29%. Researching the process of starting, we can see that the velocity of 2 m/s stabilizes when t = 2.4 s, the transition time takes up to 4 s Fig. 10. At the same time with some delay there is a jump of dynamic loadings in coupling devices of trailers. The longitudinal forces in the trailers are shifted by the values of the gaps in the road train elements, which are located between the first and second coupling device.

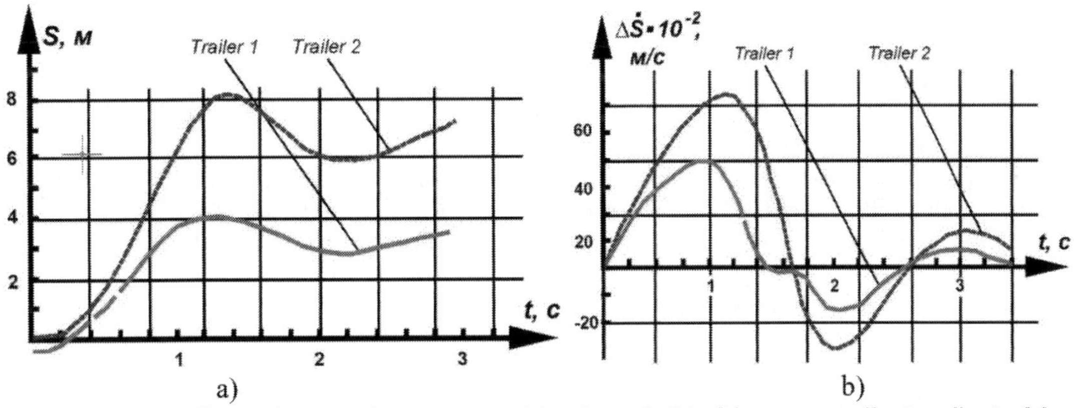

Figure 10. The difference between the movements (a) and speeds (b) of the tractor, trailer 1, trailer 2 of the three-link road train when starting.

7. Summary and conclusions

It is proposed to use graphs for developing the mathematical model of multi-link road trains with coupling devices of different designs. The calculation scheme is formed on the basis of the structure graph of the multi-link road train construction scheme. Next, the structure graph of the mathematical model generalized coordinates levels is written. This allows to build the mathematical model of appropriate complexity depending on the research objectives and operating conditions.

The mathematical motion model of a multi-link road train with an elastic-damping connection of their main elements is represented by a system of nonlinear inhomogeneous ordinary differential equations of the second order. The acceleration process for three-link road train with pneumatic coupling devices between train links was modeled. Numerical results for a specific road train were obtained.

References

[1] Aghabayk K, Sarvi M, Forouzideh N and Young W 2014 Modelling heavy vehicle car-following behaviour in congested traffic conditions *Journal of advanced transportation* **48.8** 1017-29 doi:10.1002/atr.1242

[2] Jagelčáka K J, Kiktováa M and Frančáka M 2020 The Analysis of Manoeuvrability of Semi-trailer Vehicle Combination *Transportation Research Procedia* **44** 176-181

[3] Bulgakov M, Shuklynov S, Uzhva A, Leontiev D, Verbitskiy V, Amelin M and Volska O 2019 Mathematical model of the vehicle initial rectilinear motion during moving uphill *24th Slovak-Polish International Scientific Conference on Machine Modelling and Simulations* doi.org/10.1088/1757-899X/776/1/012022

[4] Khajepour He Y, McPhee A J 2005 Dynamic modelling and stability analysis of articulated frame steer vehicles *Int J Heavy Veh Syst* **12** 28-59

[5] Cheng C, Roebuck R, Odhams A, and Cebon D 2011 High Speed Optimal Steering of a Tractor-Semitrailer *Vehicle System Dynamics* **49** 561-593

[6] Rangavajhula K and Tsao H S J 2007 Active Trailer Steering Control of an Articulated System with a Tractor and Three Full Trailers for Tractor-track Following *International Journal of Heavy Vehicle Systems* **14** 271-293

[7] Grislis A 2010 Longer combination vehicles and road safety *Transport* **3** 336-343 doi:10.3846/transport.2010.41

[8] Forkenbrock D 2003 Fatal Crash Involvement by Multiple-Trailer Trucks *Transportation Research Part A: Policy and Practice* **5** 419-433 doi:10.1016/S0965- 8564(02)00034-4

[9] Gipser M 2007 FTire the tire simulation model for all applications related to vehicle dynamics *Vehicle System Dynamics* **45** 139-151 doi:10.1080/00423110801899960

[10] Ding X, Mikaric S and He Y 2013 Design of an active trailer-steering system for multi-trailer articulated heavy vehicles using real-time simulations *Proc IMechE Part D: J Automobile Engineering* **227** 643-655

[11] Bogomolov V, Klimenko V, Leontiev D, Ryzhyh L, Smyrnov O and Kholodov M 2020 Impruving the brake control effectiveness of vehicles equipped with a pneumatic brake actuator *Science & Technique* **19** 55-62 doi:10.21122/2227-1031-202019-1-55-62
[12] MacAdam C, Hagan M, Fancher P, Winkler C, Ervin R, Zhou J and Bogard S 2000 Rearward Amplification Suppression (*Report UMTRI-2000-47*)
[13] Bruzelius F, Gomez F, Jorge A and Bruno A 2013 Basic vehicle dynamics model for driving simulators *International Journal of Vehicle Systems Modelling and Testing* **8** 364-385
[14] Georg R 2006 Vehicle modeling by subsystems J. *Braz. Soc. Mech. Sci&Eng* **24** 430-442
[15] Pozhbelko V I 2000 Structural synthesis and analysis of mechanical systems of arbitrary structure of a given level of complexity *Izvestiya vuzov* **5-6** 13-25
[16] Wulfson I 1986 *Vibroactivity of drives of machines of branched and ring structure* (Moscow: Mechanical Engineering) p 98
[17] Zinko R V 2014 *Morphological environment for technical systems investigation* (Lviv: Lviv Polytechnic Publishing) p 386
[18] Bazhenov E E 2010 Modular principle of modeling of dual-purpose articulated machines *Science and Education* **8** URL: http://technomag.edu.ru/doc/160532.html
[19] Lurie A I 1961 *Analytical mechanics* (Moscow: Fizmatgiz) p 824

Mathematical model of agricultural aggregate movement based on the Autotractor

V V Adamchuk[1], S P Pohoriliy[2, 3], V G Prysiazhnyi[2], V Yu Mirnyi[2]

[1] Director Institute of mechanics and automatics of agroindustrial production of the national academy of agrarian sciences of Ukraine
[2] Department of mechanical and energy problems of machine-tractor units Institute of mechanics and automatics of agroindustrial production of the national academy of agrarian sciences of Ukraine
[3] E-mail: pogorilyy_sergiy@ukr.net

Abstract. The paper presents theoretical studies of the movement of the agricultural unit on the basis of the MEZ-330 «Autotractor». An equivalent dynamic model of an agricultural unit on the basis of MEZ-330 «Autotractor» is presented. A system of differential equations has been compiled, which makes it possible to describe the interrelation of the main structural parameters of the unit and its operating modes. The results of modeling the motion of the MEZ-330 «Autotractor» on the implementation of technological operations are given. According to the simulation results, the dependences describing the influence of the mass of the MEZ-330 «Autotractor» on the coefficient of slipping of its wheels were established. It is established that the obtained theoretical dependences are adequate and correlate with experimental data with a 95% confidence probability.

1. Introduction

The use of agricultural units based on a car chassis to perform technological operations in the field has not yet been sufficiently studied due to the relatively recent appearance of such mobile energy means.

The creation of a mobile energy means for agro-industrial production MEZ-330 "Autotractor" [1, 2] leads to the need to conduct the oretical studies that will make it possible to establish its rational parameters and modes of operation under the conditions of technological operations.

A lot of attention was assigned to the study of the movement of the machine-tractor unit (MTA) by scientists, in particular, Chudakov E. A., Barskiy I. B., Anilovich V. A., Kutkovy H. M., Vodolazchenko Y. T., Chudakov D. A., Skotnikov V. A., Huskov V. V. [3–8] and others. The above-mentioned scientists considered the MTA movement process from the point of view of evaluating its acceleration and limiting the mathematical models with many factors and simplifications. In particular, they reduced the number of rotating masses, didn`t take into account the skidding of the engines, traction and traction properties of the unit, etc. These limitation facilitated the necessary calculations.

In view of the above, theoretical studies that will make it possible to analyze the movement of agricultural units based on the MEZ-330 "Autotractor" and determine its optimal modes are relevant.

2. Methodology

The research was carried out using methods of building mathematical models of the movement of a mobile energy means and models of the functioning of agricultural machines and tools using the basic principles of mathematics, theoretical mechanics and numerical calculations on a PC.

3. Results and discussion

In order to simulate the movement of an agricultural unit based on the MEZ-330 "Avtotractor", we will create an equivalent dynamic model (**Figure 1**).

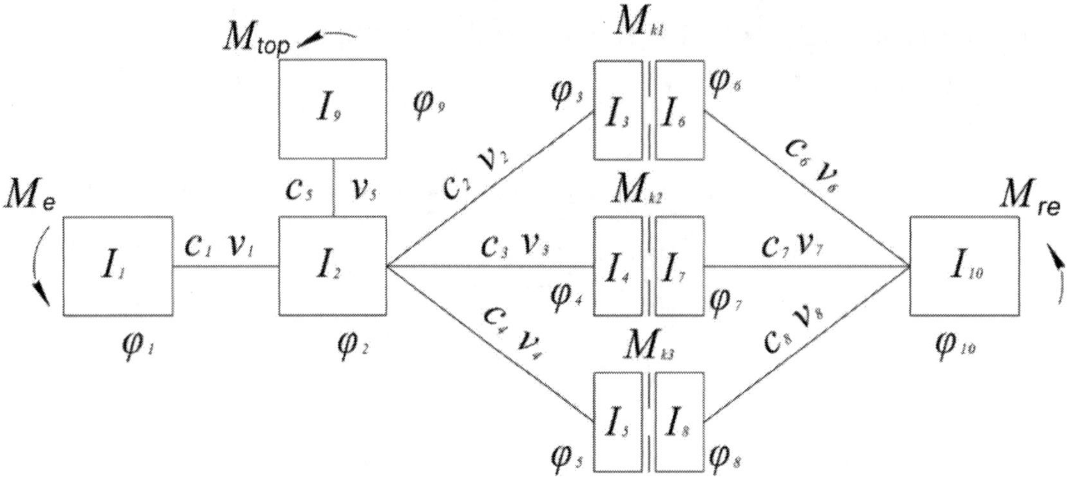

Figure 1. Equivalent dynamic model of the agricultural unit based on MEZ-330

To simplify the mathematical model, a number of assumptions were made, in particular: the tangential traction forces and the rolling resistance forces of the wheels of one axle are equal to their sum; damping of transmission parts and wheel tires is proportional to the speed of rotation. These assumptions will lead to an error in calculations of less than 3%.

D'Alembert's principle was used to compile a system of differential equations [9, 10]:

$$\begin{cases} I_1 \cdot \ddot{\varphi}_1 = M_{Д} - c_1(\varphi_1 - \varphi_2) - v_1(\dot{\varphi}_1 - \dot{\varphi}_2) \\ I_2 \cdot \ddot{\varphi}_2 = c_1(\varphi_1 - \varphi_2) + v_1(\dot{\varphi}_1 - \dot{\varphi}_2) - c_2(\varphi_2 - \varphi_3) - v_2(\dot{\varphi}_2 - \dot{\varphi}_3) - c_3(\varphi_2 - \varphi_4) - \\ -v_3(\dot{\varphi}_2 - \dot{\varphi}_4) - c_4(\varphi_2 - \varphi_5) - v_4(\dot{\varphi}_2 - \dot{\varphi}_5) - c_5(\varphi_2 - \varphi_9) - v_5(\dot{\varphi}_2 - \dot{\varphi}_9) \\ I_3 \cdot \ddot{\varphi}_3 = c_2(\varphi_2 - \varphi_3) + v_2(\dot{\varphi}_2 - \dot{\varphi}_3) - M_{к1} \\ I_4 \cdot \ddot{\varphi}_4 = c_3(\varphi_2 - \varphi_4) + v_3(\dot{\varphi}_2 - \dot{\varphi}_4) - M_{к2} \\ I_5 \cdot \ddot{\varphi}_5 = c_4(\varphi_2 - \varphi_5) + v_4(\dot{\varphi}_2 - \dot{\varphi}_5) - M_{к3} \\ I_6 \cdot \ddot{\varphi}_6 = M_{к1} - c_6(\varphi_6 - \varphi_{10}) - v_6(\dot{\varphi}_6 - \dot{\varphi}_{10}) \\ I_7 \cdot \ddot{\varphi}_7 = M_{к2} - c_7(\varphi_7 - \varphi_{10}) - v_7(\dot{\varphi}_7 - \dot{\varphi}_{10}) \\ I_8 \cdot \ddot{\varphi}_8 = M_{к3} - c_8(\varphi_8 - \varphi_{10}) - v_8(\dot{\varphi}_8 - \dot{\varphi}_{10}) \\ I_9 \cdot \ddot{\varphi}_9 = c_5(\varphi_2 - \varphi_9) + v_5(\dot{\varphi}_2 - \dot{\varphi}_9) - M_{top} \\ I_{10} \cdot \ddot{\varphi}_{10} = c_6(\varphi_6 - \varphi_{10}) + v_6(\dot{\varphi}_6 - \dot{\varphi}_{10}) + c_7(\varphi_7 - \varphi_{10}) + v_7(\dot{\varphi}_7 - \dot{\varphi}_{10}) + \\ +c_8(\varphi_8 - \varphi_{10}) + v_8(\dot{\varphi}_8 - \dot{\varphi}_{10}) - M_{re} \end{cases} \quad ,(1)$$

where I_1 – the moment of inertia of the rotating and reciprocating parts of the engine, as well as the driving parts of the clutch, brought to the crankshaft of the internal combustion engine, kg·m^2; I_2 – the moment of inertia of the rotating parts of the transmission in this gear is given (clutch drive disc, gearbox in this gear, cardan transmission between the gearbox and transfer box, transfer box in this gear), kg·m^2; I_3, I_4, I_5 – moments of inertia are given according to the shaft of the front, middle and rear axles, which take into account the mass of the wheels, the main gear and the cardan shaft, respectively, kg·m^2; I_6, I_7, I_8 – moments of inertia, which take into account the mass of the supporting surface that perceives the torque of the wheels of the front, middle and rear axles, respectively, and the mass of the part of the tires of the wheels that interacts with the supporting surface, kg·m^2; I_9 – the moment of inertia brought to the crankshaft of the diesel engine shaft, which takes into account the mass of the driven working bodies, kg·m^2; I_{10} – the moment of inertia brought to the crankshaft, which takes into account the mass of the MEZ and the technological part of the agricultural unit moving forward, kg·m^2; M_e – torque of the internal combustion engine (ICE), N·m; M_{k1}, M_{k2}, M_{k3} –– torques created on the wheels by the resistance forces of the first, second and third car axle, N·m; M_{tor} – torque applied to the crankshaft of the internal combustion engine for the drive of the machine part, N·m; M_{re} — the torque of the resistance forces of the unit's movement brought to the crankshaft of the internal combustion engine, N·m; c_1, c_2, c_3, c_4, c_5 – equivalent stiffness of transmission parts, N·m/rad; c_6, c_7, c_8 – equivalent tangential stiffness of tires of front, middle and rear wheels, N·m/rad; υ_1, υ_2, υ_3, υ_4, υ_5– equivalent damping of transmission parts, N·m·s/rad; υ_6, υ_7, υ_8– equivalent damping of tires of front, middle and rear wheels, N·m·s/rad; φ_1, φ_2, φ_3, φ_4, φ_5, φ_6, φ_7, φ_8, φ_9, φ_{10} — generalized coordinates of the equivalent dynamic model, rad.

The dependence of the torque of the internal combustion engine (M_e) on the frequency of rotation of its crankshaft, which is included in the system of differential equations (1), was obtained using the approximation of the external characteristics of the internal combustion engine. MEZ-330 "Autotractor" is equipped with a YaMZ-238DE2 engine, the nominal power of which is 243 kW, the maximum torque is 1274 N·m [11]. Approximation of the external characteristics of the diesel engine is presented in **Figure 2**. The angular speed of the diesel engine crankshaft at idle speed is ω_1 = 245 rad/s (2340 min^{-1}).

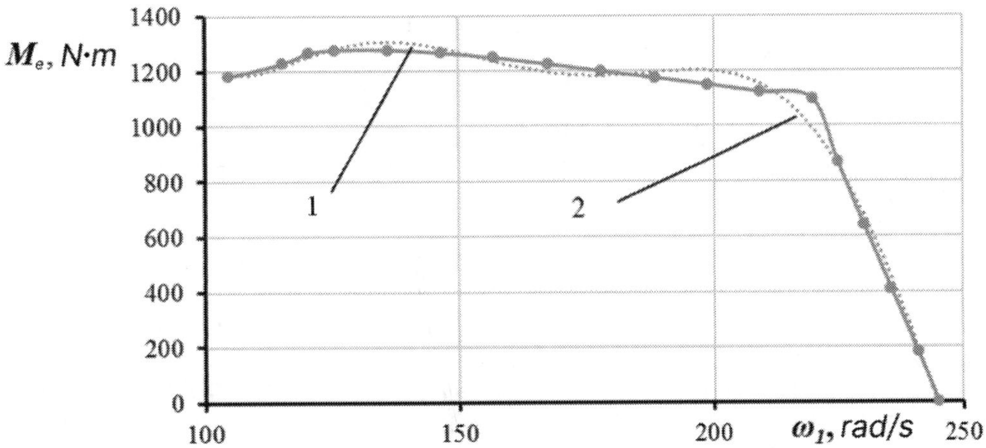

Figure 2. Dependence of the YAMZ-238DE2 engine torque on the angular velocity of its crankshaft on partial regulatory characteristics: 1 – experimental dependence [11]; 2 – the line of approximation of the experimental dependence

As a result, the regression equation was obtained:

$$M_e = 8,25 \cdot 10^{-9} \cdot \omega_1^6 - 8,75 \cdot 10^{-6} \cdot \omega_1^5 + 3,77 \cdot 10^{-3} \cdot \omega_1^4 - 0,85 \cdot \omega_1^3 + 104,34 \cdot \omega_1^2 - 6691,59 \cdot \omega_1 + 175716,04 \quad (2)$$

where ω_1 – angular speed of the crankshaft of the internal combustion engine, rad/s.

The regression equation (2) is obtained with a confidence probability of 99%.

Torques M_{k1}, M_{k2}, M_{k3}, which are created on the wheels by the resistance forces of the first, second and third axles, were determined by the formulas [5]:

$$M_{\kappa 3} = \frac{P_3 \cdot r_d}{i_t \cdot \eta_t}, \qquad M_{\kappa 1} = \frac{P_1 \cdot r_d}{i_t \cdot \eta_t}; \qquad M_{\kappa 2} = \frac{P_2 \cdot r_d}{i_t \cdot \eta_t}; \quad (3)$$

where P_1, P_2, P_3 – are traction forces created on the wheels of the front, middle and rear axles, respectively, H; r_d – dynamic rolling radius of the wheel, m; i_t – gear ratio of the transmission; η_t – the efficiency ratio of the transmission.

Traction forces P_1, P_2, P_3 are limited by dependencies [5]:

$$P_1 = \varphi_{app} \cdot R_{z1}; \qquad P_2 = \varphi_{app} \cdot R_{z2}; \qquad P_3 = \varphi_{app} \cdot R_{z3}, \tag{4}$$

where φ_{app} - the weight application factor; R_{z1}, R_{z2}, R_{z3} - there are front, middle and rear axles weight, N.

Coefficient of use of coupling weight φ was presented as a dependence $\varphi_{app} = f(\delta)$, where δ – is the coefficient of slipping of the engines. The mathematical data was obtained using the approximation of the traction characteristics of the MEZ-330 "Autotractor" (**Figure 3**).

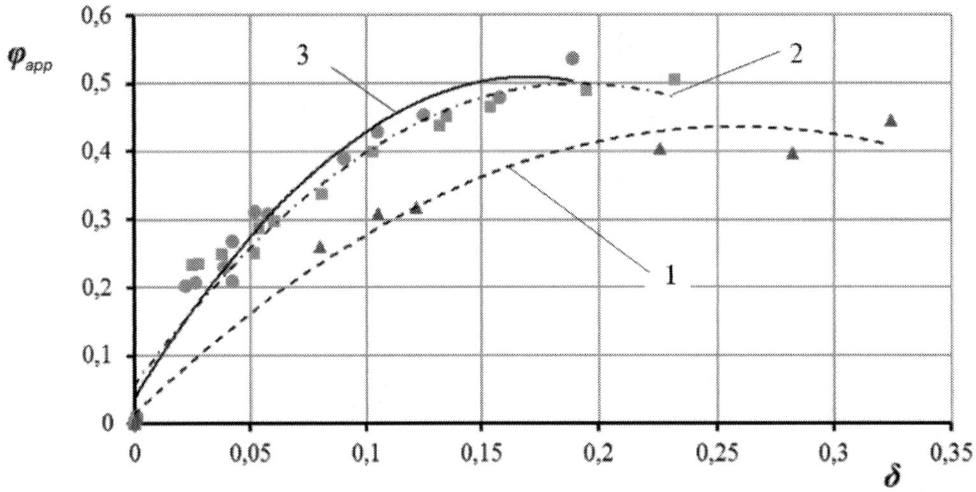

Figure 3. Dependence of the coefficient of application of the hitch weight on the skidding coefficient of the MEZ-330 "Autotractor": 1, 2, 3 - respectively, the air pressure in the tires of the wheels is 0,3; 0,1; 0,08 MPa

For example, for the MEZ-330 "Avtotractor" tire pressure of 0,08 MPa, which was chosen in order to reduce the negative effect of its motion system on the ground, the regression equation will have the followingvalue:

$$\varphi_{app} = 0{,}101 + 4{,}0445 \cdot \delta - 9{,}5898 \cdot \delta^2\,, \tag{5}$$

The regression equation (5) was obtained with a confidence probability of 0,98.

The reduced torque of the forces of resistance to the movement of the unit (M_{re}) was determined according to the known method [5], for example, for the traction unit:

$$M_{re} = \frac{(P_{re} + f \cdot G_a) \cdot r_d}{i_t \cdot \eta_t} \tag{6}$$

where P_{re} – is the resistance force created by an agricultural machin, N; f – coefficient of rolling resistance; G_a – is the weight of the agricultural unit, N.

The reduced torque on the drive of the working bodies (M_{re}) is determined similarly by the previous formula (6) in the absence of the rolling resistance component ($f{\cdot}G_a$) and instead of the dynamic rolling radius, the distance from the working bodies to the axis of the drive shaft must be used.

To substitute into the system of equations (1), the coefficient of slipping under the condition of non-locked differentials was presented in the following form:

$$\delta_{\kappa 1} = \frac{\dot{\varphi}_3 - \dot{\varphi}_6}{\dot{\varphi}_6}; \quad \delta_{\kappa 2} = \frac{\dot{\varphi}_4 - \dot{\varphi}_7}{\dot{\varphi}_7}; \quad \delta_{\kappa 3} = \frac{\dot{\varphi}_5 - \dot{\varphi}_8}{\dot{\varphi}_8}, \tag{7}$$

The solution of the system of differential equations (1) was carried out using the mathematical package Mathcad, which solves such systems of equations using the Runge-Kut method with a variable discrediting step [12, 13].

The input data were determined according to known methods [4, 5, 7, 8, 14–21] and are presented in the work [22]

The system of differential equations (1) makes it possible to simulate and investigate the movement process and operating modes of the unit. In particular, by changing the parameters of the running system (air pressure in the tires of the wheels, dynamic rolling radius of the wheels, the number of driving axles), the resistance force of the agricultural machine or tool, the gear ratio of the transmission, etc., it is possible to obtain the value of the torque at which the specified speed of movement and specific fuel consumption are ensured ICE, coefficient of slipping, etc.

According to the results of the simulation of the system of differential equations (1), the dependence of the influence of the mass of the MEZ (**Figure 4**) on the coefficient of slipping was obtained.

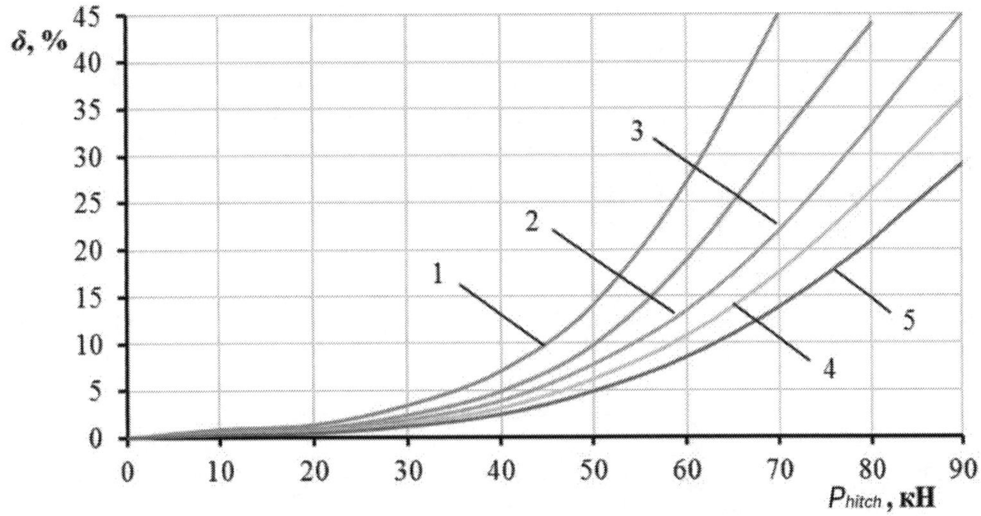

Figure 4. Dependence of the coefficient of slipping on the traction force under the condition of air pressure in the tires of the MEZ wheels of 0.3MPa: 1, 2, 3, 4, 5 – respectively, the weight of the MEZ is 11,4; 14,0; 16,0; 18,0 and 20,0 tons

From **Figure 4**, it can be seen that with an increase in the weight of the MEZ, the coefficient of slipping of its wheels decreases. So, with a traction force of, for example, 50.0 kN and MEZ masses of 11,4; 14,0; 16,0; 18,0; 20,0 t coefficient of slipping is 14,1, respectively; 9,8; 7,7; 6,2; 4,9%.

To check the correctness of the theoretical dependence obtained using the system of equations (1) with the experimental data, we will construct graphic dependences with confidence intervals of the experimental data, which were determined according to known methods [23] with a confidence probability of 95% (**Figure 5**).

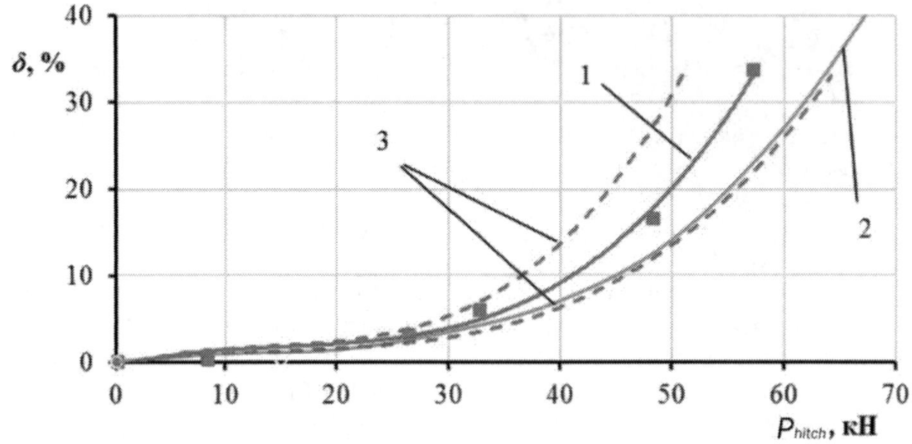

Figure 5. Dependence of the coefficient of slipping on the traction force under the condition of air pressure in the tires of the MEZ wheels of 0,3 MPa and its mass of 11,4 tons: 1 – experimental data; 2 – theoretical data; 3 – confidence intervals

From **Figure 5** it can be seen that the dependence of the coefficient of slipping on the traction force, which is obtained by modeling the system of differential equations (1) is in the confidence interval (with a confidence probability of 95%) of the experimental data. This confirms that the obtained theoretical data are adequate and correlated with the experimental data with a confidence probability of 95%.

4. Conclusion

A mathematical model of the movement of the agricultural unit based on the MEZ-330 "Autotractor" was obtained, which describes the relationship between the main design parameters of the unit and its operating modes. The dependence of the effect on the drag coefficient of the wheels of the MEZ-330 "Autotractor" and its weight has been theoretically established. It was established that the obtained theoretical dependences are adequate and correlated with experimental data with a confidence probability of 95%.

5. References

[1] Adamchuk V V, Pogorily S P 2016 The use of an automobile chassis for performing technological operations in agro-industrial production *Motrol "Commission of motorization and energetics in agriculture"* **18 (8)** 93–98.

[2] Adamchuk V V, Pogorily S P 2016 Mobile agricultural units based on automobile chassis *"Mechanization and electrification of agriculture": interdepartmental thematic collection* **50** 8–13.

[3] Chudakov E A 1950 Theory of the car. Moscow 344 p.

[4] Barsky I B, Anilovich V Ya, Kutkov G M 1973 Dynamics of the tractor. Moscow: "Mechanical Engineering" 280 p.

[5] Anilovich V Ya, Vodolazhchenko Yu T 1976 Designing and calculation of agricultural tractors. M.: "Mechanical Engineering" 456 p.

[6] Chudakov D A 1975 The fundamentals of the theory and calculation of the tractor and car. M.: Kolos 384 p.

[7] Skotnikov V A, Mashchensky A A, Solonsky A S 1986 Fundamentals of the theory and calculation of a tractor and a car. Moscow: Agropromizdat 383 p.

[8] Guskov V V, Velev N N, Atamanov Yu E, Bocharov N F 1988 Tractors: Theory: Textbook for university students by special. "Automobiles and tractors". Moscow: "Mechanical Engineering" 376 p.

[9] Targ S M 1972 A short course of theoretical mechanics. Moscow: Nauka 478 p.

[10] Komarov M S 1969 Dynamics of mechanisms and machines. Moscow: "Mechanical Engineering" 296 p.

IOP Conf. Series: Materials Science and Engineering 1277 (2023) 012028 doi:10.1088/1757-899X/1277/1/012028

[11] Technical specifications 37.319.271-96 1999 The engines are automobile YaMZ-238BE2 and YaMZ-238DE2. JSC "Avtodizel" 34 p.

[12] Vasiliev A N 2006 Mathcad 13 in the examples. St. Petersburg: BHI Petersburg 528 p.

[13] Bertyaev V D 2005 Theoretical mechanics based on Mathcad. Workshop. St. Petersburg: BHI Petersburg 752 p.

[14] Ageykin Ya S 1981 Passability of cars. M.: "Mechanical Engineering" 232 p.

[15] Aksenov P V 1989 Multi-axle cars. M.: "Mechanical Engineering" 280 p.

[16] Platonov V F 1981 All-wheel drive cars. M.: "Mechanical Engineering" 279 p.

[17] Grishkevich A I 1984 Design of transmissions of cars. Directory. Moscow: "Mechanical Engineering" 272 p.

[18] Levin M A 1989 The theory of rolling of a deformed wheel. Moscow: Nauka 272 p.

[19] Knoroz V I, Klennikov E V 1975 Tires and wheels. Moscow: "Mechanical Engineering" 184 p.

[20] Knoroz V I, Klennikov E B, Petrov I P 1976 The work of automobile tires Moscow: "Transport" 238 p.

[21] Fominykh A B, Zheglov L F 2015 Characteristics of a wide-profile automobile tire on large irregularities. *Science and Education MSTU N E Bauman* **11** 1–11.

[22] Pogoriliy S P 2018 Theoretical studies of the process of mobile energy in traction operations. *National collection "Mechanization and electrification of agriculture"* **7 (106)** 124-133.

[23] Zaks L 1976 Statistical evaluation. Moscow. Statistics 598 p.

IOP Conf. Series: Materials Science and Engineering 1277 (2023) 012029 doi:10.1088/1757-899X/1277/1/012029

Researching hydraulic shock phenomenon in complex hydraulic manifolds

R M Sheremeta

Department of Designing and Operation of Machines, Lviv Polytechnic National University, Ukraine, Lviv, S. Bandery street 12, E-mail: Roman.M.Sheremeta@edu.lpnu.ua, (ORCID 0000-0001-6558-7912)

Abstract. The proposed method describes parameters change of the working environment in any section of a complex hydraulic manifold over time at a non-stationary, arbitrarily specified flow, the manifold is bounded by elements of arbitrary stiffness. The methodology uses the theory of a real compressible fluid.

1. Introduction

A change in the pressure of the working fluid in the manifold of the hydraulic system can occur under various operational conditions. Such a change can be created when the speed of the liquid flow and the pipeline changes, when the shut-off and regulating fittings (valves) are activated quickly, when the pipelines are filled, etc. [10, 12]. Quite often, such a change in pressure can cause significant excesses in the working fluid pressures in the pipeline. The hydraulic shock phenomenon [1] can occur, which in turn can cause damage to the elements of the hydraulic manifold disabling it. The hydraulic shock phenomenon has been well researched [2-5], existing methodology allows calculating hydraulic shock parameters in pipelines [6-9]. Using the presented hydraulic shock research methods, it is possible to prevent the occurrence of destructive excess pressure in hydraulic manifolds.

2. Problem statement

Study of the parameters of the unsteady flow of the working fluid in the mains of complex hydraulic systems containing branching pipelines and connecting elements of different diameters and different stiffness is a rather difficult problem at the moment. In such complex hydraulic systems due to the superposition of reflected waves of hydraulic shock, standing wave pressure and hydraulic shock resonance can form in different parts of the hydraulic system. In the paper, the analogy between electric current and flowing liquid is used to study phenomena of resonance in the hydraulic system. [11].

3. Solution

Let's assume that an unsteady fluid flow is created at the point x_0 on the element of the pipeline, which coincides with the direction of the X axis. Let's use the wave equation (1).

$$\frac{\partial^2 P}{\partial x^2} = \gamma^2 P \quad , \tag{1}$$

where P is the amplitude of pressure waves; γ – propagation constant;

$$\gamma^2 = \frac{Fg\Omega}{c^2}\left(-\frac{\Omega}{gF} + iR\right),$$

where F is the cross-sectional area of the flow; c is the velocity of the elementary compression wave; Ω is the angular velocity of forced pressure fluctuations;

$$\Omega = \frac{2\pi}{T},$$

where $T = 4\tau$, τ – the exposure time of the impact pulse of the medium pressure in this case, the valve actuation time that changes the flow in the manifold.

Thereby:

$$\Omega = \pi/_{2\tau};$$

$$i = \sqrt{-1},$$

R – hydraulic equivalent of resistance per unit length of pipeline.

For example, for laminar fluid flow in a pipeline with a diameter d according to the Hagen-Poiseuille formula

$$R = \frac{32\nu}{gd^2F},$$

g – acceleration of free fall.

The solution of the wave equation (1) has the following form:

$$p' = (ae^{\gamma x} + be^{-\gamma x})e^{i\Omega t},$$

p' – non-stationary pressure component.

Or for the non-stationary flow component q'

$$q' = g\frac{F\Omega}{ic^2\gamma}e^{i\Omega t}\left(ae^{\gamma x} - be^{-\gamma x}\right).$$

This is a general solution to the wave equation for an arbitrary sinusoidal oscillation.

It should be noted that $e^{i\Omega t} = \cos\Omega t + i\sin\Omega t$.

The propagation constant γ is a complex value, so it can be expressed in the form

$$\gamma = \alpha + i\beta,$$

where

$$\alpha = \sqrt{\frac{Fg\Omega}{c^2}}\left[\left(\frac{\Omega}{gF}\right)^2 + R^2\right]^{\frac{1}{4}} \cdot \sin\left[\frac{1}{2}arctg\frac{RgF}{\Omega}\right], \quad (2)$$

$$\beta = \sqrt{\frac{Fg\Omega}{c^2}}\left[\left(\frac{\Omega}{gF}\right)^2 + R^2\right]^{\frac{1}{4}} \cdot \cos\left[\frac{1}{2}arctg\frac{RgF}{\Omega}\right]. \quad (3)$$

The wave resistance of the pipeline is defined as the ratio of the non-stationary pressure component to the non-stationary flow component, thus,

$$Z(x) = p'/_{q'},$$

$$Z(x) = \frac{ic^2\gamma}{gF\Omega}\left(\frac{ae^{\gamma x} + be^{-\gamma x}}{ae^{\gamma x} - be^{-\gamma x}}\right).$$

Note that the multiplier $\frac{-ic^2\gamma}{gF\Omega}=\frac{c^2\gamma}{igF\Omega}$ has a resistance dimension. It is called the characteristic resistance and is denoted by Z_c

$$Z_c=\frac{c^2\gamma}{igF\Omega}=\frac{c^2(\beta-i\alpha)}{gF\Omega}. \quad (4)$$

For the upper end of the pipeline x = 0, and for the lower x = L, forced pressure fluctuations are shown at the point x = 0. The index S denotes the upper end, R denotes the lower.

The expression that can be used to calculate the upper Z_S and lower Z_R resistance of the ends of the pipeline is as follows:

$$Z_R=\frac{Z_S-Z_c th(\gamma L)}{1-\frac{Z_S}{Z_c}th(\gamma L)}; \quad (5)$$

$$Z_S=\frac{Z_R-Z_c th(\gamma L)}{1+\frac{Z_S}{Z_c}th(\gamma L)}, \quad (6)$$

where $th(\gamma L)=\frac{e^{\gamma L}-e^{-\gamma L}}{e^{\gamma L}+e^{-\gamma L}}$, where , it is determined by the fact that:

$$\frac{e^{\gamma X}-e^{-\gamma X}}{2}=sh(\gamma X), \qquad \frac{e^{\gamma X}+e^{-\gamma X}}{2}=ch(\gamma X)$$

Determination of boundary conditions for hydraulic elements.

1. For the tank:

$$\frac{p'}{q'}=0,$$

respectively $Z=0$.

2. Absolutely rigid stopper, unfilled pipeline ends:

$$Z=\frac{p'}{q'}=\infty.$$

3. Flexible stopper (spring loaded valve).

Wave resistance in this case is determined by their characteristic resistance (4), into which expressions (2) and (3) containing the parameter Ω are substituted. In this case, Ω is the resonant circular frequency of the flexible stopper.

$$\Omega=\sqrt{\frac{c_m}{m}},$$

where m – mass of the stopper (valve mass); c_m – stopper stiffness (valve stiffness).

Valve stiffness can be defined with some assumptions as

$$c_m=\left[N(\delta)\right]'\Big|_{\delta=\delta_{\max}},$$

where $N(\delta)$ – the function of the valve drive force dependence N on the depth of valve immersion in the seat δ.

4. Complex pipe connections (figure 1).

$$\frac{1}{Z_{R1}}+\frac{1}{Z_{R4}}=\frac{1}{Z_{S2}}+\frac{1}{Z_{S3}}+\frac{1}{Z_{S5}}.$$

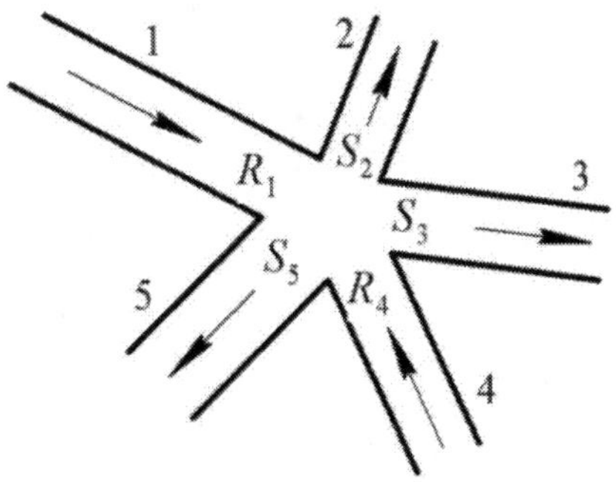

Figure 1. Generalized diagram of a complex pipeline connection.

5. Simple connection of two pipes of different diameters (figure 2):

$$Z_{R1}=Z_{S2}.$$

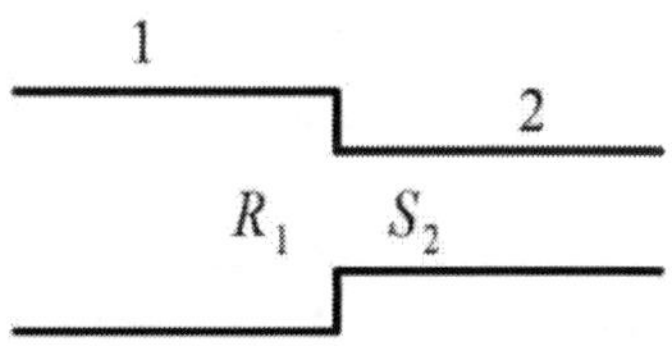

Figure 2 Two pipelines connection scheme.

Example. Analysis of transient processes in the hydraulic pipeline. Let at the input S_1 of the manifold shown in figure 3, the amplitude of the pressure fluctuation P_0 is specified.

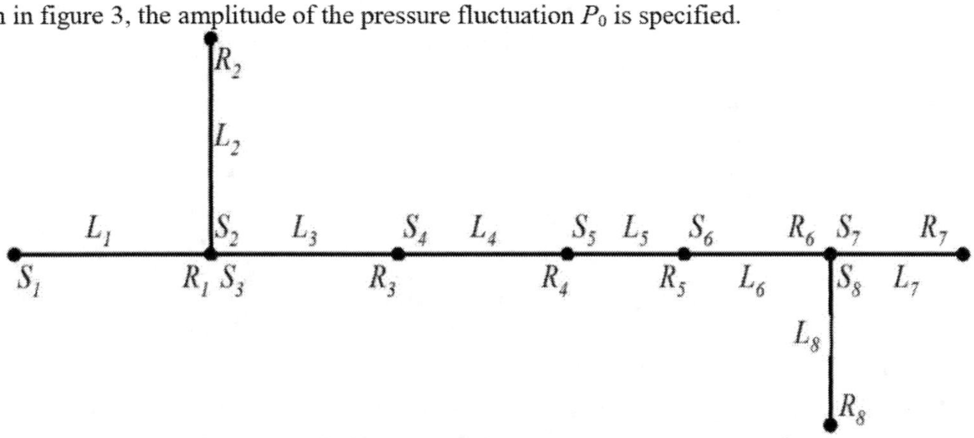

Figure 3 An example of a complex hydraulic system.

It is necessary to determine the amplitude of the medium pressure fluctuations at the point R2 of the manifold.

Boundary conditions:

$$Z_{R2} = \infty;$$

$$Z_{R7} = \infty;$$

$$Z_{R8} = \infty.$$

Let's determine the wave resistance of all elements of the manifold, expressing them in terms of boundary conditions, using equations (5) and (6):

$$Z_{S7} = \frac{Z_{c7}}{th(\gamma_7 L_7)};$$

$$Z_{S8} = \frac{Z_{c8}}{th(\gamma_8 L_8)};$$

$$Z_{S2} = \frac{Z_{c2}}{th(\gamma_2 L_2)};$$

$$Z_{R6} = \frac{Z_{S7} \cdot Z_{S8}}{Z_{S7} + Z_{S8}};$$

$$Z_{S6} = \frac{Z_{R6} + Z_{c6} \cdot th(\gamma_6 L_6)}{1 + \frac{Z_{R6}}{Z_{c6}} th(\gamma_6 L_6)};$$

$$Z_{R5} = Z_{S6};$$

$$Z_{S5} = \frac{Z_{R5} + Z_{c5} \cdot th(\gamma_5 L_5)}{1 + \frac{Z_{R5}}{Z_{c5}} th(\gamma_5 L_5)};$$

$$Z_{R4} = Z_{S5};$$

$$Z_{S3} = \frac{Z_{R3} + Z_{c3} \cdot th(\gamma_3 L_3)}{1 + \frac{Z_{R3}}{Z_{c3}} th(\gamma_3 L_3)};$$

$$Z_{R3} = Z_{S4};$$

$$Z_{S3} = \frac{Z_{R3} + Z_{c3} \cdot th(\gamma_3 L_3)}{1 + \frac{Z_{R3}}{Z_{c3}} th(\gamma_3 L_3)};$$

$$Z_{R1} = \frac{Z_{S2} \cdot Z_{S3}}{Z_{S2} + Z_{S3}};$$

$$Z_{S1} = \frac{Z_{R1} + Z_{c1} \cdot th(\gamma_1 L_1)}{1 + \frac{Z_{R1}}{Z_{c1}} th(\gamma_1 L_1)}.$$

The amplitude of the pressure wave P0 will cause a non-stationary component of pressure p at any point of the first element of the manifold:

$$p'_{X1} = e^{i\Omega t}\left[P_{S1}ch(\gamma_1 x_1) - Z_{c1}Q_{S1}sh(\gamma_1 x_1)\right];$$

$$P_{X1} = P_{S1}ch(\gamma_1 x_1) - Z_{c1}Q_{S1}sh(\gamma_1 x_1);$$

$$Q_{X1} = Q_{S1}ch(\gamma_1 x_1) - \frac{P_{S1}}{Z_{c1}}sh(\gamma_{1x_1}),$$

at the end of the first element of the manifold:

$$P_{R1} = P_{x1=L1} = P_{S1}ch(\gamma_1 L_1) - Z_{c1}Q_{S1}\cdot sh(\gamma_1 L_1),$$

at the beginning of the second element of the manifold:

$$P_{S2} = P_{R1};$$

$$Q_{S2} = \frac{P_{S2}}{Z_{S2}},$$

at the desired point of the main R2 we get the amplitude of the forced pressure fluctuation

$$P_{R2} = P_{S2}ch(\gamma_2 L_2) - Z_{c2}Q_{S2}\cdot sh(\gamma_2 L_2).$$

4. Conclusions

Proposed method of calculating the parameters of unsteady fluid flow using the wave resistance allows to determine the pressure and flow amplitude of the working fluid in any zone of the hydraulic system during hydraulic shock. It allows researching transient processes in the hydraulic manifold from the moment of injecting the liquid and optimize the structural and geometric parameters of hydraulic system elements in order to reduce the effect of hydraulic shock.

References

[1] ZHukovskij N. E. *About hydraulic shock in water pipes*. M.–L.: Gostekhizdat, 1949. – 103 s.

[2] ZHukovskij N. E. *A collection of essays. Volume 2: Hydrodynamics*. 763 s.; *Volume 3: Hydraulics. Applied mathematics*. 700 s. M.–L.: Oborongiz, 1949.

[3] Gijon M. *Research and calculation of hydraulic systems*.—M.: Mashinostroenie, 1964. — 388 s.

[4] Kalicun V. I., Drozdov E.V., Komarov A.S., CHizhik K.I. *Fundamentals of hydraulics and aerodynamics*. M.: Strojizdat, 2001. – 296 s.

[5] V. P. Sydorenko, O. M. Yakhno. *Hydraulics and hydraulic drive*. – K.: Vyd-vo Universytet «Ukraina», 2008- 163 s.

[6] Butaev D. A., Kalmykova 3. A., Podvidz L. G. i dr. *Collection of task in engineering hydraulics*., M., «Mashinostroenie» 1972, 472 s.

[7] V. V. Vakina, I. D. Denisenko, A. L. Stolyarov. *Mechanical engineering hydraulics. Calculation examples*. — K.: Vishcha shk. Golovnoe izd-vo, 1986.- 208 s.

[8] *Handbook of hydraulic calculations*. Pod redakciej P. G. Kiseleva. Izd. 4-e, pererabot, i dop. M., «Energiya», 1972. 312 s. s il.

[9] Idel'chik I. E. *Handbook of hydraulic resistances* / Pod red. M. O. SHtejnberga.— 3-e izd., pererab. i dop.— M.; Mashinostroenie, 1992.— 672 s: il.

[10] *Pipe fittings with automatic control: a Handbook* /D. F. Gurevich, O. N. Zarinskij, S. I. Kosyh i dr.; Pod obshch. red. S. I. Kosyh. — L.: Mashinostroenie, Leningr. otd-nie, 1982. — 320 s, il.

[11] Foks D. A. *Hydraulic analysis of unsteady flow in pipe networks*: Per. s angl. – M.: Energoizdat, 1981. – 248 s., il.

[12] *Probability of water hammer in the heating system, causes and consequences*. https://www.rosteplo.ru/Tech_stat/stat_shablon.php?id=2219.

Research of the dynamics of a vibrating grain cleaning separator with vertical sieves layering

V G Topilnytskyy[1,3], D P Rebot[2]

[1]Department of Designing and Operation of Machines, Lviv Polytechnic National University, Stepan Bandera Street 12, Lviv, Ukraine
[2]Computer-Aided Design Department, Lviv Polytechnic National University, Stepan Bandera Street 12, Lviv, Ukraine
[3]E-mail: topilnvol@gmail.com

Abstract. The construction of a vibrating grain cleaning separator is proposed and its nonlinear mathematical model with vertical layering of sieves is built. The system of analytical expressions obtained on the basis of asymptotic methods of nonlinear mechanics and the Lagrange equation of the second kind is taken as the model. The model makes it possible to describe the movement of the oscillating part of the grain cleaning separator. The non-linearity of the obtained model will enable its further use for a number of similar separators with different design features. With its help, the impact of the main design parameters of the separator and the kinematic indicators of its drive on the characteristics of its movement and the intensity of the separation process is determined.

1. Introduction

With a constant deficit in providing third world countries with grain crops, the question arises of raising their productivity and maximally reducing losses during storage. The storage of grain crops is affected by several factors, in particular, the presence of inclusions in the volume of grain during its collection, insufficient drying of grain crops, and inappropriate climatic conditions of the storage itself. Freshly harvested grain crops, the so-called grain heap, contain particles of straw, their ears, chaff, and seeds of other plants from the field (weeds), sand, pieces of soil and other inclusions. The presence of such inclusions may vary in percentage, but they will always be there. It is obvious that such inclusions worsen the quality of harvested grain in the aspect of its storage as food material and seed fund material. In fact, untimely and poor cleaning of grains from such impurities leads to an increase in its moisture content, the occurrence of self-heating, mold growth, breeding of pest populations, etc.

The equipment used for cleaning and sorting grain crops from impurities is called separators. There are many types and sizes of them. An important task today, in terms of energy-saving technologies and environmental protection, is the design and operation of such separators, which would: a) performed its functions fully and qualitatively; b) was simple, reliable and maintainable; c) consumed a minimum of energy resources; d) had a minimal negative impact on the environment. How to achieve these requirements? Modern design is automated using CAD – CAE – CAM systems. These systems, in particular CAE calculation systems (including equipment), are based on nonlinear mathematical models of the dynamics of such equipment during its operation. That is, having a system of analytical dependencies, which includes all parameters of this or that separator on the one hand (kinematic, geometric, force), and parameters that determine the level of efficiency of its functioning on the other, it is possible to clearly monitor the influence of the first factors on the second, and determine the conditions of the optimal shape, design and operation of the designed separators.

The need to apply the phenomenon of separation determines the use of a wide range of separators, which can be classified into separators with movable and stationary sieves [1]. Actually, separators of the first type have significantly higher productivity and, accordingly, the level of application due to the better level of interaction of the screen surface with the mixture being separated. The level of rationality of the separation process also increases with an increase in the number of sifting surfaces (sieves) in the design of the separator [2], [3].

Concentric [4] placement of sieves in drum-type separators, or sequential placement of round or U-shaped cross-sectional sieves [5] in separator structures enable their effective use as separators - conveyors with simultaneous provision of separation phenomena and transfer of separated fractions in space.

Separators with moving sieves are quite complex dynamic systems, the effective design of which and subsequent operation are possible only on the basis of their thorough theoretical research, in particular, with the use of mathematical modeling. There are a number of attempts to theoretically describe the dynamics of separators, but all of them have their own limited application. Thus, in [6], [7] only the dynamics of individual parts of the separator is theoretically studied, there are no studies of the entire separator as a single dynamic system. In [8], the movement of oscillating surfaces is studied only in the linear formulation of the problem, and the obtained theoretical models are linear, which narrows the possibility of their application. In [9], the dynamics of separators is studied by numerical methods. By authors of the article, on the basis of asymptotic methods of nonlinear mechanics [10], is developed a number of models of vibration processing systems [4, 5, 11, 12], including separators, in the form of sets of analytical expressions, which included the key parameters of these systems: kinematic, geometric and power. These models are non-linear and make it possible to describe the dynamic processes occurring in one or another researched oscillating system. According to this method the presented vibration separator with vertical layering of sieves was studied.

2. Materials and Method

Having a non-linear mathematical model of a grain cleaning separator of a certain type, it is possible to calculate quickly it under certain operating conditions (the type of grain array to be cleaned, possible types of impurities in it, the required separation performance, etc.), develop its design and manufacture the separator itself, and at the stage operation of such equipment is to choose the necessary separation modes. In this regard, a mathematical model describing the dynamics of a grain cleaning separator, which must be developed in order to use it further for the calculation and design of a number of many similar or simpler separators, must meet the following requirements:

- it should reflect the unified, universal and complicated design of the grain cleaning separator;
- it must be parametric, that is, all the parameters of the separator must be included in it in a symbolic format with the possibility of varying them both in their ranges of change and among themselves in arbitrary combinations;
- it must be non-linear, to adequately ensure the display of the operation of such a non-linear mechanical system;
- it should be non-linear, to adequately ensure the display of the work of such a non-linear mechanical system; it should be simple (relatively), to enable its application, in particular, inclusion in the automated design and calculation system. To fulfill this condition, the model should be presented in the form of a set of analytical dependencies of the factors of the efficiency of the separator (the trajectory of the sieves, the amplitude and frequency of their oscillations) on its parameters.

Based on the analysis of the available grain cleaning equipment, we choose an air-sieve type separator with a vibration drive of a flat system of parallel sieves for modeling. In particular, the following are located on the frame of the separator: a system of parallel sieves (attached to it through an elastic suspension), a drive for generating vibrating oscillations of the sieves, and the same elastic suspension. In Figure 1a schematic diagram of the oscillating block of such a separator is presented.

The diagram (Figure 1) shows the main components of the grain cleaning separator, the direction of movement of the grain mass through the sieves, the direction of loading of unrefined grain into the body with sieves and unloading of the separated fractions. The sieve system drive consists of an electric motor, an elastic coupling, an imbalance shaft and the imbalance itself. As an alternative, the drive can be a motor-vibrator, which is directly attached to the body with sieves and oscillates together with them.

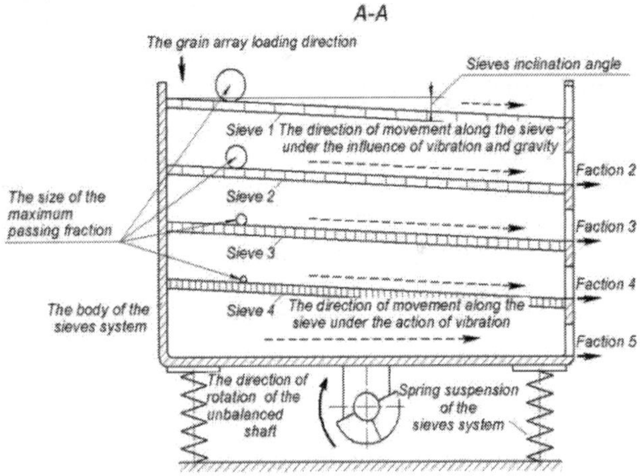

Figure 1a. The main type of grain cleaning separator.

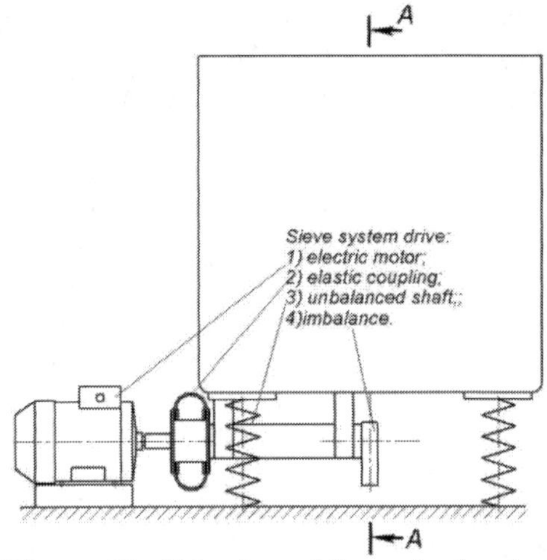

Figure 1b. Side view of the grain cleaning separator.

Figure 1. Schematic diagram of the selected for modeling vibrating grain cleaning separator with vertical layering of sieves.

To build a mathematical model of the grain cleaning separator based on its principle scheme (Figure 1), its calculation scheme (Figure 2) was developed.

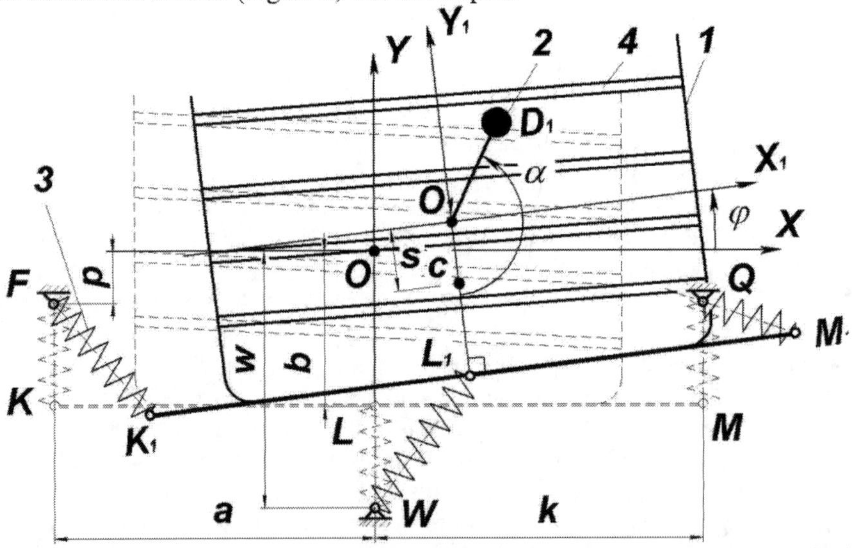

Figure 2. Calculation scheme of the vibrating grain cleaning separator with vertical layering of sieves: 1 – oscillating body with sieves; 2 – center of mass imbalance; 3 – elastic suspension; 4 - separating sieves.

The calculation scheme is shown in Figure 2 is a diagram of the movement of an arbitrary vertical section of the oscillating mass of the separator (sieve system) in the plane of rotation of the unbalanced mass, i.e. imbalance. Therefore, it is assumed that the grain cleaning separator is in flat motion. Oscillations in other planes were neglected due to the small amplitude of their oscillations compared to the amplitude of the sieve oscillations in the plane of the rotary motion of the vibration drive. Based on the principles that the model should be unified and complicated, three pairs of springs were introduced into the separator suspension system (one spring from each pair is visible in the diagram), which can be arbitrarily placed in the plane of the separator's plane movement.

In the diagram, the screen system (body with sieves) of the separator at the initial moment of time (the drive is turned off) is depicted by a dashed line, and by a solid line, respectively, the body with sieves at an arbitrary time of grain mass separation.

On the diagram (Figure 2) and for the construction of the mathematical model, the following notations are adopted:

1. The oscillating part of the separator is marked as M_{os} kg. It includes: the weight of the sieve body - M_{body}, the mass of the sieves themselves - M_{sieve}, (it can be divided into masses of separate sieves of the separator), the mass of grain array on sieves - M_{gr} (similarly, it can be presented as the sum of the array masses on separate sieves). Accordingly $M_{os} = M_{body} + M_{sieve} + M_{gr}$;

2. I - the moment of inertia of the oscillating part of the grain cleaning separator relative to the axis that crosses through its center of mass (point c);

3. D_1 - the center of mass of the rotating unbalanced mass of imbalance M_D. Its distance to the axis of rotation $O_1D_1 = r$ - radius of imbalance. Point O_1 - the actual axis of rotation of the unbalanced shaft. The model provides that the unbalanced unit of the separator drive can be located at an arbitrary point of the oscillating movement of the separator. $\alpha = \omega t$ - point D_1 rotation angle at an arbitrary moment in time relative to the initial position, ω - angular velocity of rotational movement of the unbalanced shaft;

4. Coordinate systems: XOY - motionless, $X_1O_1Y_1$ - the mobile system, tied immovably to the oscillating part of the separator. When the drive is turned off, these systems match;

5. φ - the angle of rotation of the grain cleaning separator during its oscillating motion (coordinate system $X_1O_1Y_1$ relative to motionless XOY);

6. The distance from the axis of rotation of the unbalanced shaft to the center of mass of the oscillating part of the separator is denoted by $s = cO_1$;

7. The total stiffness of the elastic suspension will be denoted as $C = C_1 + C_2 + C_3$. It consists, respectively, of the sum of stiffness of three pairs of springs. The springs in the complicated model of the separator can have an arbitrary position in the plane of its oscillations, which are determined by the coordinates of their ends, set by the user of the model;

8. FK_1, QM_1, WL_1 – the length of the suspension springs of the separator during its operation; FK, QM, WL - the length of the suspension springs when the separator is not working;

9. The distance from the attachment of the left spring to the axis OY - a; distance from the attachment of the right spring to the axis OY - k; w, p, b - the corresponding vertical coordinates of the suspension mounting and the geometric parameters of the body with sieves.

The mathematical model of the vibrating grain-cleaning separator will be a system of analytical expressions that will make it possible to describe the movement of the oscillating part of the grain-cleaning separator (sieve system) in the plane of the rotational movement of the drive depending on the parameters of the separator itself. Having the laws of motion of the oscillating part of the separator, it is possible to find the amplitude of its oscillations and investigate the efficiency of the separation process. The law of the oscillating part of the grain cleaning separator is expressed through the laws of the oscillating motion of its individual points. The oscillating part of the separator is absolutely rigid, therefore, knowing the law of motion of one (base) point of it depending on the parameters of the separator, we can find and use the law of motion of any other point of the movable part of the separator. For the convenience of modeling and based on the practical experience of developing mathematical models for studying the dynamics of vibration processing systems, we will take the center of rotation of the unbalance shaft as the base point – the point O_1 . This point can move (constructively) arbitrarily in the plane of vibration of the separator. Thus, the laws of changing the Cartesian coordinates of this point, depending on the parameters of the separator, will determine its movement. Another law of movement of the separator, which is necessary for studying its dynamics, will be the angle of rotation of the sieve system in the plane of oscillations relative to its original position. Let's find these laws and, accordingly, build a model for studying the dynamics of a vibrating grain cleaning separator with vertical layering of sieves. Building the model consists of several stages.

The Lagrangian of the oscillating part of the grain cleaning separator will have the form:

$$\begin{aligned}L = {} & \frac{\left(M_{body} + M_{sieve} + M_{gr}\right)}{2}\left(\left(\dot{x}_{o1} + S\cos\varphi\cdot\dot{\varphi}\right)^2 + \left(\dot{y}_{o1} + S\sin\varphi\cdot\dot{\varphi}\right)^2\right) + \frac{1}{2}I\dot{\varphi} + \frac{M_D}{2}\times \\ & \times\left(\left(\dot{x}_{o1} + r\cos(\omega t+\varphi)\cdot(\omega+\dot{\varphi})\right)^2 + \left(\dot{y}_{o1} + r\sin(\omega t+\varphi)\cdot(\omega+\dot{\varphi})\right)^2\right) - \\ & -\left(\frac{C_1}{2}\left(\left(x_{o1} - a\cos\varphi + b\sin\varphi + a\right)^2 + \left(y_{o1} - a\sin\varphi - b\cos\varphi - p\right)^2 - (b+p)^2\right) + \frac{C_2}{2}\times\right. \\ & \times\left(\left(x_{o1} + b\sin\varphi\right)^2 + \left(y_{o1} - b\cos\varphi + w\right)^2 - (w-b)^2\right) + \frac{C_3}{2}\times \\ & \times\left(\left(x_{o1} + k\cos\varphi + b\sin\varphi - k\right)^2 + \left(y_{o1} + k\sin\varphi - b\cos\varphi - p\right)^2 - (b+p)^2\right)\Big) + \\ & + \left(\left(M_{body} + M_{sieve} + M_{gr}\right)g\left(y_{o1} - S\cos\varphi\right) + M_D g\left(y_{o1} + r - r\cos(\omega t+\varphi)\right)\right).\end{aligned} \tag{1}$$

The following system will be the primary differential equation of the mathematical model of the oscillatory motion of the grain cleaning separator with vertical layering of sieves:

$$\begin{cases} \ddot{x}_{o1} + \dfrac{C}{M_{body} + M_{sieve} + M_{gr} + \mathrm{M_D}}x_{o1} = \varepsilon f_x(\varphi,\dot{\varphi},\ddot{\varphi},\omega t); \\ \ddot{y}_{o1} + \dfrac{C}{M_{body} + M_{sieve} + M_{gr} + \mathrm{M_D}}y_{o1} = \varepsilon f_y(\varphi,\dot{\varphi},\ddot{\varphi},\omega t); \\ \ddot{\varphi} + \dfrac{b_1}{a_1}\sin\varphi = \varepsilon' f_\varphi(\varphi,\dot{\varphi},\ddot{x}_c,\ddot{y}_c), \end{cases} \tag{2}$$

here $\varepsilon = \dfrac{1}{M_{body} + M_{sieve} + M_{gr} + M_D}$, $\varepsilon \ll 1$,

$\varepsilon' = a_1 = I + \left(M_{body} + M_{sieve} + M_{gr}\right)S^2 + M_D r^2 \approx \varepsilon$, $\sqrt{\dfrac{C}{M_{body} + M_{sieve} + M_{gr} + M_D}}$ – natural frequency of oscillations of the sieve system, $b_1 = C_1\left(a^2 - pb\right) + C_2 wb + C_3\left(k^2 - pb\right) - M_k gS$.

The solution of system (2) will be in the first approximation:

$$x_{01}(t) = x_0\sin\left(\sqrt{\frac{C}{M_{body} + M_{sieve} + M_{gr} + M_D}}t + \alpha_x\right) +$$

$$+\varepsilon\int_0^t [(k1^2(L_1\sin k1u+L_2\cos k1u)((M_{body}+M_{sieve}+M_{gr})S\cos(L_1\sin k1u+L_2\cos k1u)+$$

$$+M_Д r\cos(\omega\omega+L_1\sin k1u+L_2\cos k1u))+$$

$$+k^2(L_1\cos k1u-L_2\sin k1u)^2\times((Mкуз+Mсит+Mзер)S\times$$

$$\sin(L_1\sin k1u+L_2\cos k1u)+M_Д r\sin(\omega\omega+L_1\sin k1u+L_2\cos k1u))+M_Д r\times$$

$$\sin(\omega\omega+L_1\sin k1u+L_2\cos k1u)(\omega^2+2\omega k1(L_1\cos k1u-L_2\sin k1u))-b\times$$

$$\sin(L_1\sin k1u+L_2\cos k1u)(C_1+C_2+C_3)+(1-\cos(L_1\sin k1u+L_2\cos k1u))\times$$

$$(C_3k-C_1a)]\cdot\sin\left(\sqrt{\frac{C}{Mкуз+Mсит+Mзер+M_Д}}(t-u)\right)du; \quad (3)$$

$$y_{01}(t)=y_0\sin\left(\sqrt{\frac{C}{Mкуз+Mсит+Mзер+M_Д}}t+\alpha_y\right)+$$

$$+\varepsilon\int_0^t [(k1^2(L_1\sin k1u+L_2\cos k1u)((Mкуз+Mсит+Mзер)S\sin(L_1\sin k1u+L_2\cos k1u)+$$

$$+M_Д r\sin(\omega\omega+L_1\sin k1t+L_2\cos k1t))-k^2(L_1\cos k1t-L_2\sin k1t)^2(M_K S\times$$

$$\cos(L_1\sin k1t+L_2\cos k1t)+M_Д r\cos(\omega\omega+L_1\sin k1t+L_2\cos k1t))-M_Д r\times$$

$$\cos(\omega\omega+L_1\sin k1t+L_2\cos k1t)(\omega^2+2\omega k1(L_1\cos k1t-L_2\sin k1t))+b\times$$

$$\cos(L_1\sin k1t+L_2\cos k1t)(C_1+C_2+C_3)-(\sin(L_1\sin k1t+L_2\cos k1t))(C_3k-C_1a)+$$

$$+p(C_1+C_3)-C_2w+(Mкуз+Mсит+Mзер+M_Д)g]\times\sin\left(\sqrt{\frac{C}{Mкуз+Mсит+Mзер+M_Д}}(t-u)\right)du;$$

$$(t)=L_1\sin k1t+L_2\cos k1t.$$

The system of analytical expressions (3) will be a mathematical model of the vibrating grain-cleaning separator, which will make it possible to describe the movement of the oscillating part of the grain-cleaning separator (sieve system) in the plane of the rotary motion of the drive depending on the parameters of the separator itself.

The expressions that will make it possible to find the law of oscillatory motion and, accordingly, the amplitude of oscillations of any point of the oscillating part of the grain cleaning separator will be the following:

$$x_{i1}=x_{o1}+x_i\cos\ (t)-y_i\sin\ (t),$$
$$y_{i1}=y_{o1}+x_i\sin\ (t)+y_i\cos\ (t). \quad (4)$$

In the system of expressions (4) x_{i1} y_{i1} are the Cartesian coordinates of the separator point, the movement of which should be investigated relative to the system XOY (motionless); x_i , y_i are the Cartesian coordinates of the separator point, the movement of which should be investigated relative to the system $X_1O_1Y_1$ (moving system, which is rigidly connected to the sieve system of the separator).

Thus, system (3) makes it possible to determine the trajectory of the base point oscillating movement of the grain cleaning separator oscillating part, and system (4) makes it possible to move to any other point of the oscillating part.

Knowing the trajectory of movement of the points of the separator, it is possible to find the amplitude of their oscillations depending on the change in the kinematic, mass-inertial and geometric parameters of the separator both at the stage of its design and when determining the operating modes, because the amplitude and frequency of the oscillations of the separator are key factors in determining the efficiency of the separator.

The next step of this work is to study the influence of the geometric, kinematic, mass-inertial parameters of the separator and the time parameter on the efficiency of its operation. These studies were carried out on the basis of the use of the systems of analytical expressions obtained above (3, 4) and the CAE system of mathematical calculations MahtCad.

What exactly was studied? The subject of study is the influence of the described above parameters of the separator on the amplitude of its vertical oscillations as a factor in determining the effectiveness of the separator. Why? Because oscillations of the separator are created by a vibration drive, in this case, an unbalanced one.

Based on this, the investigated separator can be classified as one of the types of vibrating technological machines, the efficiency of which is determined by the dependence: $q = k \cdot A^2 \cdot \omega^3$, in which q is the factor of the efficiency of the vibrating technological machine, A (mm) - the magnitude of the oscillation amplitude of the vibrating technological machine's executive body, ω (s^{-1}) - oscillation frequency of the executive body, k - empirical coefficient, which depends on the type of machine and its parameters.

Let's analyze the components of the expression to determine efficiency. Coefficient k - empirical, established experimentally and determined by the type of vibration machine. It is not unified, therefore, for the consideration of a whole series of separators, in the aspect of studying their dynamics, its importance is insignificant. Frequency of oscillations of the executive body ω - is determined by the frequency of disturbance of vibration oscillations, which is set by the vibration drive, more precisely, it is the frequency of the drive, for example, the frequency of the rotational movement of the imbalance. That is, this indicator is an input parameter of one or another vibration technological machine (it is not affected by other parameters of the machine) and is set by the user when simulating the work process of this vibration technological machine.

Amplitude of oscillations A of the executive body (in this case, the sieve system) directly depends on the geometric, kinematic (including frequency of the drive oscillations ω), mass-inertial parameters of the investigated vibration machine and the time parameter.

Based on these considerations, the amplitude of oscillations of the executive body of one or another vibrating machine is a key factor in determining the efficiency of the machine and is the subject of research on the influence of the parameters of the vibrating technological machine (in this case, the grain cleaning separator).

But there is a certain nuance here. Executive bodies of vibrating technological machines can be in three types of movements: rectilinear-reversible (oscillating along a certain straight line), flat (oscillations in a plane), complex (oscillations in space). In this case, the grain cleaning separator has oscillations of the sieve system in the plane. That is, the amplitude of the sieve system oscillations can be divided into two components: Ay (vertical) and Ax (horizontal). With complex motion, we have three component amplitudes along all Cartesian axes of the coordinate system, with one-dimensional, respectively, one component.

Horizontal component Ax - directed parallel to the sieve system of the grain cleaning separator and is responsible for the intensity of transportation of the grain array along the sieve, the vertical component Ay is directed perpendicular to the sieves. It determines the level of mixing of the grain array on the sieve and the level of interaction of its parts (cultured grain and impurities) with the separating sieves. In other words, it determines the intensity of the separation process. Of course, it can be said that the intensity of movement of the grain array along the sieve also affects on the intensity of separation of the grain array by the sieve, however, from a practical point of view, the vertical level of interaction of the grain array with the sieve (its vertical movement), and, accordingly, the vertical component of the amplitude of oscillations Ay will be dominant in determining the efficiency of the separation process.

Summarizing the above, in this work a study of the influence of the parameters of the grain cleaning separator on the amplitude of vertical oscillations Ay of its sieve system was carried out.

Now let's determine which parameters influence the vertical amplitude of oscillations to be investigated. From the point of view of the complexity of changing the parameters of a real grain cleaning separator, they can be divided into: parameters that are difficult to change in the finished design of the separator; parameters that the design of the separator allows us to change.

The geometric dimensions of the separator are unambiguously assigned to the first group of parameters (see the calculation diagram and comments on it), to the second group, respectively, the parameters of the unbalance drive (the mass of the unbalance, its radius and the frequency of rotational motion), the parameters of the oscillating mass (the mass of the sieves, the mass of the body sieves i.e sieve fastening systems, the mass of the grain array on the sieves), suspension parameters i.e stiffness values of individual pairs of springs.

The following parameters of the grain cleaning separator were used for substitution in the model:

1. Parameters, the change of which is permissible during the exploitation of the finished design of the grain cleaning separator: r – the radius of imbalance varies from 0 to 0.1 m; ω = 100 s^{-1} – angular velocity of the unbalanced drive M_D – the weight of the imbalance varies from 1 to 4 kg; $M_{os} = M_{body} + M_{sieve} + M_{gr}$ – the oscillating part of the separator varies from 50 to 450 kg; $C = C_1 + C_2 + C_3$ – the total stiffness of the suspension springs varies from 10 to 50 kN/m; $s_{max} = cO_1 = 0.3m$ – the largest value of the distance from the axis of rotation of the unbalanced shaft to the center of mass of the oscillating part of the separator.

2. Parameters that are difficult (or unrealistic) to change during the operation of the finished design of the grain cleaning separator (see Figure 2): $a = 0.6m$ – distance from the attachment of the left spring to the axis OY; $k = 0.6m$ – distance from the attachment of the right spring to the axis OY; $p = 0.15m$, $b = 0.75m$, $w = 1.05m$ – the corresponding vertical coordinates of the suspension mounting and the geometric parameters of the body with sieves.

Having a set of parameters listed above, a study of their influence on the vertical amplitude of the oscillating motion of the grain-cleaning separator was conducted in various combinations of them, and a corresponding analysis was made. First of all, the dependence of the vertical coordinate y of the movement trajectory of the sieve system from the time parameter t was obtained. This was done for different values of the separator parameters listed above, and already from the maximum values of the vertical coordinate, the corresponding values of the vertical amplitude were determined depending on the time parameter.

3. Results

The obtained research results are displayed on the graphs (Figuren 3 - Figure. 8). In Figuren 3 the dependence of the vertical amplitude of oscillations of the grain cleaning separator on its oscillating mass at different imbalance radii is shown. From the obtained graph (Figure 3) it can be seen that the value of the vertical amplitude of oscillations of the grain cleaning separator is nonlinearly dependent on the oscillating mass and decreases with the growth of the latter. In particular, for a separator with an imbalance radius of 0.1 m and an angular speed of the drive of 100 s^{-1} with an increase in the oscillating mass of the grain cleaning separator from 50 to 300 kg (by 6 times), the amplitude of oscillations decreased from 8.1 to 4 mm (by 4.1 mm), that is, approximately by 2 times. As it was emphasized above, the oscillating mass includes the mass of the sieves, the mass of the body and the mass of the grain array on the sieves. The amount of the latter can be easily changed, while changing the value of the oscillating mass. It is also possible to add unbalance mass to the value of the oscillating mass, but with large values of the oscillating mass, the influence of the unbalance mass will be insignificant. A relatively small change in the amplitude of the separator vibrations with a significant change in the oscillating mass (ratio 1 to 3) can be explained, in this case, by the high stiffness of the suspension springs.

IOP Conf. Series: Materials Science and Engineering 1277 (2023) 012030 doi:10.1088/1757-899X/1277/1/012030

Also, the size of the amplitude is affected by the size of the imbalance radius. As it increases, the value of the amplitude also increases. In particular, with an increase of the radius from 0.02 m to 0.1 m i.e in 5 times for the oscillating mass of the separator of 200 kg, the amplitude of oscillations increased from 3.8 mm to 6.9 mm (in 1.8 times).

Let's analyze how the actual stiffness of the suspension springs affects the vertical amplitude of vibrations of the grain cleaning separator. This dependence is shown in Figure 4 for different values of the imbalance radius. It can be seen from the graph (Figure 4) that the value of the vertical amplitude of oscillations of the grain cleaning separator depends nonlinearly on the stiffness of the suspension and decreases with the growth of the latter. In particular, for a separator with an imbalance radius of 0.06 m and an angular speed of the drive of 100 s^{-1} with an increase in the stiffness of the suspension of the grain cleaning separator from 10 to 50 kN/m (by 5 times), the amplitude of oscillations decreased from 8.2 to 3 mm (by 5.2 mm), that is, by approximately 2,7 times. The stiffness of the suspension of the oscillating mass of the grain cleaning separator can be changed relatively easily by adjusting the number of springs in the suspension, or by installing springs with different levels of stiffness.

The next step of the research was to determine how the radius of imbalance directly affects the amplitude of vertical oscillations of the grain cleaning separator. This dependence is shown in Figure 5 at different angular speeds of rotation of the drive motor. As can be seen from the obtained dependence, an increase in the radius of imbalance leads the vertical amplitude of oscillations of the grain cleaning separator increases non-linearly. Thus, at an angular speed of the drive motor of 70 s^{-1}, with an increase in the radius of imbalance from 1 cm to 10 cm (by 10 times), the amplitude of oscillations increased from 0.4 to 6.2 mm, i.e. by 15.5 times.

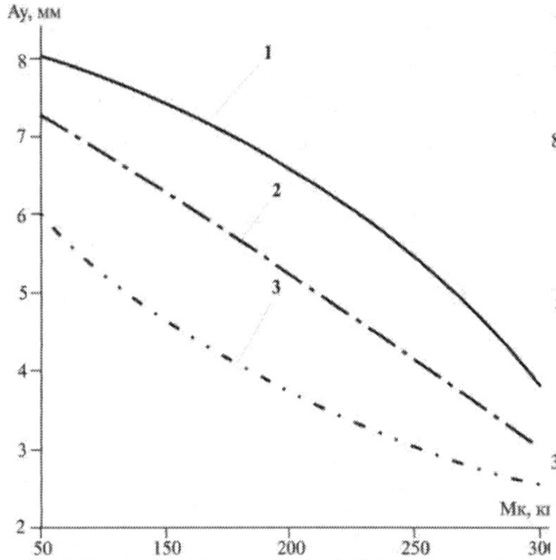

Figure 3. Dependence of the vertical amplitude of vibrations of the separator on its oscillating mass: 1 – radius of imbalance 0.1m, 2 – radius of imbalance 0.06m, 3 – radius of imbalance 0.02m.

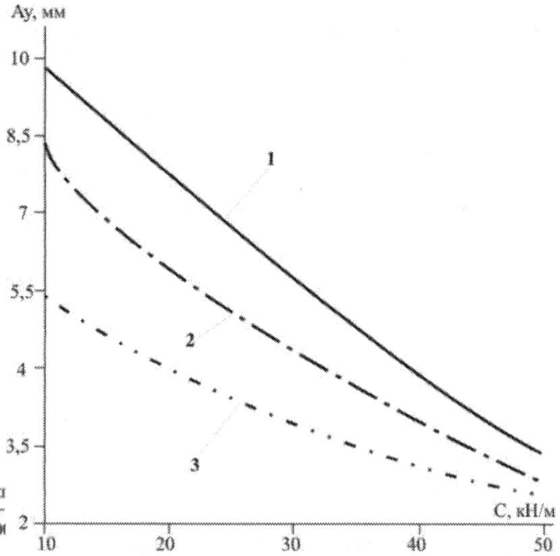

Figure 4. Dependence of the vertical amplitude of vibrations of the separator on the stiffness of the suspension: 1 – radius of imbalance 0.1m, 2 – radius of imbalance 0.06m, 3 – radius of imbalance 0.02m.

It is not difficult to change the radius of the imbalance during the operation of the separator, since it is constructively possible to foresee its assembled structure of individual masses that move (rotate) independently of each other around the axis of the imbalance shaft. That's why this parameter can be considered key when setting the required amplitude of oscillations of the grain cleaning separator. As can be seen from the graph shown in Figure 5, an increase in the angular speed of the drive of the grain cleaning separator leads the amplitude of vertical oscillations of its moving part decreases. More detail, the dependence of the amplitude of oscillations on the angular speed of the drive motor, the unbalanced shaft or the imbalance itself is shown in Figure 6. The graph is built for different stiffness values of the separator suspension. As can be seen from the obtained dependence, an increase in the angular speed of the drive leads the amplitude of the vibrations of the moving part of the separator decreases non-linearly. Thus, for a separator with a suspension spring stiffness of 40 kN/m with an increase in angular velocity from 80 s^{-1} to 160 s^{-1} (in 2 times), the amplitude of vertical oscillations decreased from 5.1 to 3.7 mm, that is by 1.38 times.

It is not difficult to change the radius of the imbalance during the operation of the separator, since it is constructively possible to foresee its assembled structure of individual masses that move (rotate) independently of each other around the axis of the imbalance shaft. That's why this parameter can be considered key when setting the required amplitude of oscillations of the grain cleaning separator. As can be seen from the graph shown in Figure 5, an increase in the angular speed of the drive of the grain cleaning separator leads the amplitude of vertical oscillations of its moving part decreases. More detail, the dependence of the amplitude of oscillations on the angular speed of the drive motor, the unbalanced shaft or the imbalance itself is shown in Figure 6. The graph is built for different stiffness values of the separator suspension. As can be seen from the obtained dependence, an increase in the angular speed of the drive leads the amplitude of the vibrations of the moving part of the separator decreases non-linearly. Thus, for a separator with a suspension spring stiffness of 40 kN/m with an increase in angular velocity from 80 s^{-1} to 160 s^{-1} (in 2 times), the amplitude of vertical oscillations decreased from 5.1 to 3.7 mm, that is by 1.38 times.

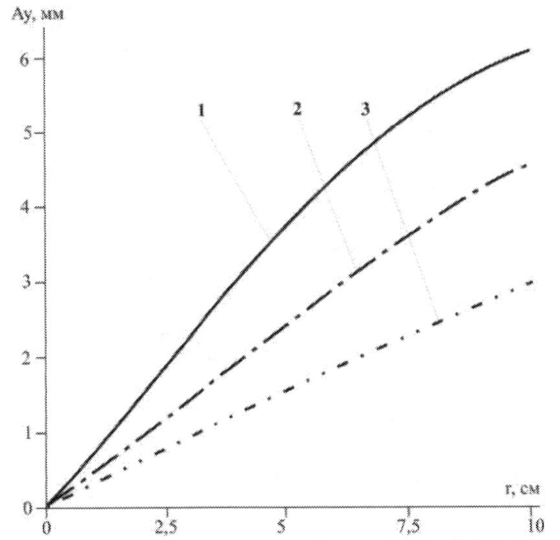

Figure 5. Dependence of the vertical amplitude of the separator oscillations on the imbalance radius: 1 – angular speed of rotational movement of the drive motor 70 s^{-1}, 2 – angular speed of rotational movement of the drive motor 110 s^{-1}, 3 – angular speed of rotational movement of the engine 150 s^{-1}.

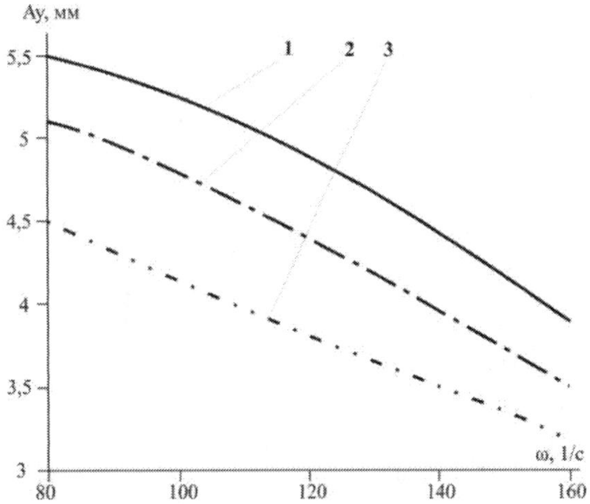

Figure 6. Dependence of the vertical amplitude of vibrations of the grain cleaning separator on the angular speed of the drive: 1 – suspension stiffness 30 kN/m, 2 – suspension stiffness 40 kN/m, 3 – suspension stiffness 50 kN/m.

A real available option for adjusting the amplitude of grain cleaning separator oscillations during its operation is the use of unbalance structures with the possibility of changing its unbalanced rotating mass M_D. This variant of the possibility of changing the amplitude of oscillations in terms of availability and speed takes the second place after changing the radius of imbalance. So, Figure 7 shows the dependence of the vertical amplitude of oscillations of the grain cleaning separator on the imbalance mass for different values of the oscillating mass of the separator M_{os}.

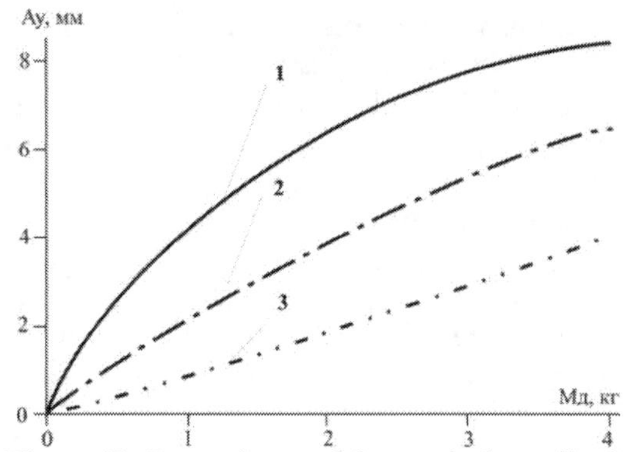

Figure 7. Dependence of the vertical amplitude of oscillations of the moving part of the grain cleaning separator on the imbalance mass M_D : 1 – oscillating mass 135 kg, 2 – oscillating mass 235 kg, 3 – oscillating mass 335 kg.

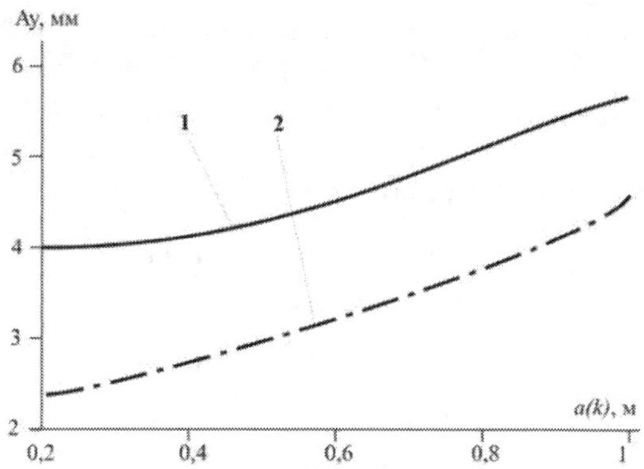

Figure 8. Dependence of the vertical amplitude of oscillations of the grain cleaning separator on half the distance between the supports of the outermost springs of the suspension: 1 – oscillating mass of 100 kg, 2 – oscillating mass of 175 kg.

As can be seen from the obtained dependence, the increase in the value of the unbalanced rotating mass, with other parameters constant leads the value of the vertical amplitude of oscillations of the grain cleaning separator increases non-linearly. Thus, for the oscillating moving mass of the separator of 235 kg with an increase in the imbalance mass from 1 to 4 kg (by three times), the amplitude of the oscillations of the moving part of the separator increased from 0.8 to 3.5 mm (by 4.4 times). It is obvious that the value of the oscillating mass of the moving part of the separator will influence the value of the amplitude and the range of its change. As the value of the oscillating mass increases, the value of the corresponding range of changes in the vertical amplitude of oscillations will decrease non-linearly.

The final stage of studies of the influence of the separator parameters on its performance indicators is the determination of the influence of the geometric parameters of the separator on the vertical amplitude of vibrations of its moving part. These geometric parameters are laid down in the design of the separator and it is very difficult to change them during its operation. In particular, Figure 8 shows the influence of the distance between the supports of the moving part of the separator a anad k to the vertical axis of symmetry of the separator OY, this is half the distance between the supports of the outermost springs of the suspension (see Figure 2).

From the obtained dependence, it follows that with an increase in the distance between the extreme supports of the separator suspension springs, the amplitude of the vertical oscillations of the grain cleaning separator increases non-linearly. Thus, with an increase in the indicated distance from 0.2 to 1 m - by 5 times, for the oscillating mass of the separator of 175 kg, the amplitude value increased from 2.4 to 4.5 times - by 1.9 times. This dependence should be taken into account at the separator design stage. If it is necessary to ensure larger amplitude of oscillations of its moving part, and thereby increase the performance of the separation, then the distance between the supports of the elastic suspension in the design of the separator should be increased as much as possible.

4. Conclusions

Thus, the subject of research in this work is a separator for cleaning and sorting grain crops after its harvesting. This separator has a system of vertically layered sieves, which are in an oscillating sign-changing motion, i.e. they are vibrated. The source of generation of this mode of movement is the unbalanced drive and placement of the screen system on an elastic suspension. Having chosen the type of separator for research, which is noted for its universality and reliability of operation, according to practical experience of use, a nonlinear, parameterized, unified model of its movement (study of its dynamics) was built. Further, on the basis of the obtained model, the influence of the key parameters of the separator on the efficiency of its operation was investigated and the corresponding analysis was carried out. The non-linearity of the model, with the use of appropriate mathematical apparatus and methods during its construction, made it possible to use it for a number of similar separators with different design features. Also, this model is easy to automate with the appropriate CAD-CAE system. Carrying out a complex of such research works is relevant from both a theoretical and an applied point of view and will have practical application.

The studies carried out in the work showed the qualitative and quantitative influence of the parameters of the grain cleaning separator with vertical layering of sieves and the efficiency of its operation. As can be seen from the obtained results, some of the parameters are quite significant based on the position of their influence on the amplitude of oscillations and, first of all, the possibility and complexity of their change during the operation of the finished design of the grain cleaning separator. These parameters include the radius of imbalance and its mass. They will be basic when setting this or that value of the amplitude of oscillations of the grain cleaning separator, depending on the type of grain array, which should be cleaned of impurities and divided into fractions during the required period of time of the separator.

References:

[1] Johnson E 2004 Low profile vibratory screen separators. Increasing capacity for tough materials *Powder and bulk engineering* **18(12)** 17–23.

[2] Singh R 2004 Vibratory separators still make the grade for screening dry bulk powders Oxford Eng. Elsevier Advanced Technology *Filtration and separation* **41(1)** 20–24 https://doi.org/10.1016/S0015-1882(04)00107-7

[3] Hosking J and Shrgold F 1962 *The effect of some operating conditions on the efficiency of separation of a vibratory screen* London: Quarry Managers Journal 140

[4] Topilnytskyy V, Rebot D, Sokil M, Velyka O, Liaskovska S, Verkhola I et al 2017 Modeling the dynamics of vibratory separator of the drum type with concentric arrangement of sieves *Eastern–European Journal of Enterprise Technologies Applied Mechanics* **2.7(86)** 26–35 https://doi.org/10.15587/1729-4061.2017.95615

[5] Stotsko Z A, Topilnytskyi V G, Kusyi Y M and Rebot D P 2018 Mathematical model for studying the dynamics of a vibrating separator with sequential placement of sieves *Vibration in engineering and technology* **2(89)** 49–57

[6] Lawinska K and Modrzewski R 2017 Analysis of sieve holes blocking in a vibrating screen and a rotary and drum screen *Journal Physicochemical Problems of Mineral Processing* **53** 812–828.

[7] Zhou N 2015 Dynamic characteristics analysis and optimization for lateral plates of the vibration screen *Journal of Vibroengineering* **17(4)** 1593–1604.

[8] Subach A P 1991 *Dynamics of processes and machines of volumetric processing: textbook* Zinatne Riga

[9] Ivanov K and Vaisberg L 2015 New Modelling and Calculation Methods for Vibrating Screens and Separators *Journal Lecture Notes in Control and Information Sciences* **22** 55–61 https://doi.org/10.1007/978-3-319-15684-2_8

[10] Mitropolskii Yu A 1997 *Nonlinear Mechanics Single-Frequency Oscillations: Textbook* Institute of . mathematic NAS of Ukraine, Kyiv

[11] Stotsko Z, Sokil B, Topilnytskyj V and Rebot D 2013 The influence of the loose medium parameters on the process of vibratory separation *Journal of Manufacturing and Industrial Engineering,* **12(1–2)** 17–19 https://doi.org/10.12776/mie.v12i1-2.179

[12] Topilnytskyy V G, Stotsko Z A, Kysyj J M and Rebot D P 2014 Investigation of the dynamics of vibratory separator with unbalanced drive *Journal Automation of manufacturing processes in machine tooling and instrument making* **786** 53–61.

Analysis of exploration parameters of industrial car jack used in car services in Kazakhstan

Saltanat Nurkusheva[1], Michał Bembenek[2]

[1] S.Seifullin Kazakh Agrotechnical University, Kazakhstan, The Department of Transport Equipment and Technologies, Zhenis 62B, Nur-Sultan, Kazakhstan, saltanat.nurkusheva@kazatu.kz
[2] AGH University of Science and Technology, Faculty of Mechanical Engineering and Robotics, A. Mickiewicza 30, 30-059 Krakow, bembenek@agh.edu.pl

Keywords: car jack, lifting equipment, elevator at the car service, equipment for cars, classification of car jacks.

Abstract. This article deals with the problem of the effective use of car jacks for passenger cars. The aim of the study is to create a theoretical basis for improving the efficiency and safety of the application of car jacks in car service enterprises. Based on the analysis of the review of existing designs, which are used in the technological process of car service, the universal classification of the car jack is modernized. As a result of the study, a comparison of technical indicators of the types of car jacks used in car service centers in Kazakhstan has been carried out. We propose an optimal range of load capacity. The results of the work, when implemented in the practice of design, will provide a reasonable reduction in the present costs of equipping the lifting equipment in car service stations and the operation of car jacks at the expense of more advanced methods of calculation and design.

1. Introduction

An increase in the number of passenger cars, the complexity of their design, and the presentation of stringent requirements for the technical condition of the car cause the need to improve the system of car maintenance [1-3]. The main structure of the service system is a car service center with its hierarchy of enterprises: service stations, car centers, repair plants, and workshops that perform rather complex repairs for cars. The development of quality auto maintenance is possible under the condition of improving the efficiency of workers, which is largely facilitated by the full mechanization and automation of technological processes [4]. Significantly, the high volume of supply and demand in the market of car maintenance, repair, and service, united in the work by the single term "automobile maintenance", defines high competition in this field [5,6].

The use of lifting equipment in auto maintenance facilitates access to the units and parts of motor vehicles, as well as reduces the labor intensity of maintenance works during their implementation, which leads to increased labor productivity of service employees. The efficiency of the use of elevators for vehicles at service stations depends primarily on the adaptability and their design to carry out maintenance work on them in specific technological areas, as well as indicators of the degree and rationality of their use depending on the time. The higher adaptability of an elevator design to carry out technical influences on the car, the less labor intensity and cost of their performance, and the higher productivity of a post, a site zone, and service stations as a whole.

The main characteristics of car jacks: lifting capacity, lifting speed and height, ground clearance, and type of construction. Lifting capacity characteristics depend on the weight of the vehicle, and the equipment must safely lift vehicles to a certain height. The allowed weight for the repair and maintenance of cars and small trucks is 1.5 to 5 tons. In most cases, the lifting speed is from 1 to 2 minutes. Electromechanical equipment has a higher lifting speed of the machine. The height of the elevators is generally not more than 2 meters. The total height does not exceed 4 meters. In terms of lifting height, the units for tire shops can lift the machine to a small height. The clearance of the machine elevators is the minimum height when the legs are lowered. Often, for many types of equipment, this parameter is in the range from 95 to 110 millimeters. If low-profile mechanisms, the size of the clearance ranges from 75 to 90 millimeters. Such systems are relevant for cars with low clearance or body kit. According to the type of drive, equipment is divided into: electromechanical, hydraulic, electro-hydraulic, pneumo-hydraulic, and pneumatic. According to the way the support elements are made, one can distinguish rack and plunger lift, as well as scissor units. Each type of equipment has its own design features [1]. Analysis and review of existing designs showed that the most popular are electro-hydraulic and -mechanical car jacks. The predominant majority of car jacks at large car service centers are: two-column, four-column, and scissor jacks [7-15].

Experience in using the car jacks at service stations showed undeniable advantages in comparison with the inspection trenches, as they take up little space, provide better working conditions, and maximum accessibility to the serviced units and assemblies of vehicles from the bottom and side. There is a problem with safety in the operation of the jack at car service stations in Kazakhstan. Often mechanics lift cars that have a lifting capacity greater than the car jack's ability to support the weight. For an increase in high profits, the same car jack can lift a car over a dozen times a day. So one of the important things is to create a safe working environment in car repair shops. When selecting lifting equipment to perform maintenance and current repair works, only the design or operational indicators of the elevator are analyzed. And the economic performance indicators are considered separately, not in conjunction with other indicators, and also do not take into account the measurement of technological equipment parameters such as wear and tear and aging during operation, as well as the associated increase in operating costs.

2. Methodology

To determine the dependence of lifting capacity and lift height on time, the corresponding calculations were carried out based on the methods of mechanics, analytical and numerical methods of analysis are used. Excel 2010 was used for data processing and solving mathematical problems [16-18].

The subject of the study is the operational characteristics of the car jack. The object of the study are the car jacks of Heshbon, Norberg, Oma, Ravaglioli, Rotary jack, Sivik, and Trommelberg, which are used in Nur-Sultan. All technical parameters are taken from the catalog of jack passports. The main objectives of the study are: analysis of the use of car jacks at auto service enterprises; a collection of technical data on the car jack used at auto service enterprises of Nur-Sultan; development of methods for calculating the parameters of lifting equipment. In the study of literary sources, we offer a generalized classification, **Figure 1** is based on existing classifications of car jacks [1].

3. Results

The problem of selection and placement of the jack at service stations is quite complicated since its solution requires taking into account many different factors. However, the solution to these issues will not only reduce capital investments in the development of the material base of service enterprises, but also reduce the cost of maintenance and repair work, as well as reduce transport costs.

Consider solving this problem using the example of the city of Nur-Sultan. At service stations for passenger cars, different types of foreign-made car jacks are used.

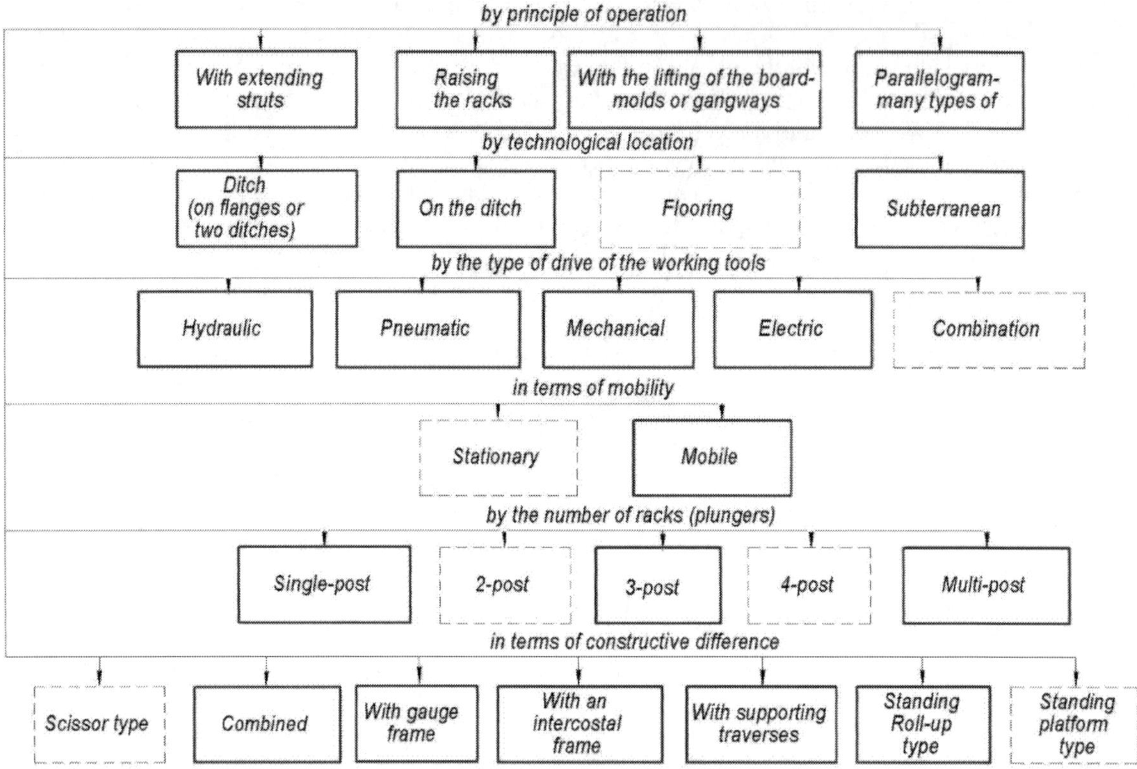

Figure 1. Classification of car jacks.

Technical characteristics of the jacks are given in **table 1** below [19-32].

Table 1. Technical characteristics of the car jacks

	Brand	Model	Capability, kg	Lifting time, sec	Lowering time, sec	Max height, mm
1	Heshbon	HL-51G	3500	60-70	30-100	1780
2	Heshbon	HL-27M	6000	50-60	30-40	1800
3	Nordberg	N412A-4T	4000	45	22	1800
4	Nordberg	N631L-3	3000	55	22	1850
5	Nordberg	4122A-4T	4000	52	22	1900
6	Nordberg	4455	5500	45	30	1750
7	Oma	526B-450AT	4000	50	45	1750
8	Ravaglioli	RAL	4500	35	35	2000
9	Ravaglioli	RAV 4655L	6500	75	60	2012
10	Ravaglioli	RAV 4505L	5000	43	30	2050
11	Ravaglioli	KPS244HEK	3200	50	50	2040
12	Rotary lift	ATO77	3500	45	40	1880
13	Rotary lift	SPOA10	4500	25	19	2195
14	Sivik	ПГА-5000	5000	60	60	1858
15	Trommelberg	tst45sw	4500	50	40	1850
16	Sivik	ПГА-3500-E	3500	50	75	1750

The graph **figure 2-3** show a clearer view of the lifting capacities and lifting times of different car lifts.

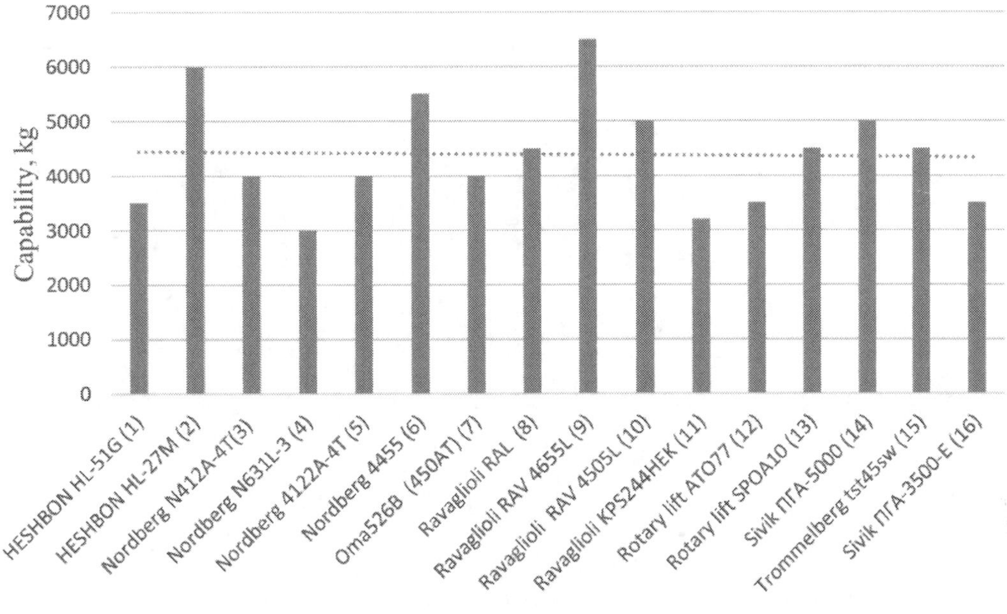

Figure 2. Lifting capacities of analyzed car jacks.

Based on the statistical analysis data obtained, the average value payload was 4387.5kg, the variance was 952343.75, the standard deviation was 1007.89, and the coefficient of variation was 22.97.

The Ravaglioli RAV 4655L brand had the highest increase in relative lifting capacity among the car jack, which was 6,500 kg, and the Nordberg N631L-3 brand had the lowest increase of 3,000 kg. The brands Nordberg N412A-4T (4000kg) and 4122A-4T (4000kg), Ravaglioli RAL (4500kg), Oma526B (4000kg), Rotary lift SPOA10 (4500kg), Trommelberg tst45sw (4500kg) were close to average values.

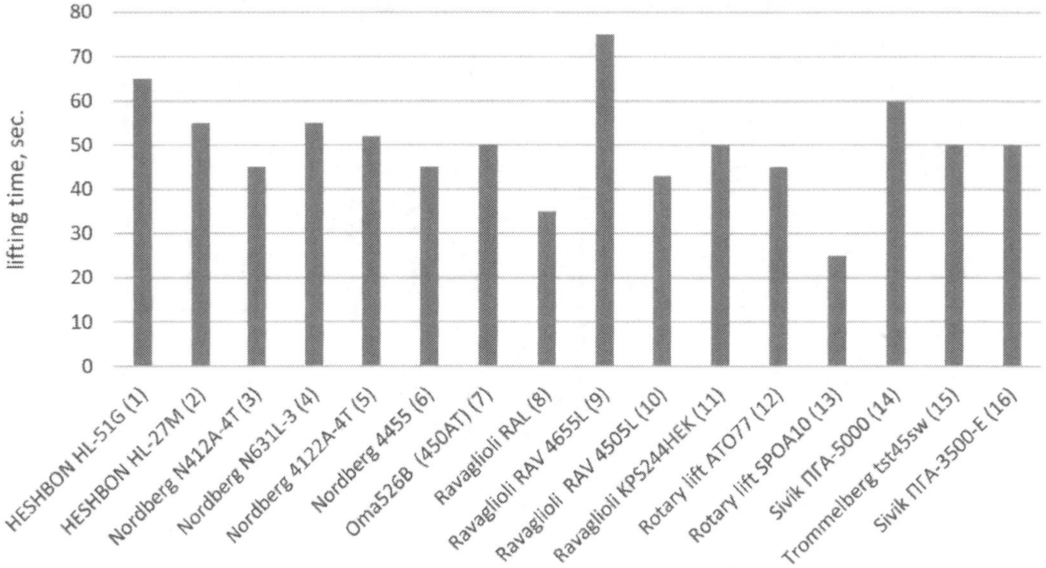

Figure 3. Lifting time of analyzed car jacks.

Among the car jack models under consideration there is a variety of values of lifting capacities (from 3000kg to 6500kg) and lifting speeds (from 25sec to 75sec). Based on the statistical analysis data obtained, the average value lifting time was 50sec, the variance was 123.63, the standard deviation was 11.48, and the coefficient of variation was 22.97. The Ravaglioli RAV 4655L had the highest lift time, which was 75 seconds, and the Rotary lift SPOA10 had the lowest, which was 25 seconds. The average value was 50 seconds, among them brands Oma526B and Ravaglioli KPS244HEK, Trommelberg tst45sw, and Sivik PGA-3500-E. The brands HESHBON HL-51G (65sec.), Sivik PGA-5000 (60sec.), HESHBON HL-27M, and Nordberg N631L-3 (55sec.), Nordberg 4122A-4T (52sec) were above average. Below average values were for the brands Nordberg N412A-4T and 4455, Rotary lift ATO77 (45sec.), Ravaglioli RAV 4505L (43sec.) Ravaglioli RAL (35sec.).

Sampling distribution functions or histograms play an important role in estimating the degree of dispersion of random variables. The Excel program calculated the number of hits in each of the intervals, the integral probability distribution function in %, which is shown in **Figure 4**.

The entire range of load capacity values is divided into intervals equal to each other. The entire sample is determined by the number of hits of the values in each interval, for which the relative frequency is calculated - the ratio of the number of hits in the interval to the total number of data. In our case, there are 7 intervals equal to a step of 500 kg (from 3000kg to 3500kg, from 3500kg to 4000kg, from 4000kg to 4500kg, etc.).

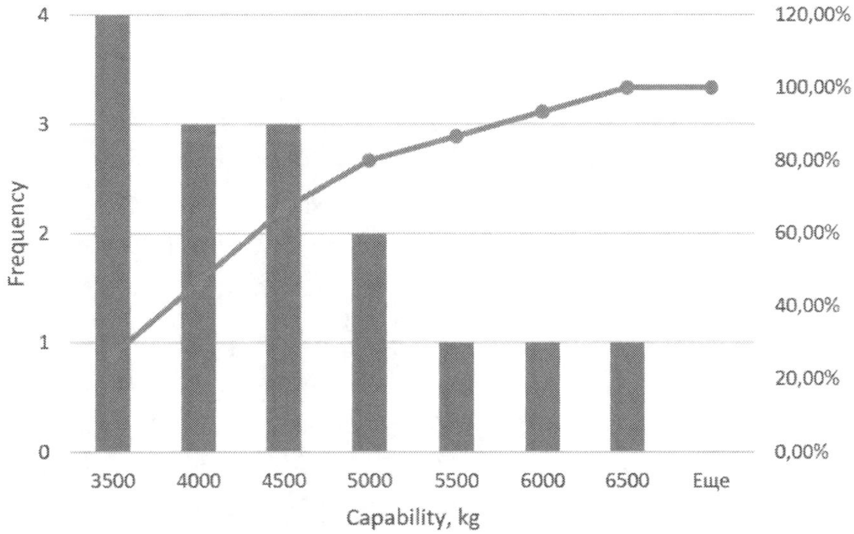

Figure 4. Histogram and integral curve of load capacity distribution.

In probability theory, there are "goodness-of-fit criteria" for assessing the validity of a distribution law fitted to the data. One of the main criteria of agreement is the Pearson criterion χ^2 (chi-square), which in this case was 0.77.

Most jacks operate automatically. Changing the lifting speed of the vehicles is done using automatic devices. These devices are included in the engine control system, according to a pre-calculated tachogram, which is repeated every cycle of the jack. As a consequence, the automatic operation of the jacks is carried out according to a planned program [33].

The lifting speed is a function of time

$$\vartheta = f(t) \tag{1}$$

We determine the value, which depends on the lifting capacity, lifting height, and lifting time. Taking into account the three factors, the mathematical model of the problem is as follows:

$$Z = \frac{Q \cdot H}{t} \to \min. \tag{2}$$

The lifting capacity of the hoist should not be less than 3000 kg $Q \geq 3000$

The lifting speed must be safe when the elevator is in operation $\vartheta = \frac{H}{t} \to A$.

As a result of the calculations according to the above, we obtain the graph in **Figure 5.**

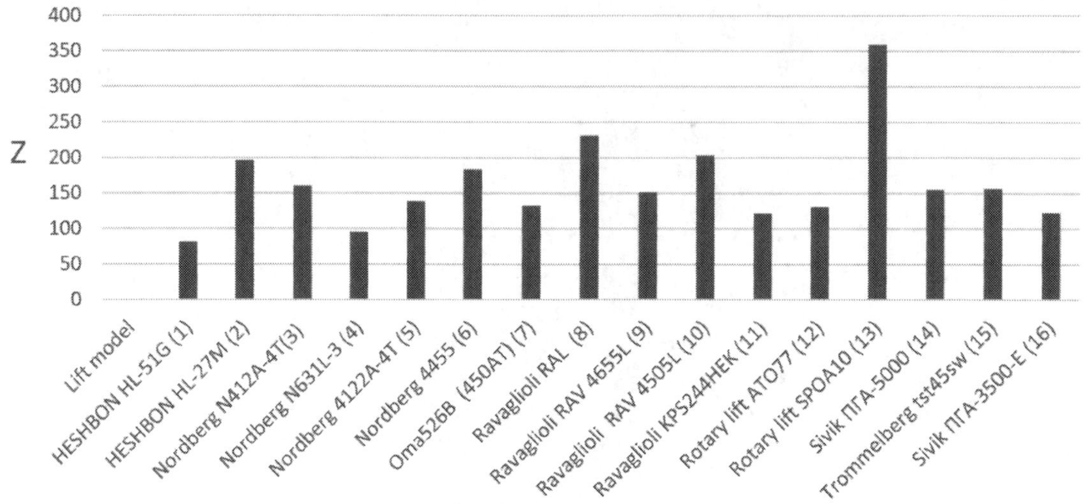

Figure 5. Z-value of analyzed car jacks.

The Rotary lift SPOA10 had the highest Z-value of 359.1kg·m/sec and the HESHBON HL-51G had the lowest of 81.61kg·m/sec. The average was 163.71kg·m/sec, among them the closest brand Nordberg N412A-4T, which was 1601kg·m/sec. Relative to the maximum value, the Z value for the model HESHBON HL-27M was 54.7%, Nordberg 4455 - 51%, Ravaglioli RAL - 64.3%, Ravaglioli RAV 4505L - 56.7%. Relative to the average value of the Z value for the model Nordberg N631L-3 was 58.1%, the model Nordberg 4122A-4T - 84.6%, the model Oma526B - 80.9%, Ravaglioli RAV 4655L - 92, 6% of the model Ravaglioli KPS244HEK - 74,5% of the model, the model Rotary lift ATO77 - 79.8% of the model Sivik PGA-5000 - 94.6% of the model Trommelberg tst45sw - 95.6% of the model Sivik PGA-3500-E - 74.8%.

4. Conclusions

The relevance of the work performed is presented. Based on the results of the literature study and studies, the classification of car jacks has been developed, which allows not only to critically assess the existing car jack, but also makes it possible to structurally supplement the proposed classification to modernize and create new types of car jacks.

Statistical analysis was performed based on the technical characteristics of the car jack. Analyzing the technical characteristics of a car jack can determine the best option for three factors, namely lifting capacity, time and height of lifting.

Results of statistical analysis: the mean value of the payload was 4387.5kg, the variance was 952343.75, the standard deviation was 1007.89, the coefficient of variation was 22.97, the average value lifting time was 50sec., the variance was 123.63, the standard deviation was 11.48, and the coefficient of variation was 22.97. Z-value less than 100 was obtained in two jack models (1 and 4), from 100 to 150 - five jack (5, 7, 11, 12, 16), from 150 to 200 - six jacks (2, 3, 6, 9, 14, 15), over 200 - in three models (8, 10, 13).

The purpose of the further study is a detailed analysis of the car jack design for strength and stability.

5. References

[1] Figenbaum, E., Fearnley, N., Pfaffenbichler, P., Hjorthol, R., Kolbenstvedt, M., Jellinek, R., Emmerling, B., Bonnema, G. M., Ramjerdi, F., Vagane, L. and Iversen, L. M. (2015) 'Increasing the competitiveness of e-vehicles in Europe', *European Transport Research Review*, 7(3). doi:10.1007/s12544-015-0177-1

[2] Wang, C. F. and Wang, F. L. (2013a) 'The Design of the Dual Power mechanical simple garage lifting platform', *Advanced Research on Mechanical Engineering, Industry and Manufacturing Engineering Iii*, 345, pp. 54-+. doi:10.4028/www.scientific.net/amm.345.54

[3] Xing, X. and Lee, G. (2017) 'The Design of Goods Transport and Boosting-device Equipment', *2017 4th International Conference on Manufacturing Engineering and Technology For Manufacturing Growth (Metmg 2017)*, pp. 109-114.

[4] Kostyuchenkova ON, Nurkusheva SA, Review of the use of lifting and transport equipment at a car service, *Publishing house: Kurgan State Agricultural Academy named after TS Maltseva*, Lesnikovo, 2020. - 686-689 p.

[5] Howell, J., Windsor, S. and Passmore, M. (2021) 'Some Observations on Shape Factors Influencing Aerodynamic Lift on Passenger Cars', *Fluids*, 6(1), p. 16. https://doi.org/10.3390/fluids6010044

[6] Petru, M., Sevcik, L., Masin, I., Lepsik, P., Lufinka, A. and Novak, O. (2014) 'Dynamic Analysis of Lifting Platform Construction for Car Relocation', *Modern Methods of Construction Design*, pp. 517-524.

[7] Kostyuchenkova O.N., Nurkusheva S.A., To the question of justification of classifications of motor transport enterprises and technological equipment of car service enterprises, *Publishing house: Kurgan State Agricultural Academy named after TS Maltseva*, Lesnikovo, 2021 .- 371-374 p.

[8] Napolskiy G.M., Zenchenko V.A. Justification of the demand for car service services and technological calculation of the passenger car service station. - *M .: Publishing house "Moscow Automobile and Highway State University (MADI)"*, 2000. - 83p.

[9] Balabaev O.T., Sarzhanov D.K., Abishev K.K., Kostyuchenkova O.N., Kozhukhova M.V., Improvement of the lift balcony type for vehicles, *Republican scientific-theoretical conference "SEYFULLIN READINGS - 12: Youth in Science - the Innovative Potential of the Future "S. Seifullin KazATU*, Nur-Sultan, 2016.

[10] Alexandris, N. A., Barney, G. C. and Harris, C. J. (1979) Multi-car lift system-analysis and design, *Applied Mathematical Modelling*, 3(4), pp. 269-274. doi:10.1016/s0307-904x(79)80057-8

[11] Burian, M., Havlik, J., Folta, Z., Trochta, M. and Marsalek, P. (2014) 'Solution the Drive Lifting Scissor Platforms', *Modern Methods of Construction Design*, pp. 19-26. doi:10.1007/978-3-319-05203-8_3

[12] Takesue, N., Komoda, Y., Murayama, H., Fujiwara, K. and Fujimoto, H. (2016) 'Scissor lift with real-time self-adjustment ability based on variable gravity compensation mechanism', *Advanced Robotics*, 30(15), pp. 1014-1026.

[13] Tao, L., Jian, S. and Ieee (2009) 'Simulative Calculation and Optimal Design of Scissor Lifting Mechanism', *Ccdc 2009: 21st Chinese Control and Decision Conference, Vols 1-6, Proceedings*, pp. 2079-+. doi:10.1109/ccdc.2009.5192393

[14] Zhang, W., Wang, X. X. and Wang, L. W. (2012) 'Strength Optimization Design of Box-shape Arms of a Scissor Lift Mechanism with Single Hydraulic Cylinder', *Functional Manufacturing and Mechanical Dynamics Ii*, 141, pp. 513-+. doi:10.4028/www.scientific.net/amm.141.513

[15] Zhang, W., Zhang, X. F., Yan, C., Xiang, S. J. and Wang, L. W. (2015) 'A characteristic triangle method on input vectors of scissor lift mechanism and its applications in modeling and analysis', *Journal of Advanced Mechanical Design Systems and Manufacturing*, 9(3). doi:10.1299/jamdsm.2015jamdsm0042

[16] Gong, J., Zhang, D. Q., Guo, Y., Tang, Z. Y., Liu, C. S., Hu, P., Zhao, Y. M., Quan, W. C. and Jin, Y. P. (2020) 'Potential energy recovery method based on alternate recovery and utilization of multiple hydraulic cylinders', *Automation in Construction*, 112. doi:10.1016/j.autcon.2020.103105

[17] Ma, Y. L., Lin, L., & Liu, X. W. (2011). Kinematics and Dynamics Analysis of the Lifting System of a Welding and Tube-Transferring Car for Deepwater Pipe-Laying Vessels. *Advanced Materials Research*, 199-200, 32–40. doi:10.4028/www.scientific.net/amr.199-200.32.
[18] Kostyuchenkova O.N., Valge A.M., The use of Excel and MathCAD when conducting research on the mechanization of agricultural production. *Training manual*. Astana 2017, 168 p.
[19] https://en.heshbon.com/index.php?mid=Down_1&page=2&document_srl=1967
[20] https://en.heshbon.com/index.php?mid=Down_1&page=2&document_srl=1951
[21] https://nordberg.ru/catalog/product/?cat_id=844
[22] https://nordberg.ru/catalog/product/?cat_id=849
[23] http://www.proalign.ru/images/pdf/manuBERG-4455.pdf
[24] https://www.mkslift.ru/catalog/product/1688/
[25] https://dm-astana.kz/p56653200-podemnik-dvuhstoechnyj-45t.html
[26] https://ravaglioli.com/wp-content/uploads/2019/03/RAV_4505L_4655L_02.pdf
[27] https://ravaglioli.com/wp-content/uploads/2017/02/RAV-PT-2C-KPN-306HE-04-RUS.pdf
[28] https://rotarylift.com/wp-content/uploads/2019/11/Rotary-ATO77.pdf
[29] https://rotarylift.com/product/spoa10/
[30] https://kz.sivik.ru/catalog/avtopodyemnoe_oborudovanie/dvukhstoechnye/pga_5000249/
[31] https://trommelberg.ru/fcatpic/man_TST45SW.pdf
[32] https://kz.sivik.ru/catalog/avtopodyemnoe_oborudovanie/dvukhstoechnye/pga_3500/
[33] Z.M. Fedorova, I.F. Lukin, A.P. Nesterov, Lifts. *Vyshcha Shkola Publishers' Association*, 1976, 296 p.

Comparative assessment of the efficiency of some car drives

Y Rudyk[1, 4], A Gavryliyk[1], V Kuts[2], V Yatsuk[2], S Vinogradov[3]

[1] Lviv State University of Life Safety
[2] National University Lviv Polytechnic
[3] National University of Civil Defence of Ukraine
[4] Author to whom any correspondence should be addressed
yurudra@gmail.com

Abstract. The fundamental basis for the scientific progress is the cheaper technology. Systems analysis and engineering, decision-making systems are based on data completeness and their validation. The study aims to provide the following grounds for a systematic comparison of economic and environmental performance of vehicles with different fuel drives. The issue of logistical attractiveness and compatibility for Ukraine is especially relevant today. Thus, specific fuel consumption is calculated by determining the cost of electricity that goes to charge the traction battery from the mains and is equivalently converted into the cost of fuel. The paper presents a method of calculating the equivalent fuel consumption for electric vehicles according to European (l per 100 km) and American standards (miles per 1 gallon). According to the results of fuel efficiency studies, it is established that the equivalent fuel consumption on the territory of Ukraine of the studied electric vehicles is 8-10 times lower than the fuel consumption of cars of the same class with internal combustion engines.

1. Introduction

The increase in the world cars fleet, which exceeded 1 billion units, caused the huge emissions of flue gas into the atmosphere, which greatly worsened the environmental situation.

The imposition of increasingly stringent environmental standards on internal combustion engines (ICEs) has led world engineering to seek alternatives to the use of ICEs on vehicles that have become electric motors.

The use of electric motors in vehicles requires an electricity source. Lithium-ion batteries are the most widely used because of a number of advantages, such as high power capacity, specific power, and an enough resource compared to lead-acid, nickel-cadmium or sodium-metal-chloride batteries. At the same time, this type of energy elements is capable of igniting or even exploding in mechanical damage or recharging [1, 10].

2. Method of calculating the equivalent fuel consumption development

The undoubted advantage of electric vehicles (EV) is not only their environmental friendliness, but also their cost-effectiveness. The constant increase in the cost of petroleum products outlines the reasons why consumers choose low-fuel-efficient cars without reducing technical performance. For electric vehicles, the economic component can be described by the equivalent fuel savings.

The equivalent fuel savings will be different in different countries for one brand of electric vehicle. This is due to the different cost of electricity and fuel. Thus, calculating and estimating the equivalent fuel consumption of electric vehicles in different countries is an actual task. Methodological bases for estimating the fuel economy of electric vehicles have been developed. This will allow potential buyers, owners or economists of the trucking industry to objectively estimate the equivalent fuel consumption and to successfully select a particular brand of electric vehicle.

According to [4], as well as the most environmentally friendly electric vehicle (the lowest energy consumption per unit of distance) in 2018, Hyundai Ioniq Electric is equipped with an electric motor of 120 hp. (88 kW), which develops torque of 295 Nm, consuming combined (55% of highway, 45% of city) 25 kW / 100 miles or equivalent in the US fuel market is 136 MPGe, lithium energy capacity - ionic battery is 28 kWh. The claimed mixed cycle run on a single full charge of the battery under the EPA tests is 124 miles (198 km) [5,6].

The second place is occupied by Tesla Model 3 Long Range, equipped with an electric motor with a capacity of 283 hp. (211 kW) with a maximum torque of 510 Nm, which uses in the combined mode 26 kWh / 100 miles or 130 MPGe, the energy capacity of the lithium-ion battery is 75 kWh. The declared mileage in the mixed cycle on one full charge of the battery according to EPA tests is 325 miles (520 km) [7,8].

Closes the top three BMW i3, equipped with an electric motor with 168 hp. (125 kW) with a maximum torque of 250 Nm, which in combined mode uses 27 kWh / 100 miles or 124 MPGe, the energy capacity of the lithium-ion battery is 42 kWh. The declared mileage in the mixed cycle on one full charge of the battery according to EPA tests is 160 miles (256 km) [9-11].

According to the American Automobile Internet Resource [3], various ratings have been compiled, including the most saving electric vehicles, citing the latest trends in world automotive production. There is a difference between the European and American units. For convenience, we will hereafter indicate both dimensions of quantities. According to American standards, the mileage (1 miles = 1.609 km) that a car is capable of overcoming one gallon of fuel (1gal lig = 3.755 l) is taken per unit of fuel economy (MPG).

Instead, the economy of an electric vehicle is determined by the number of miles the vehicle can overcome using energy equivalent to that contained in a gallon of gasoline. This technique is used for plug-in hybrid electric vehicles (PHEVs) and alternative fuel vehicles (NGV-natural gas vehicles; FCV-fuel cell vehicles). That is, one gallon of gasoline equivalent indicates the number of kilowatt-hours of electricity, the volume of natural gas, or the mass of hydrogen, which is equivalent to the energy of a gallon of gasoline (1 MPGe = 33.7 kW · h = 121 MJ) at which the vehicle walks a distance in one mile. In vehicles that use two or more fuels (PHEV, NGV, FCV), they indicate the consumption of each fuel in gallons of gasoline equivalent.

3. Analysis the different between two measurement units systems

However, all vehicles in the lineup since 2013, by decision of the United States Environmental Protection Agency (EPA) and the National Highway Traffic Safety Administration (NHTSA), in addition to the equivalent fuel consumption, indicate the amount of energy required to overcome 100 miles of path.

The purpose of this work is to evaluate the efficiency of electric vehicles by developing methods for estimating their specific fuel consumption and the relative cost of 100 km mileage in different countries. The results can be used by potential consumers of electric vehicles, as well as economists of motor transport companies.

The aim of the study is to analyze the fuel economy of the above mentioned electric vehicles and determine the specific fuel consumption in different countries, taking into account electricity and fuel prices, drawing conclusions about the economical use of these electric vehicles in different countries.

Problem solving. Specific fuel consumption is calculated by determining the cost of electricity that goes to charge the traction battery from the mains and is equivalently converted into the cost of fuel.

The efficiency of transportation by car through the working process of a reciprocating internal combustion engine in terms of energy efficiency and highlighting such processes are considered. By combining the transformation of the physical process of evaporation (change in the aggregate state), the chemical process of combustion, and the physical and mechanical process of heat generation under variables in a wide range of temperatures and pressures, with a difficult-to-control variability of the working fluid state are obtained [13, 14]. Heat transfer inside the cylinder and through its walls to the cooling system is of great importance in the operation of the ICE. Extraction of heat from the working medium (gas) causes energy losses. Hence, the evaluation of the efficiency of devices with internal combustion engines as the degree of use of thermal energy determines the level of ecological impact on the environment. Internal combustion engines as part of vehicle drives, in electric energy cogeneration systems create a significant ecological pressure on the environment, burn a large volume of fuel and air with relatively low productivity [15, 16]. Among the various indicators of efficiency of internal combustion engines as fuel filling factors λpv [kg/kg)] , as air filling λpl [kg/kg)], average effective pressure pe [Pa], specific fuel consumption ge [g/(kW·h)], the effective efficiency factor ηe [J/J] are considered.

Determine the specific fuel consumption for the above mentioned electric vehicles. Taking into account the energy capacity of the battery of the electric car, as well as the efficiency of the cycle "charge-discharge" and the charger, the amount of electricity that will be consumed to fully charge the battery (subject to full discharge) will be:

$$Q = \frac{c}{\eta}; \quad (1)$$

where Q - the amount of energy consumed when charging the battery, kWh; η - total efficiency (taking into account the efficiency of the charger and the efficiency of the "charge-discharge" cycle), c - the energy capacity of the battery, kWh.

Given the maximum mileage on a fully charged battery and the cost of one kilowatt-hour of electricity, the expression for determining the specific value of 100 km of mileage will look like:

$$C_1 = \frac{c \cdot a}{\eta \cdot l} \cdot 100; \quad (2)$$

where C1 - the cost of electricity consumed to overcome 100 km of road; a - the cost of one kilowatt-hour of electricity; lkm - maximum mileage on a fully charged battery, km.

Then the equation for determining the specific fuel consumption will look like:

$$C_2 = \frac{c \cdot a}{\eta \cdot l \cdot b_l} \cdot 100; \quad (3)$$

where C2 - equivalent fuel consumption of the electric car l / 100km; bl - the cost of one liter of fuel.

Or in the case of calculating the specific fuel consumption in MGPe (miles / 1 gallon) the equation will look like:

$$C_3 = \frac{l_m \cdot b_{gal} \cdot \eta}{c \cdot a}; \quad (4)$$

where C3 is the equivalent fuel consumption of the electric vehicle in MPGe (miles / 1 gallon); bgal is the cost of one gallon of fuel; lm is the maximum mileage on a fully charged battery, miles.

It is clear that the specific fuel consumption and the specific cost of 100 km run directly depends on the cost of energy, which are different in different countries. Therefore, it is important to study these indicators.

4. Results and Discussion

Figure 1 shows the cost of a kilowatt-hour of electricity and liters of gasoline in certain countries as of the last quarter of 2019 in UAH.

According to the results of EPA research and calculations, we obtain in the form of a graphical dependence of the relative mileage for Hyundai Ioniq Electric, Tesla Model 3 Long Range and BMW i3 in different countries.

The analysis of graphic dependence shows that in Ukraine the relative cost of mileage for electric cars Hyundai Ioniq Electric is 26.35 UAH / 100 km, for Tesla Model 3 Long Range 27.28 UAH // 100 km and BMW i3 28.52 UAH / 100 km, which is the lowest among all European countries. However, the lowest relative cost of 100 km is observed in Kazakhstan, as electricity prices in this country are the lowest.

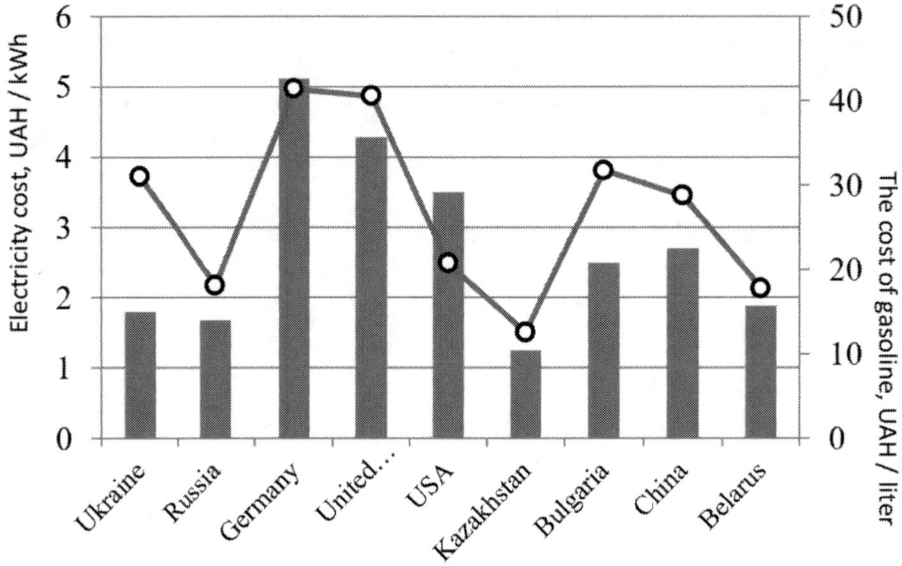

Figure 1 - The cost of a kilowatt-hour of electricity and gasoline in different countries as of 2019 in UAH

Instead, the highest equivalent fuel consumption of these vehicles in the United States, which is three times higher than the equivalent fuel consumption in Ukraine. At the same time, this creates preconditions for the expediency of using electric cars in Ukraine.

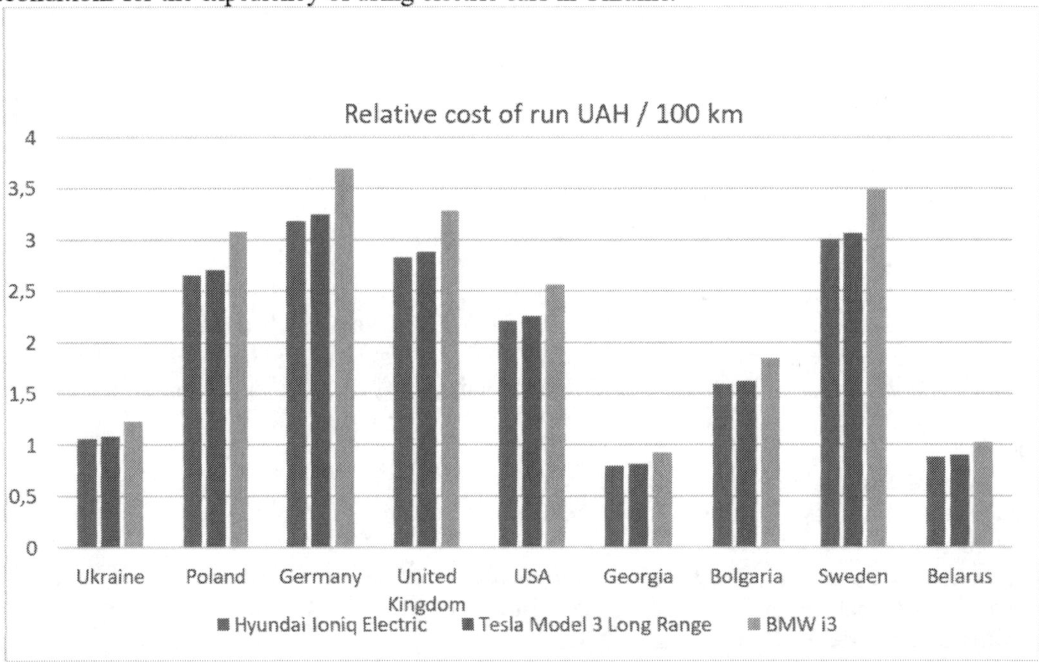

Figure 2 - Relative mileage of Hyundai Ioniq Electric, Tesla Model 3 Long Range and BMW i3 in different countries

Taking into account fuel prices, the specific fuel consumption of these electric vehicles is shown in Figure 3. The average cost of electricity in the United States as of August 2019 is $ 0.14 per 1 kWh, and the cost of a gallon of gasoline - $ 3.1. In Ukraine, the cost of electricity is 1.68 UAH / kWh, and gasoline A-95 - 31 UAH / liter. The results are shown in table 1.

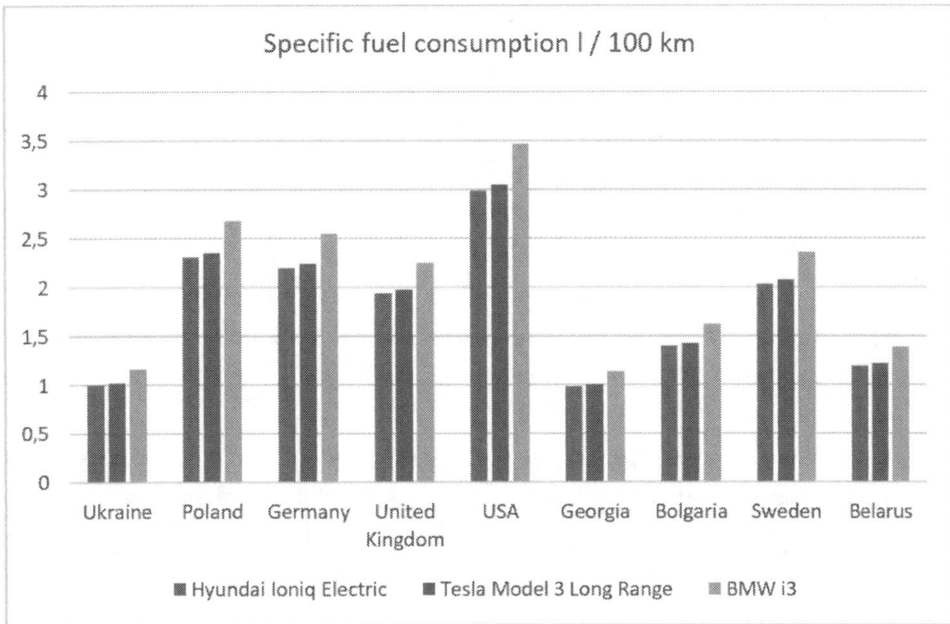

Figure 3 - Specific fuel consumption of Hyundai Ioniq Electric, Tesla Model 3 Long Range and BMW i3 in different countries

Taking into account the equivalent fuel consumption, it can be argued that in Ukraine the use of these electric vehicles is 8-10 times more economical than cars with internal combustion engines. Overall goal of any activity is gaining success, which often be estimated by economy values. Thus, fuel and costs saving are important not only for consumer and producer by direct impact in the price of production, including its modern very advanced electronic part, but also delayed effects on the environment and operational safety in general.

Table 1.Fuel economy of Hyundai Ioniq Electric, Tesla Model 3 Long Range and BMW i3 electric cars

Parameter	Dimensionality	Hyundai Ioniq Electric	Tesla Model 3 Long Range	BMW i3
Fuel economy (according to EPA)	MPGe (mpg)	136	130	124
Electricity consumption	kWh / 100 miles	25	26	27
Specific fuel consumption	l / 100 km	0,85	0,88	0,92
Relative mileage 100 km (mixed mode)	UAH / 100 km	26,25	27,3	28,35

5. Conclusion

The paper presents a method of calculating the equivalent fuel consumption for electric vehicles according to European (l / 100 km) and American standards (miles / 1 gallon).

It is established that the equivalent fuel consumption in different countries may vary and depends on the cost of energy.

It is investigated that in Ukraine the specific fuel consumption for electric cars Tesla Model 3 Long Range is 0.88 l / 100 km, for Hyundai Ioniq Electric 0.82 l / 100 km and for BMW i3 0.92 l / 100 km, which is the lowest among all European and other countries that have been studied, thus creating the preconditions for the development of electric vehicles in Ukraine.

According to the results of fuel efficiency studies, it is established that the equivalent fuel consumption on the territory of Ukraine of the studied electric vehicles is 8-10 times lower than the fuel consumption of cars of the same class with internal combustion engines.

References

[1] Gavrilyuk A and Lin S 2017 Protipozhezhnij zahist kolisnih transportnih zasobiv ta shlyahi jogo pidvishchennya *Pozhezhna bezpeka* **31** pp. 11-17

[2] Maxwell J. Clerk 1892A Treatise on Electricity and Magnetism, 3rd ed., vol. 2. Oxford: Clarendon, pp.68–73.

[3] Ramchandani M and Whitehouse N 1976 Heat transfer in a piston of a four stroke diesel engine *SAE Prepr.* **760007** 9 p.

[4] Kelley Blue Book [Online resource] / The Trusted Resource - Access Mode: https://www.kbb.com

[5] United States Environmental Protection Agency and U.S. Department of Energy (2016-11-18). "Compare Side-by-Side: 2017 Hyundai Ioniq Electric." fueleconomy.gov. Retrieved 2016-11-19.

[6] Szostech, M Hyundai IONIQ Electric Specifications. *My Electric Car Forums*. Retrieved 2016-11-25.

[7] Kane, Mark (2016-03-02). "Hyundai IONIQ Electric & IONIQ Plug-in At The Geneva Motor Show (Gallery, New Stats)." InsideEVs.com. Retrieved 2016-03-02.See more details in the official press release.

[8] Rteslamotors - FW 2019.8.3 actually increased Model 3 AWD peak power by 8% above 45 mph (70 km / h) ". Reddit. Retrieved July 14, 2019.

[9] Powell, Derek (May 22, 2019). "Tesla Model 3 vs. BMW 330I vs. Genesis G70 Comparison Test." MotorTrend. US Retrieved May 25, 2019.

[10] U. S. Environmental Protection Agency and U.S. Department of Energy (July 25, 2014). "Most Efficient EPA Certified Vehicles." fueleconomy.gov. Retrieved 13 June 2015.Current Model Year excludes all-electric vehicles

[11] Gavriluk A, Rudyk Y, Kuts V and Naumchuk R 2020 Required safety component of automotive cyber-physical systems PiCS&T IEEE Conference Proc.

[12] U. S. Environmental Protection Agency and U.S. Department of Energy (November 16, 2016). "Top Fuel Sippers (EPA Ratings, 2017 Model Year)." fueleconomy.gov. Retrieved November 21, 2016. The 2014–16 BMW i3 BEV (60 Amp-hour battery) was the most efficient EPA-certified vehicles considering all fuels and of all years until November 2016, when it was surpassed by the 2017 Hyundai Ioniq Electric

[13] Haschuk P and Nikipchuk S 2018 Peculiarities of heat generation in an internal combustion engine *Automotive Transport* **42** P. 12-21. DOI: 10.30977/AT.2219-8342.2018.42.0.12

[14] Decan G, Broekaert S, Lucchini T, D'Errico G, Vierendeels I and Verhelst S 2018 Evaluation of wall heat flux calculation methods for CFD simulations of an internal combustion engine under both motored and HCCI operation *Applied Energy* **232** P. 451-461. DOI: 10.1016/j.apenergy.2018.09.214

[15] Hashchuk P 1992 *Car energy efficiency* (Lviv: Svit) 208 p

[16] Yatsuk V and Klinkovska O 2017 Peculiarities of quality assurance in the European automotive industry. *8th International science and technology conf. in memory of Professor Ihor Kisil* November 14-15

Simulation of the Production and Transport Problem in the FlexSim Environment

O T Velyka[1], E V Martyn[2], S E Liaskovska[3]

[1] Department of Designing and Operation of Machines, Lviv Polytechnic National University, Ukraine
[2] Lviv State University of Life Safety, Ukraine
[3] Department of Artificial Intelligence Systems, Lviv Polytechnic National University, Ukraine
[1] E-mail: veloks@ukr.net

Abstract. The method of building a simulation model of the production and transportation problem is described. In the FlexSIM environment, a simulation model of a part of the production process was built and the impact of changing input parameters on output parameters and the behavior of the technological system as a whole was investigated and analyzed.

1. Introduction

Modern industrial productions are complex dynamic systems characterized by a high level of uncertainty of initial information and the complexity of their behavior. To analyze and solve many problems associated with the management of such systems, it is advisable to use simulation modeling. Simulation modeling in the research of technological processes is a research method in which the system under study is replaced by a model that describes the real system with sufficient accuracy. Computer studies are conducted with it to obtain information about this system [1-3]. Simulation modeling is the most universal method of system research and quantitative assessment of their functioning characteristics. Modern simulation modeling of production processes[4] allows you to reproduce the sequence of operations or stages of the production process by simulating its individual elements, as well as to determine the best type of production line with optimal performance, the ability to perform testing with a variety of production parameters, including flow capacity and quantity necessary equipment. This allows you to determine the type of production that will be most effective for the production of the product, and the factors that affect the final indicators of the quality of the product being created. Thanks to simulation modeling and conducting experiments with the use of the model, it is possible to achieve an increase in the efficiency of the use of equipment, optimization of production, reduction of production cycle time, increase in production volumes and improvement of product quality[5]. New developments and technologies in simulation modeling, namely their appearance, make it possible to increase the efficiency of the technological process and bring it to the required production level [6].

2. Materials and Method

The method of simulation modeling is widely used in production to solve various problems from optimization of intermediate processes to strategic management. Such modeling allows analyzing not only a specific process, but also the production system as a whole, which makes it possible to check the capital intensity of a particular management strategy. A typical experiment of a simulation model includes the following stages:
— construction of the created model;
— analysis of the constructed model;
— implementation of a simulation model and conducting research;

— analysis and evaluation of the obtained results.

In the process of simulation modeling, it is possible to reproduce not only the relationship between the objects of the system, but also to simulate the development of the system over time, which is very important for the enterprises of the machine-building industry. After all, the study of their industry structure requires a systemic approach and requires a comprehensive analysis of eco- nomic, statistical and other sources of information both at the level of a separate production process and at the level of the enterprise as a whole. The production and transport task is to ensure consumer demand with the lowest possible total costs. In addition to information about demand, not only transport tariffs for transporting a unit of products from suppliers to consumers and the production capacity of suppliers are taken into account, but also the selling prices of products from each of the suppliers. In a one-stage transport problem, products from suppliers go directly to consumers. On the contrary, in multi-stage transport tasks, products from suppliers first arrive at intermediate points (transport network nodes, distribution centers, warehouses), where, if necessary, they are reloaded or unloaded and stored for a certain time. That is, the final consumers receive products not from suppliers, but from the specified intermediate points of transport networks. There are two-stage transport tasks, when the transportation of products is carried out in two stages: first from the supplier to an intermediate point, then from this intermediate point to the consumer (**Figure 1**) and multi-stage - when there are several intermediate points on the way from the supplier to the consumer (**Figure 2**).

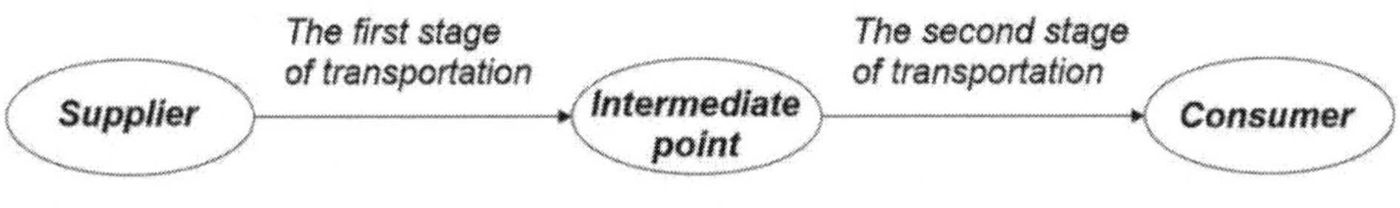

Figure 1. The structure of the two-stage transport problem

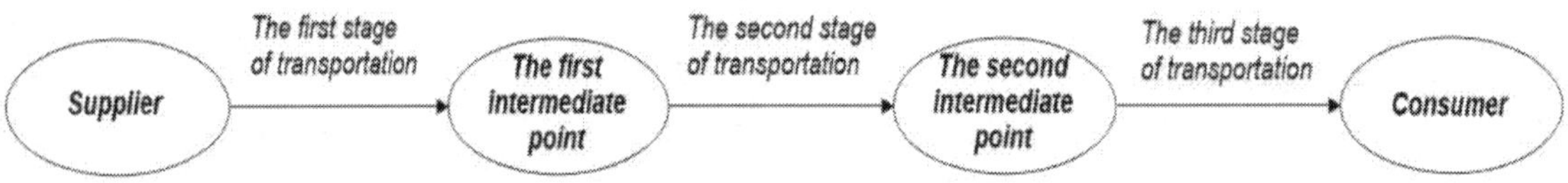

Figure 2. The structure of the multi-stage transport

The economic-mathematical model of the production-transport problem has the following form:

$$\begin{aligned} & z = \sum_{i=1}^{n} s_i y_i + \sum_{i=1}^{m} \sum_{j=1}^{n} c_{ij}\, x_{ij} \to min \\ & \sum_{i=1}^{m} x_{ij} \geq b_j \; (j = \overline{1,n}), \\ & 0 \leq y_i = \sum_{i=1}^{m} x_{ij} \leq N_i (i = \overline{1,m}), \\ & x_{ij} \geq 0 \; (i = \overline{1,m},\; j = \overline{1,n}) \end{aligned} \qquad (1)$$

where: m - the number of enterprises - producers of products (suppliers); i – individual supplier number ($i=1, m$); N_i– production capacity of the i-th producer;

s_i– sell-ing price of a product unit from the i-th supplier;

n - the number of consumers; – individual consumer number ($j=1,m$);

b_i – demand for products from the j-th consumer;

c_{ij} – costs for transportation of a unit of products along the route $i \to j$ $(i = \overline{1,m},\; j = \overline{1,n})$.

x_{ij} – the volume of transportation along the route $i \rightarrow j$ $(i = \overline{1,m},\ j = \overline{1,n})$.

y_i – volume of products manufactured at the i-th enterprise and sent to the consumer;

z – total costs of purchasing products from suppliers and transporting these products to consumers.

The objective function of the problem reflects the requirement to minimize the total costs of purchasing products and transporting them to consumers. The limitations of the task take into account the requirement to satisfy the demand of each of the consumers, the condition that the purchase volumes correspond to the levels of the production capacities of each of the suppliers, and the balance equations for the distribution of the products manufactured by each of the supplier enterprises among the consumers. Requirements on the non-negativity of production volumes and transportation of products are also taken into account. Mathematically, the model (1) of the production and transport problem is a problem of linear programming of the transport type. It can be solved by the simplex method. It is also possible to reduce this problem to a transport problem and solve it using the method of potentials. But first you should check whether the condition for the existence of a solution to this problem is met. The production and transportation problem has a solution if and only if the total production capacity of all producers corresponds to the aggregate demand for products from all consumers:

$$\sum_{i=1}^{m} N_i \geq \sum_{j=1}^{n} b_j \quad (2)$$

Under condition (2), the production-transport problem can be reduced to a classical transport problem. And then such a problem is solved by the simplex method or the method of potentials.

In modern conditions, the transportation of products from the producer to the consumer is carried out in two ways: from the supplier to the consumer (the most economically advantageous) and from the supplier-base to the consumer (requires more transport and other costs). The delivery of products through the base to the consumer requires the construction of a model of a multi-stage transport problem, in which the minimum value of the total transport costs is usually taken as the optimality criterion. The transportation plan between suppliers and warehouses and the transportation plan between warehouses and consumers are independent of each other. Two independent transport tasks are solved separately and in any order. If the total capacity of the warehouses is greater than the total capacity of the suppliers, then it is necessary to make a single calculation in order to obtain a more economically efficient multi-stage transportation plan.

Sequential processes of production of various types of products, their delivery to processing points and other types of production of these types of products and their delivery to end consumers are the task of production and transport logistics. They are reflected in the setting of a multi-stage transportation problem, the objective function of which is to minimize all total costs for the production and transportation of raw materials and finished products. Solving such problems is problematic, as it requires the development of interconnected models, methods, and algorithms. Thus, the task of developing high-quality tools for building models of multi-stage production and transport problems is urgent.

One of the modern simulation tools is FlexSIM multifunctional simulation environment. With the help of the FlexSIM system, it is possible to build a model of the production and transportation task at the enterprise, as well as individual parts manufacturing sites and transportation lines by simulating all elements of technological processes even before the start of their real production. The advantage of the FlexSIM software tool is that this simulation environment supports design, model development, allows you to perform computer experiments with the model, including various types of analysis.

3. Results

Let's consider one of the variants of the two-stage production and transportation problem. The enterprise uses five types of resources to manufacture seven types of products. At each of the five sites, a certain type of resource (parts) is produced and processed. Further, these resources are transported to an intermediate point (warehouse), from where they arrive at branches (shops) for the manufacture of specific equipment (product). The network model of this problem is shown in **Figure 3**. The classic transport problem is one-stage in the sense that in it the products from suppliers come directly to consumers. However, in practice it is quite common for products from suppliers to first arrive at intermediate points (transport network nodes, distribution centers, warehouses), where, if necessary, they are reloaded or unloaded and stored for some time. That is, the final consumers do not receive products from suppliers, but from these intermediate points of transport networks.

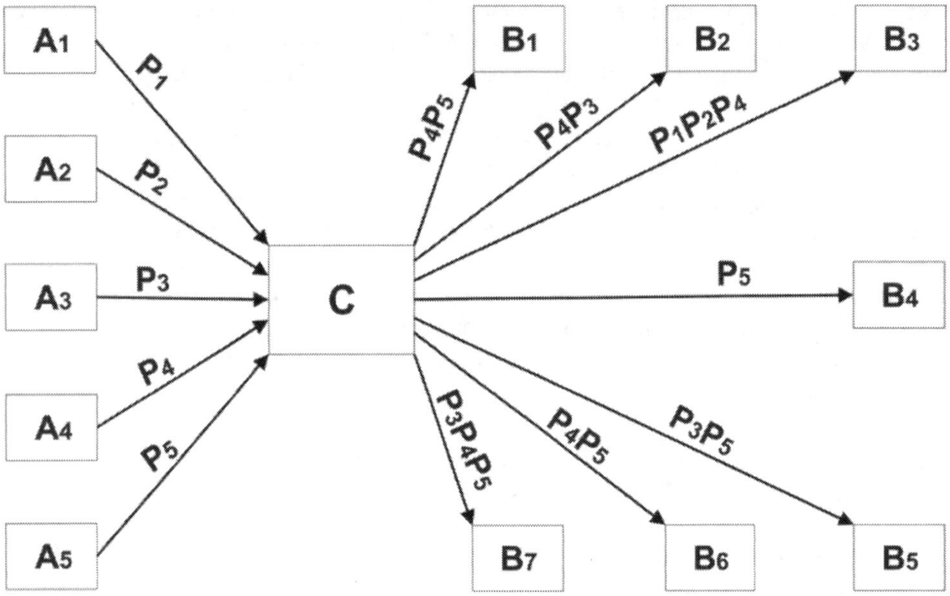

Figure 3. The transport network of the several stages transport task

We consider FlexSim software tools as an effective software to analyze transport tasks. As FlexSim is useful tool for solving transport tasks and effective for different technical tasks [1] , economical tasks[3], industry tasks[6]. Principle of simulation modeling, gave as a chance to analyze every part of process. We used the transport network (**Figure 3**) for creating the simulation process in FlexSim (**Figure 4**). We can see all stages are connected between each other. We used basic elements: sources (Source), processes (Process) and queues (Queue). In our case, sources are the elements from which information or objects enter the product model it is point A1 from **Figure 3**. Queues 1, 2, 3, 4, Conveyors, Separator, Racks 1, 2, 3, 4, 5, 6 Combiner in FlexSIM are intermediate stages of model (**Figure 3**). Sink 1 and Sink 2 we take as points B1 and B2 of our transport network. The algorithm of creating imitation model in FlexSim software is the next:

Step 1. Input, the first element is Source, where we specify the following logic:
Source1 – Source – FlowItemClass -Box.
Source1 – Flow – Send To Port – First Available.
Step 2. The next stage is the generation of products of different types (four types):
Triggers – On Creation – Set Item Type and Color-Item Type – duninform (1,4, get stream(current)).
Step 3. We have to divide each type of product into a different port.
We have four ports to each conveyor.
Separator – Flow – Send to Port – Port by Case.
Step 4. Create conveyors for distributing products to next stage.
Step 5. Create Racks and put the following logic for each rack:
Maximum content, Number of Bays, Number of Levels
Step 6. Create separator and combiner like intermediate points.
Step 7. Create Sink1 and Sink2 like points B1 and B2 (like destination points)
Step 8. Put connections between elements of the simulation model.

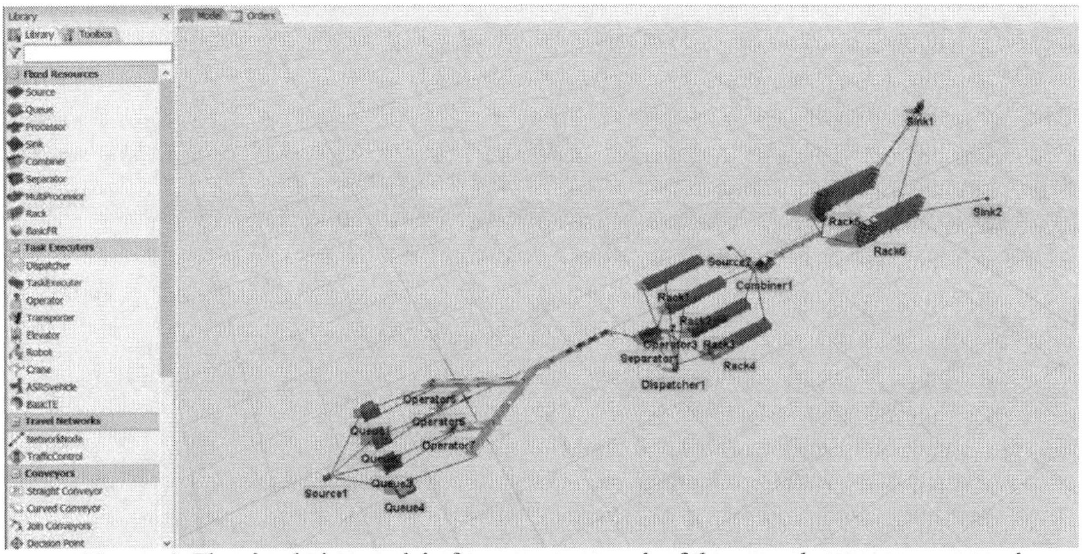

Figure 4. The simulation model of transport network of the several stages transport task.

The developed simulation model in the FlexSim environment allowed seeing the process of transportation. The study allowed us to set the parameters of transportation, for example, the moving products through intermediate stages to destination points (B_1 and B_2) FlexSim tools allowed to visualize the transport network of the optimal support plan (**Figure 5, Figure 6**).

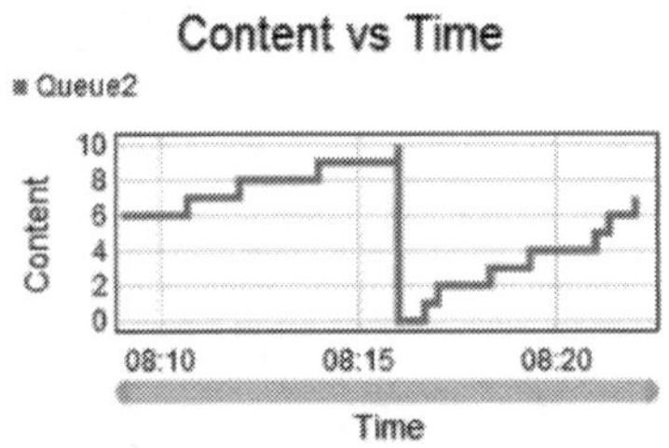

Figure 5. Dependence of the content and time for Queue2

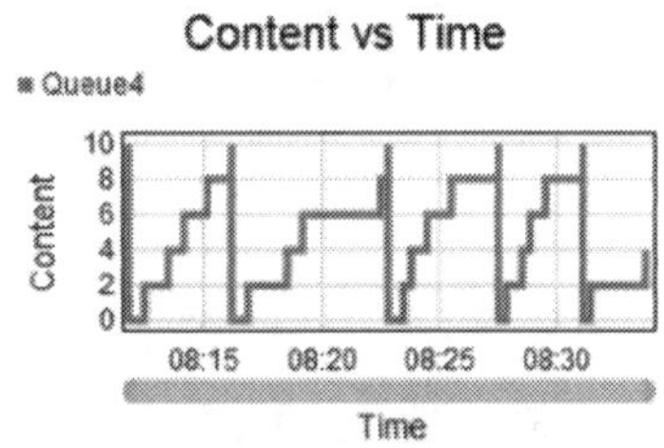

Figure 6. Dependence of the content and time for Queue4

The analysis of the study conducted a mathematical analysis of the transport problem is the basic step of creation the simulation model of this process in FlexSim simulation software. The result is the optimal path of movement of products from the producer through intermediate points to the consumer. The model of process in the FlexSim environment is created allows to analyze each stage of delivery. For example, the number of each type of product delivered to the consumer, stops of products in warehouses (Rack 1, 2, 3, 4), the number of surplus products for each warehouses.

4. Conclusions

The process of building a simulation model of the production and transportation problem in the FlexSim environment is demonstrated. A diagram of relationships between objects in a two-stage transport problem was created, which clearly demonstrated the impact of input data both on each object separately and on the technological system. The sequence of setting the logic for the interaction of the objects of this technological process is proposed.

5. References

[1] Mustafa Fatih Yegul, Fatih Safa Erenay, Soeren Striepea and Mustafa Yavuza 2017 Improving configuration of complex production lines via simulation-based optimization *Computers & Industrial Engineering* **109** 295-312. https://doi.org/10.1016/j.cie.2017.04.019

[2] Manavalan E and Jayakrishna K 2019 A review of Internet of Things (*IoT) embedded sustainable supply chain for industry 4.0 requirements* **127** 925-953. https://doi.org/10.1016/j.cie.2018.11.030

[3] Sornettea D, Maillart T and Kröger W 2013 Exploring the limits of safety analysis in complex technological systems *International Journal of Disaster Risk Reduction* **6** 59-66. https://doi.org/10.1016/j.ijdrr.2013.04.002

[4] Ljaskovska S, Martyn Y, Malets I and Prydatko O 2018 Information Technology of Process Modeling in the Multiparameter Systems *IEEE Second International Conference on Data Stream Mining & Processing (DSMP)* 177-182. https://doi.org/10.1109/DSMP.2018.8478498

[5] Ljaskovska S, Martyn Y, Malets I and Velyka O 2020 Optimization of Parameters of Technological Processes Means of the FlexSim Simulation Simulation Program *IEEE Third International Conference on Data Stream Mining & Processing (DSMP)* 391-397. https:// doi.org/ 10.1109/DSMP47368.2020.9204029

[6] Hao Peng, Qiushi Zhu 2017 Approximate evaluation of average downtime under an integrated approach of opportunistic maintenance for multi-component systems *Computers & Industrial Engineering* **109** 335-346. https://doi.org/10.1016/j.cie.2017.04.043.11

[7] Yang T, Hou Z, Liang J, Gu,Y and Chao X 2020 Depth Sequential Information Entropy Maps and Multi-Label Subspace Learning for Human Action Recognition *IEEE Access* **8** 135118–135130. https://doi.org/10.1109/access.2020.3006067

[8] Sena Daş G 2017 New Multi objective models for the gate assignment problem *Computers & Industrial Engineering* **109** 347-356. https://doi.org/10.1016/j.cie.2017.04.042

[9] Riko I Made, Chee Lip Gan, Liling Yan, Katherine Hwee Boon Kor, Hong Ling Chia, Kin Leong Pey and Carl V. Thompson 2012 Experimental characterization and modeling of the mechanical properties of Cu–Cu thermocompression bonds for three-dimensional integrated circuits *Acta Materialia* **60 (2)** 578-587. https://doi.org/10.1016/j.actamat.2011.09.038

Computation of Tense-Deformed State of the Die Mold Matrix

O.T. Velyka[1], M. V. Bojko[1], N.-T. I. Velykyj[1]

[1] Department of Robotics and Integrated Mechanical Engineering Technologies, Lviv Polytechnic National University, 12 S. Bandera Street, Lviv, 79013, Ukraine

Corresponding author's e-mail address: oksana.t.velyka@lpnu.ua

Abstract. A method of computation of the tense-deformed state of the die mold matrix is described in the given article. Using the criterion of the same durability an algorithm of minimization of its weight is built after the structural parameters. The stress-strain state of the matrix was calculated in the AUTODESK INVENTOR . Research and analysis of the stress-strain state of the matrix was carried out and its optimal parameters were determined. As a result, it makes it possible to optimize the structure of the matrix in terms of stiffness and stability and ensure its optimal performance
Keywords: Press-form; matrix; finite element method.

1. Introduction

One of the main tasks in creating CAD press forms for molding plastic products and optimizing the design process is the development of a formalized methodology for technological and design computation [2]. The design computation method includes calculation of all component parts of the mold (Fig.1). Matrix is one of the main parts where the formation of plasma products occurs. Matrix is a part of the mold taking place in the formation of the outer surface of the product afterward. It is under the variable load (which occurs at the moment of injection of plastic into the mold), that increased requirements are placed on matrix design, so its computation is of practical interest [4-6].

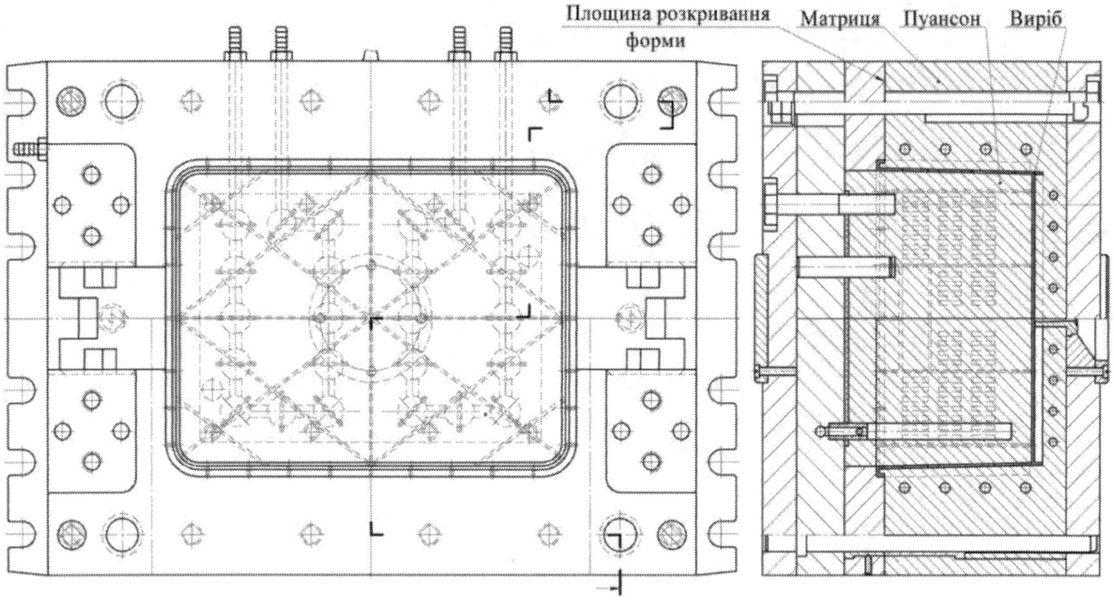

Figure 1. Press-form for plastic injection molding

2. Materials and method

Let's look at the matrix structure as a box structure made of five rectangular plates (Fig. 2). Because the rigidity of each plate in its area significantly exceeds its bending rigidity, the deformations of the plates due to forces acting in the plane of the plate will be much smaller than the deformations due to bending moments. Thus, we will not take them into account. For the same reason, we will consider that each plate is resting on the adjacent plates as a rigid support, that is, to say, the lines of conjugation of the plates (ribs of the matrix) are elastic. We also assume that the angle between elements perpendicular to their common side remains straight during deformation for adjacent plates

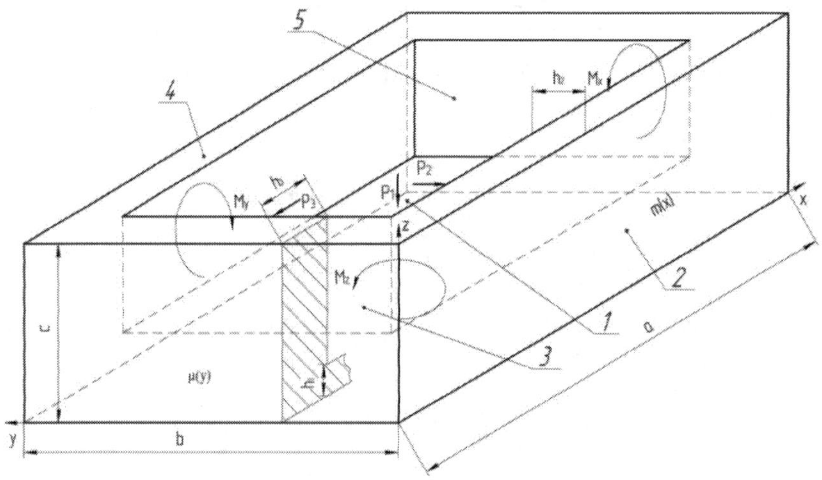

Figure 2. Matrix. The design diagram .

Considering The effort of the injection molding machine, which ensures that the mold does not open during filling with plastic is reduced to a certain moment load, which changes along the upper ends of the side walls.

This moment load is further determined by Fourier's coefficients, which are included as parameters of the sought solution. When filling the forming cavity matrix with plastic the internal pressure of plastic injection acts on the bottom and walls of the structure.

According to the design loading scheme, the task is limited to determining the stress-strain state of the matrix, composed of five plates, which are under the action of static loads of different intensity, as well as bending moments, which are applied to the ends of side plates 2-5.

Due to the geometric and force symmetry – the progins, the corners of the turn, the bends, turning angles, torques, and forces in plates 4, and 5 will be similar for plates 2. Thus, it is enough to limit ourselves only to the calculation of plates 1–3 (Fig. 2)

Let's introduce the following representations:

a, c, c – length, width and height of the matrix; h1, h2, h3 – thicknesses of plates 1–3;

E – moduls of elasticity; ν – Poisson's ratio; P_1, P_2, P_3 – the intensity of static loads acting on the plates; 1–3; w (x, y), u (x, z), v (y, z) – plate deflections;1–3; $M_x^{(1)}, M_y^{(1)}, M_x^{(2)}, M_z^{(2)}, M_y^{(3)}, M_z^{(3)}$ – bending moments of plates; 1–3; D_1, D_2, D_3 – cylindrical stiffness of the plates; 1–3: $D_i = \frac{Eh_i^2}{12(1-\nu^2)}$, $(i = 1,2,3)$; $\Delta_1 = \frac{\partial^2}{\partial x^2} + \frac{\partial^2}{\partial y^2}$, $\Delta_2 = \frac{\partial^2}{\partial x^2} + \frac{\partial^2}{\partial z^2}$, $\Delta_3 = \frac{\partial^2}{\partial y^2} + \frac{\partial^2}{\partial z^2}$ - two-dimensional Laplace operators.

We use the classic plate theory to calculate plates [1]. .

The plates should be made to fit in areas that occupy the plates, not uniform

:

$$\Delta_1\Delta_1 w = \frac{\partial^4 w}{\partial x^4} + 2\frac{\partial^4 w}{\partial x^2 \partial y^2} + \frac{\partial^4 w}{\partial y^4} = \frac{P_1}{D_1},$$
$$\Delta_2\Delta_2 u = \frac{P_2}{D_2}, \quad \Delta_3\Delta_3 v = \frac{P_3}{D_3}. \tag{1}$$

By setting the values of the intensities in a certain way, it is possible to simulate the corresponding types of loads.

Thus, with a vertical load we have:

$P_1 = \text{const} \neq 0, \quad P_2 = P_3 = 0,$

With horizontal transverse load:

$P_2 = \text{const} \neq 0, \quad P_1 = P_3 = 0,$

When the load is horizontal:

$P_3 = \text{const} \neq 0, \quad P_1 = P_2 = 0,$

With a full load:

$P_1 = \text{const} \neq 0, \quad P_2 = \text{const} \neq 0, \quad P_3 = \text{const} \neq 0.$

According to the assumptions made above, the deflections of the plates must also satisfy the conditions of bearing on a rigid contour

$$w(0,y) = w(a,y) = w(x,0) = w(x,b),$$
$$u(0,z) = u(a,z) = u(x,0) = u(x,c), \tag{2}$$
$$v(0,z) = w(b,z) = v(y,0) = v(y,c),$$

conditions for the angle of rotation of the adjacent plates:

$$w'_y(x,0) = -u'_z(x,0), \quad w'_x(0,y) = -v'_z(y,0), \quad u'_x(0,z) = -v'_y(0,z), \tag{3}$$

the conditions for the equines of the bending moments along the direction line of the plates are as follows:

$$M_y^{(1)}(x,0) = M_z^{(2)}(x,0), \quad M_x^{(1)}(0,y) = M_z^{(3)}(y,0), \quad M_x^{(2)}(0,z) = M_y^{(3)}(0,z), \tag{4}$$

and also, the equality of the bending moments applied to the ends of the side walls should be ensured:

$$-M_z^{(2)}(x,c) = m(x), \quad -M_z^{(3)}(y,c) = \mu(y). \tag{5}$$

Thus, the task is reduced to determining the deflections w, u, v that satisfy the above equations and boundary conditions.

After finding the deflections, all values characterizing the stressed state are determined by the formulas (1). So for plate 1, we will have:

bending and twisting moments

$$M_x^{(1)} = D_1\left(\frac{\partial^2 w}{\partial x^2} + \nu\frac{\partial^2 w}{\partial y^2}\right), \quad M_y^{(1)} = D_1\left(\frac{\partial^2 w}{\partial y^2} + \nu\frac{\partial^2 w}{\partial x^2}\right), \quad H_{xy}^{(1)} = -D_1(1-\nu)\frac{\partial^2 w}{\partial x \partial y}, \tag{6}$$

tension

$$\sigma_x^{(1)} = \frac{12z}{h_1^3}M_x^{(1)}, \sigma_y^{(1)} = \frac{12z}{h_1^3}M_y^{(1)}, \tau_{xy}^{(1)} = \frac{12z}{h_1^3}H_{xy}^{(1)}. \tag{7}$$

The solution of the closed system of equations (1) - (5) lies in Leur's method in relation to the theory plates using trigonometric Fourier's series.

Finally, the task is to solve the endless system of linear algebraic equations to find three groups of the unknown: A_m, B_n, C_k , coefficients resulting in the solution for deflections

Because the distribution law of of bending moments acting on the upper ends of the side plates is unknown, it is advisable to study the limiting cases, namely:

a) moments $m(x)$ and $\mu(y)$ are absent, the upper edge of the structure is; $z = c$ freely supported;

б) the upper edge of the structure is tightly clamped, which corresponds to the rigidly closed position of the mold.

In the case of free resistance $m(x) = \mu(y) = 0$. In the case of hard clamping of the upper end of the side walls if $z = c$ turn angle equals zero:

$$u'_z(x,c) = v'_z(y,c) = 0 \tag{8}$$

Stress on the outer surfaces of the plates is calculated by the formulas (7)

The task of the rational use of the material in the design of the matrix, based on the conditions of strength and geometric limitations, is considered. We will assume that the structure under consideration can be conditionally divided into N elements (substructures) of a given geometry with N constructive controlled parameters that determine the consumption of material. Separate parts of the structure (bottom, side walls) are considered as elements, and their thicknesses are considered as defining structural parameters $h_j (j = \overline{1,N})$.

Due to linear weight dependence on controlled parameters h_j the target function (weight function) looks like:

$$F(\overline{h}) = \sum_{j=1}^{N} G_j \cdot h_j, \quad h_j > 0 \tag{9}$$

So, for example, considering the matrix as a five-plate system, we will have the following parameters of the objective function: $N = 3;$ $G_1 = ab\rho$, $G_2 = 2ac\rho$, $G_3 = 2bc\rho$, де ρ – acceleration of earth gravity.

The task of weight optimization of the matrix design considering the conditions of strength and geometrical limitations can be formulated as a problem of mathematical programming:

it is necessary to find the optimal vector of controlled parameters $\overline{h}_{opt}$, where

$$F(\overline{h}_{opt}) = \min_{h \in D} F(\overline{h}) \tag{10}$$

$$D = \left\{ \overline{h} : \max_{V_j} \sigma_j(\overline{h}) \le [\sigma]_j, \quad h_{\min j} \le h_j \le h_{\max j}, \quad j = \overline{1, N} \right\} \tag{11}$$

$$\text{where } [\sigma]_j = \begin{cases} [\sigma_+]_j, \sigma_j \ge 0 \\ [\sigma_-]_j, \sigma_j < 0 \end{cases} \tag{12}$$

$[\sigma_+]_j$, $[\sigma_-]_j$ – permissible limits of tension and compression in j plate;

V_j – volume j - the element of its system.

The task (10) of minimizing the weight of the plate system is equivalent to the inverse problem of creating a discrete uniform-strength deformed plate system:

We have to find vector $\overline{h}^*$ to fulfill the following condition:

$$\max_{V_j} \sigma_j(h_1^*, \ldots, h_N^*) \equiv [\sigma]_j, \quad h_{\min j} \le h_j^* \le h_{\max j}. \tag{13}$$

If the task (13) has a solution, then each j element will have a specified margin of safety, and the system as a whole will have minimal weight.

The model of the optimum distributing h_j of a discrete uniform-strength plate system can be implemented according to the following algorithm

$$h_j^{(i+1)} = \frac{1}{2} h_j^{(i)} \left[1 + K_j^{(i)}(h_j^{(i)})\right], \quad K_j^{(i)} = \frac{\sigma_j^{(i)}}{[\sigma]_j} \tag{14}$$

where i –iteration number; $h_j^{(i)}$, $h_j^{(i+1)}$ – value of thickness j⁻plate on i i $(i+1)$ iteration; $\sigma_j^{(i)}$ –ultimate stress in j plate.

The calculation process ends when the condition is simultaneously fulfilled in each extreme node of the plate system:

$$\left| h_j^{(i+1)} - h_j^{(i)} \right| < \varepsilon h_j^{(i)}, \quad \forall\, j = \overline{1, N} \tag{15}$$

where ε is a small given positive number that characterizes the accuracy of conditions (13).

If the values of the thicknesses in some nodes are $h_j^{(i+1)}$ less h_{min} or exceed h_{max}, then we use $h_j^{(i+1)} = h_{min}$ or $h_j^{(i+1)} = h_{max}$.

3. Results and Discussion

To calculate the stress-strain state of the mold matrix, an automated calculation algorithm was used for designing by the CAE finite element method [1,3]. This method is implemented in the AUTODESK INVENTOR software environment. The given method is implemented through the AUTODESK INVENTOR software environment. The choice of this software is due to wide possibilities for conducting studies of the stress-strain state of the mold matrix; in particular, it allows to automatically calculate the strength of the part model according to the input data (the system selects nodes in the volume of the part, divides the part into finite elements, numbers the nodes, builds a matrix of elements, makes calculation equations). The calculation results are displayed as tabular data and visual notes of the searched parameters.

For calculation by the method of glass elements we used the "Stress Analysis" module of AUTODESK INVENTOR. This module allows you to find the best options for the design of the part. For this, the "Stress Analysis" module provides tools for express analysis of the stress-strain state with the determination of stresses, strains, displacements, safety margin.

Performing stress analysis includes the following steps:

- Enter the "Stress Analysis" module
- Create the "Stress analysis" model and set its parameters
- Exclude elements from the analysis that do not significantly affect the result in this simulation
- Check whether the materials used are correct
- Set structural dependencies (supporting surfaces and type of fastening)
- Apply a load to the faces, edges or vertices of the part
- Adjust the grid if necessary
- Run the simulation
- Perform review and interpretation of results
- Perform report generation

A 3-D matrix for plastic injection model was designed for the research (Fig.3)

Figure 3. Matrix designed in the AUTODESK INVENTOR environment.

The next step is to set the boundary conditions and applied forces that correspond to the loads. (Fig.4) And then we bulid a mesh of finite elements on the model (Fig. 5). The density of the mesh is adjusted if necessary.

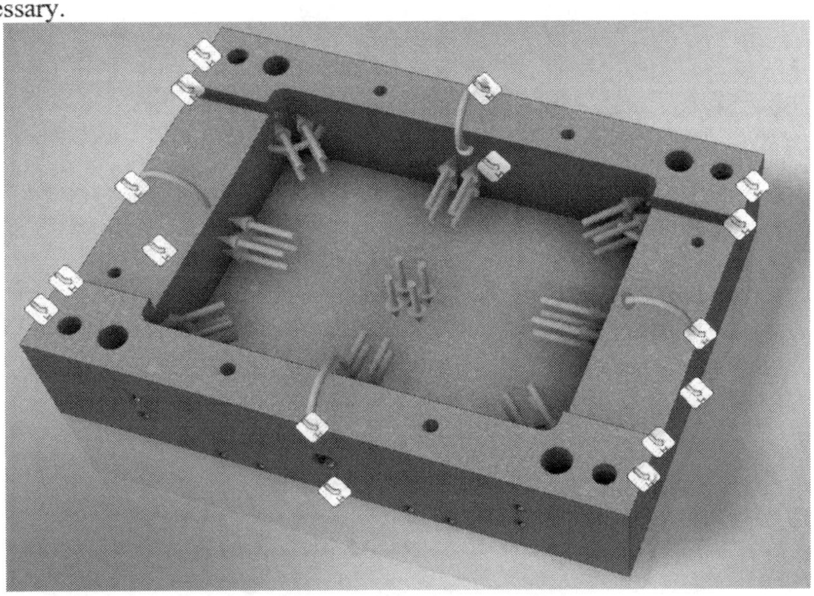

Figure 4. Setting of forces and torques acting on the matrix

Figure 5. Mesh construction of finite elements.

Then we perform calculations for various mold parameters and choose the optimal model. Analyzing results of the calculation of the stress-deformed state, critical areas were identified in which the greatest loads occur, and, therefore, deformation is also possible. (Fig. 6).

Figure 6. Detection of critical areas

4. Conclusions

As a result of studies on the stress-strain state of a 3-D model for a plastic injection model in the AUTODESK INVENTOR environment, we determined its optimal design and the parameters that would ensure its operational efficiency

References

[1] Dubenets V. H., Khilchevskyi V. V., Savchenko O. V., Osnovy metodu skinchennykh elementiv: Navchalnyi posibnyk, Chernihiv: ChDTU, 2007., 288 s [In Ukrainian]

[2] Dobrianskyi, S. S. Tekhnolohichni osnovy mashynobuduvannia. Kyiv : KPI im. Ihoria Sikorskoho, 2020. – 379 s. [In Ukrainian].

[3] Petrenko A. M., Semenov O.Y. Osnovy postroenyia system avtomatyzyrovannoho proektyrovanyia. -K.: Vyshcha shkola, 1984. -296 s. [In Russian]

[4] Boiko M. V., Velyka O.T., Topilnytskyi V.H., Lampika R.V. Rozrakhunok priamokutnoi matrytsi pres-formy na mitsnist i zhorstkist, Avtomatyzatsiia vyrob. protsesiv u mashynobud. ta pryladobuduvanni, NULP 2010. — Vyp. 44. -s. 74-79. . [In Ukrainian]

[5] Velyka O., Boiko M., Rozrakhunok priamokutnoi matrytsi pres-formy na mitsnist i zhorstkist Teoriia ta praktyka ratsionalnoho proektuvannia, vyhotovlennia i ekspluatatsii mashynobudivnykh konstruktsii, Materialy konferentsii NULP, Lviv., 2010. — s. 38 . [In Ukrainian]

[6] M. Boiko, O. Velyka, V. Topilnytskyi, Rozrakhunok ta optymizatsiia matrytsi vyrubnykh shtampiv z umovy mitsnosti ta zhorstkosti, *9-y Mizhnarodnyi sympozium ukrainskykh inzheneriv-mekhanikiv u Lvovi: prysviach. 165-richchiu Nats. un-tu «Lviv. politekhnika»*, 20-22 trav. 2009 r. : pratsi / Nats. un-t «Lviv. politekhnika» [ta in.]. — L., 2009. — S. 155-156. [In Ukrainian]

IOP Conf. Series: Materials Science and Engineering 1277 (2023) 012035 doi:10.1088/1757-899X/1277/1/012035

Study of microclimate parameters in the bus driver's cabin in the warm season

Yurii Voichyshyn[1], **Kostyantyn Holenko**[2], **Dmytro Roman**[1], **Orest Horbai**[1]

[1] Department of Equipment Design and Operation, Institute of Mechanical Engineering and Transport, Lviv Polytechnic National University, Lviv, Ukraine
[2] Department of Tribology, Automobiles and Materials Science. Khmelnytsky National University. Khmelnytsky, Ukraine

Abstract. Work describes an approach to complex thermal analysis of buses salons starting from engine heat radiation and with its further distribution along the internal bus volume. Provided workflow chain in the article could be used as the base for any bus type climate and thermal investigation with the next optimization events. Heat analysis with materials thermal conductivity parameters was proceeded using Ansys Fluent and Ansys Steady State Thermal addons. Based on the obtained results, it is possible to draw conclusions about the impact of different parameters (temperature, heat radiation, humidity, etc) on the buses salons microclimate. Presented work could be much actual in practical plane like methodology used for efficient bus bodies development by engineers who could apply suggested materials for calculating the distribution of heat from the engine in the passenger compartment of the bus and studying the microclimate in different operating conditions. That could be especially valuable at the stage of the new bus models design and modelling.

1. Introduction

The topic of passenger transportation comfort does not stand till today and is extremely relevant both in the industrial and scientific world. One of the important points in design of public transport vehicles is the creation of comfortable conditions for passengers and the driver of bases for today.

Many factors shape comfort during passenger transportation, such as smoothness of movement, manoeuvrability, noise insulation, environmental safety, as well as thermal comfort in the passenger compartment and the driver's cabin and bus salon in common. In this regard, the microclimate should be considered one of the main factors of high-quality and comfortable transportation, and accordingly, is actual among the modern buses development.

It is quite difficult to determine heat losses in the passenger compartment during research because the city bus makes frequent stops during transportation on the route, during which air enters the passenger compartment from the environment.

Another important question is bus layout diagram: engine location (front, middle or rear), type of the bus (tourist, city or shuttle), which defines the microclimate conditions inside the bus salon as well.

2. Problem statement

The layout of bus bodies buses determines many indicators of their operation efficiency, like:

- **Weight distribution** along the axles, which depends on the mass of the corresponding units and aggregates of the bus (points C on fig.1). For example, such types like school buses have to match Regulation No 94 of the Economic Commission for Europe of the United Nations (UN/ECE) — uniform provisions concerning the approval of vehicles with regard to the protection of the occupants in the event of a frontal collision [1]. That's why school buses are mostly equipped with the front location of engine under the hood. At the same time mass points of engine with gearbox (C5 and C7 on fig.1), which are located in the rear overhang, is quite positive solution for the better road adhesion coefficient, because the rear drive axle additionally loaded with the weight of powertrain [2-5].
- **Salon space square** is extremally important for the efficient commercial exploitation of buses in conditions of high-populated cities, as each square inch of space determines the bus usage profit. For example, point C1 with the 2 rows of seats places above the engine provides extra passengers capacity even for Low-entry type bus body (fig.1). Places for standing (C6, C7) play an important role in filling the bus with passengers during rush hours.
- Finally, the **heat (thermal) distribution** has an import role in all of this as well – efficient thermal energy radiated by the engine must reach from point C5 to the front of the bus (point C4 on fig.1) through the whole volume of salon. Besides, we have to pay attention, that being located on the different height those points C1-C7 will have different level of temperature due to the fact that the heat always tends to the top, so our research should cover 3D investigation of heat radiation.

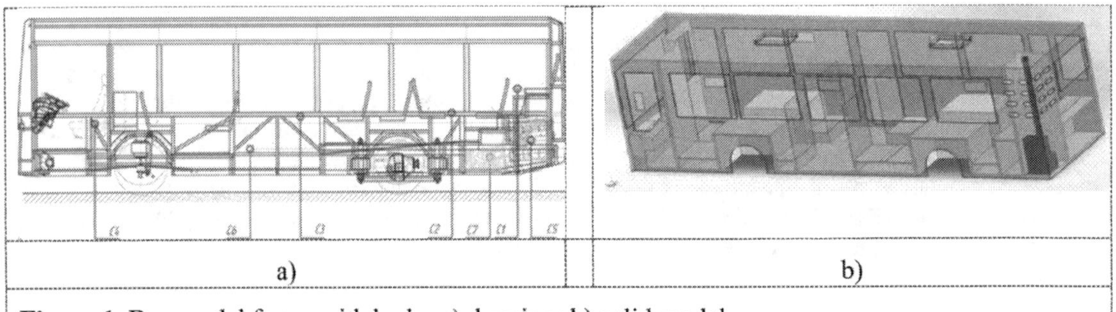

Figure 1. Bus model frame with body: a) drawing; b) solid model.

Our task is to form a methodology for calculating the distribution of heat from the engine in the passenger compartment of the bus and studying the microclimate in different operating conditions.

Besides our research is conducted in the areas of improving air exchange and heating systems, taking into account the requirements of state and international regulatory documents. Air exchange systems components (fans, elements of air ducts, placement of inlets) and their influence on the efficiency of the air circulation are studied in this article as well. Theoretical part of presented methodology uses materials of system analysis, natural experiment and mathematical modeling with the solution based on differential equations, techno-economic analysis and heat balance methods, which were previously researched [6].

There is no general normative documentation for ensuring microclimate in the world for today. It's because the climate is different in different countries [7]. Regulatory documents in various countries of the world include the following documents: DIN 1946-3, DIN ENISO 7730 1995 (Germany, EU), EN ISO 7730:2005, IDT (EU), ASHRAE Standard 55 (USA), GOST 30393-2015 (Russia) and outdated GOST R 50993-96 (Ukraine) [8-11].

3. Methods and results

Being the heart of the bus, engine generates heat in the bus salon through ventilation ducts that must be properly managed inside the bus cabin. It is obvious that modern buses have efficient air conditioning systems to create a favourable microclimate, especially if they were designed by the biggest players on the market (MAN, Mercedes-Benz, Volvo, Scania, etc), but our goal is to investigate the heat distribution from engine throughout the passenger compartment, which could be actual for small bus building companies, who are using existing aggregates, but develop own bodies.

Let's investigate the theoretical part of the thermal conductivity of materials and convention. The amount of energy conducted through a body per unit area and unit thickness during the unit time when the difference in temperature between the faces causing heat flow is unit temperature difference.
Thermal conductivity depends essentially upon the following factors [12]:
1) Material structure; 2) Density of the material; 3) Moisture content; 4) Pressure and temperature.

Fourier's low of heat conduction is an empirical law based on observation and states as follows: "The rate of heat flow through a simple homogeneous solid is directly proportional to the area of the section at right angles to the direction of heat flow, and to change of temperature with respect to the length of the path of the heat flow".

Mathematically, it can be represented by the equation:

$$Q = -\lambda(t)\frac{dt}{dx}A, \tag{1}$$

where:

Q – heat flow through a body per unit time, W;
$\lambda(t)$ – thermal conductivity coefficient;
A – surface area of heat flow, m^2;
dt – temperature difference of the block faces through which heat flows, C^0 C0 or K;
dx – thickness of body in the direction of flow, m.

Thus,

$$Q = -kA\frac{dt}{dx} \tag{2}$$

where:

k = constant of proportionally and is known as thermal conductivity of the body.

Let's get acquainted with the parameters definitions, which we will use then in Ansys Steady State Thermal based on the uniform wall investigation. Consider a one-sided wall of thickness δ, the thermal conductivity coefficient of which is constant and equal to λ. Constant temperatures t_{c1} and t_{c2} are maintained on the outer surfaces (fig.2a). Temperature changes only in the direction of the x-axis. In this case, the temperature field is one-dimensional, the isothermal surfaces are flat and located perpendicular to the x axis.

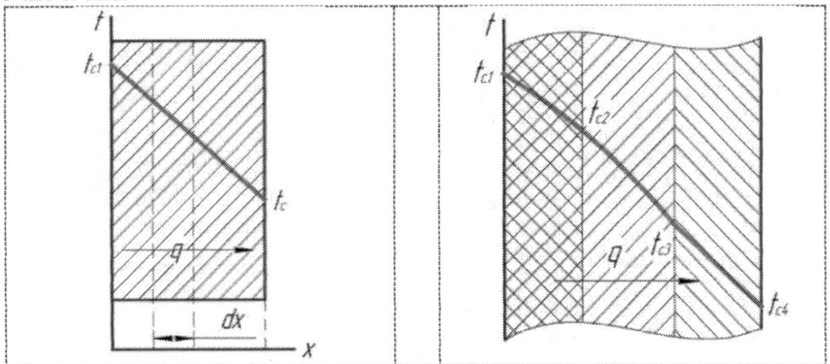

a) b)

Figure 2. Flat wall types: a) uniform, b) multi-layered wall

We select a plane of thickness dx inside the wall at a distance x, bounded by two isothermal surfaces. Based on the Fourier law, we can write:

$$q = -\lambda\frac{dt}{dx} \text{ or } dt = -\frac{q}{\lambda}dx \tag{3}$$

Quantity q in the stationary thermal regime is constant in each section, therefore:

$$t = -\frac{q}{\lambda}x + C \tag{4}$$

Integration constant C is determined from the boundary conditions, namely, when $x = 0$, $t = t_{c1} = C$, and when $x = \delta, t = t_{c2}$. Substituting these values into equation (4), we have:

$$t_{c2} = -\frac{q}{\lambda}\delta + t_{c1} \qquad (5)$$

Unknown value of the specific heat flux is determined from equation (5), namely:

$$q = \frac{\lambda}{\delta}(t_{c1} - t_{c2}) = \frac{\lambda}{\delta}\Delta t \qquad (6)$$

Therefore, the amount of heat transferred through 1 m^2 of the wall per unit of time is directly proportional to the thermal conductivity coefficient λ and to the temperature difference of the outer surfaces Δt but inversely proportional to the wall thickness δ.

Equation (6) is the calculation formula for thermal conductivity of the flat wall. It connects four quantities: $q, \lambda, \Delta t, \delta$

Multi-layered wall case is more complicated (fig.2b), but it's typical for bus body elements, so we have to get acquainted with process inside to have the complete understanding: let the wall consist of three heterogeneous, but closely adjacent layers. The thickness of the first is δ_1, the second is δ_2, and the third is δ_3. Accordingly, the thermal conductivity coefficients of the layers are equal to λ_1, λ_2, λ_3. In addition, the temperatures of the wall outer surface is t_{c1} and t_{c4} are known. We consider the thermal contact between the surfaces to be ideal, and denote the temperature at the contact points by t_{c2} and t_{c3}.

Flux density q is constant and the same for all layers in the stationary regime, therefore, on the basis of (6), we can get:

$$\left\{\begin{matrix} q = \frac{\lambda_1}{\delta_1}(t_{c1} - t_{c2}) \\ q = \frac{\lambda_2}{\delta_2}(t_{c2} - t_{c3}) \\ q = \frac{\lambda_3}{\delta_3}(t_{c3} - t_{c4}) \end{matrix}\right\} \qquad (7)$$

Then we can write the calculation formula for defining an n-wall with several layers:

$$q = \frac{t_{c1} - t_{cn+1}}{\sum_{i=1}^{n}\frac{\delta_i}{\lambda_i}} \qquad (8)$$

3.1 Engine heat radiation analysis.

Now we can move to the practical part of experiment – let's formulate an approach how to analyse temperature and heat flux on the surface of engine based on solid-model of engine of internal combustion (fig.3) to understand the typical calculation scenario, which could be used for any type of engine. Boundary conditions are presented in Table 1.

Table 1. Boundary conditions of engine heat distribution analysis.

Boundary conditions location	Meaning	Tags on fig.3
Combustion chamber	2300 C	A
Convection of the outer surface (air flow)	25 $W/m^2\ C$	B
Convection of coolant	1000 $W/m^2\ C$	C

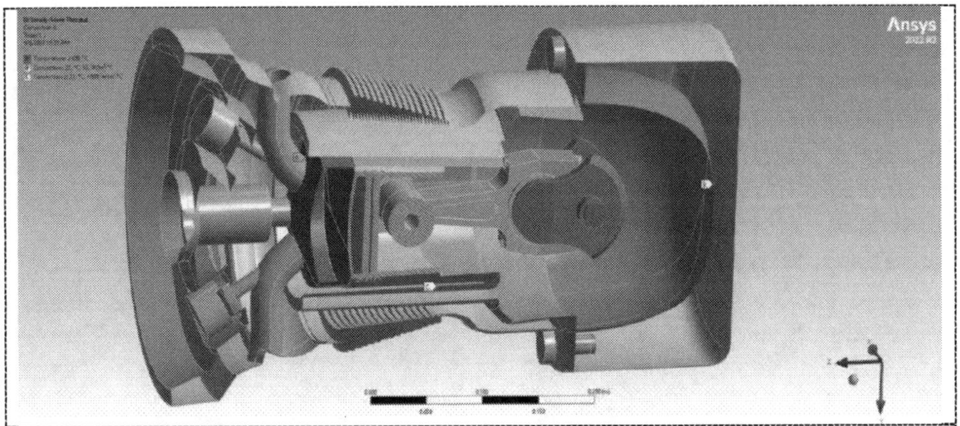

Figure 3. Boundary conditions of the engine thermal analysis.

Results of the Ansys Steady State Thermal calculation are presented on Fig.4 and allow us to determine temperature in the necessary surface location. Our interest is focused around the outer surface of the engine cylinder cooling fins – temperature is around 80-120 C *(Fig.4)* which was noticed during the most hot 3rd stroke of engine (working stroke).

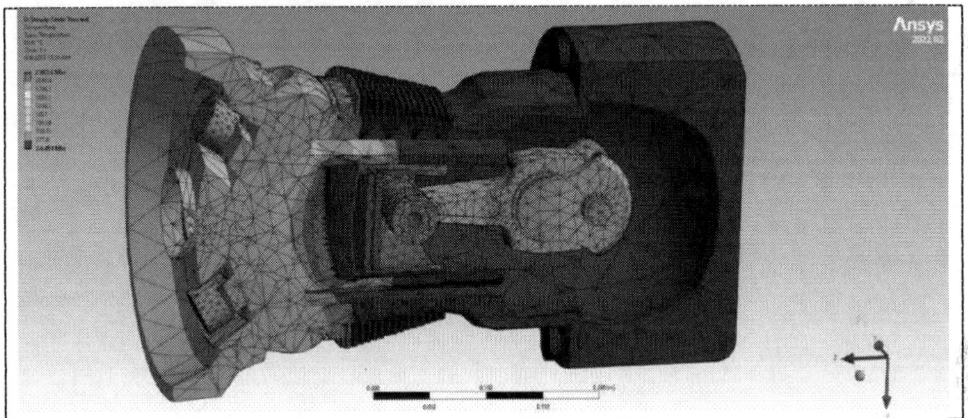

Figure 4. Thermal map of engine during the working stroke.

3.2 Bus body structure heat radiation analysis

By understanding the range of temperature radiation in appropriate locations we can move forward with applying them to complex model of the bus body (Low-entry type) with rear location of engine (fig.5) according to the Table 2.

Table 2. Formatting sections, subsections and subsubsections.

Boundary conditions	Meaning
Engine temperature	90 C
Convection of the outer surface (air flow)	25 $W/m^2\,C$
Convection of the salon surfaces (static air)	5 $W/m^2\,C$

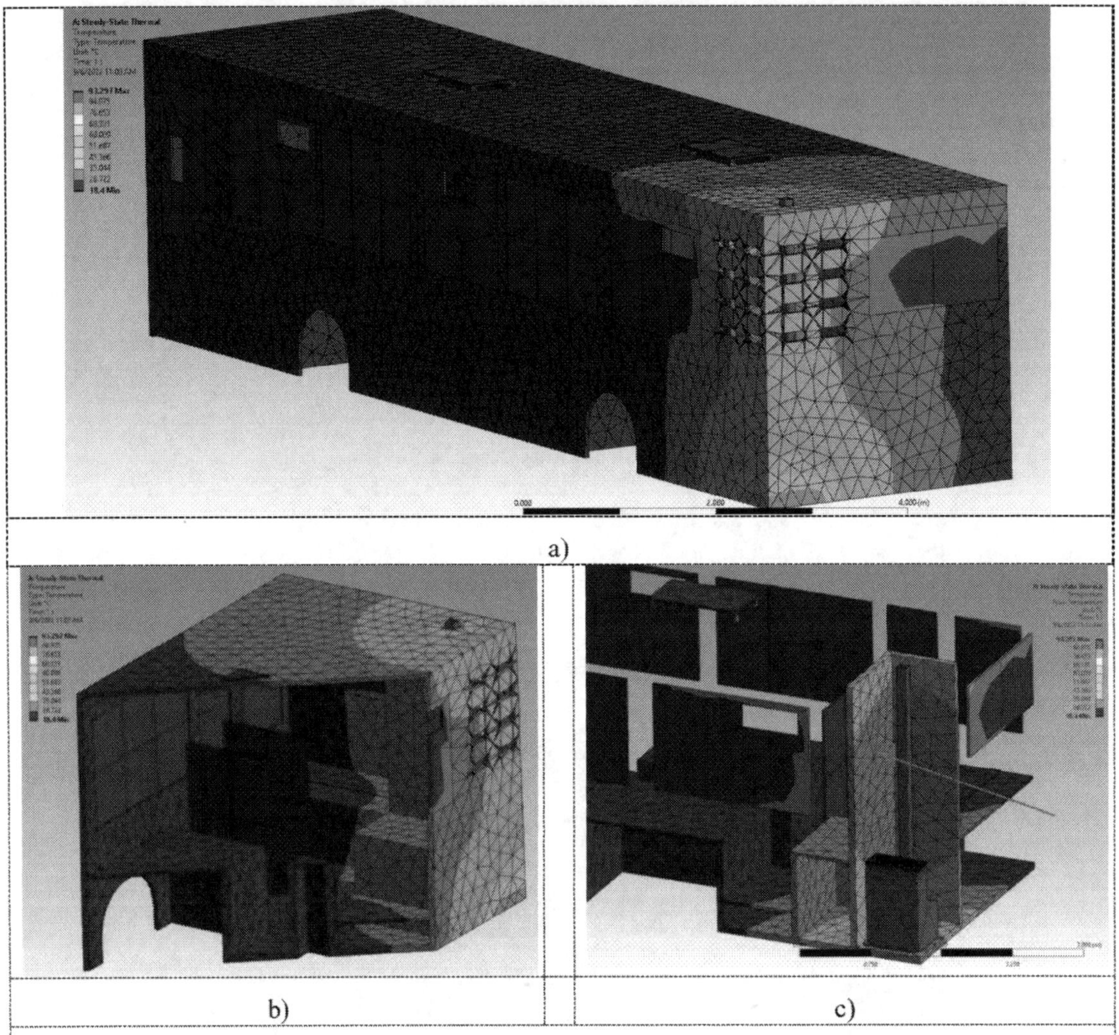

Figure 5. Temperature map along the bus body and salon structures: a) whole bus body; b) section of body by plane; c) windows, frame and engine view (without body).

Outer body surfaces convection is 5x higher, than internal one for emulating air flow in real bus operating conditions. Physical and mechanical properties of the materials suit to appropriate ones in Ansys Engineering: Steel 20 for the frame and body elements: Glass for the windows, Carbon fiber panels.

The goal of heat radiation analysis is the next: we need to know the meaning of temperature on the vertical wall (marked area on fig.5c), which separates engine section with salon. Therefore, the range of temperature in the specified location fluctuates around 30-50 C. By finding the necessary meaning of temperature we can move to the final step now – emulate heat flow distribution along the bus salon thinking that the mentioned wall (fig.5c) is equipped with some heating fan for getting the warm air masses from engine and spreading them along the salon to heat the internal volume. Before we proceed to calculations it makes sense to get acquainted with theory [13].

Navier-Stokes equations are used to describe the flow of a non-ideal gas or liquid for mechanical systems. First, let's consider the essence of the viscosity concept for a (non-ideal) gas, which can be easily illustrated by Couette's experiment. Let the flow be given between two very long parallel plates, one of which (for example, the lower one) is stationary, while the other moves in its own plane with a constant speed. Let us denote the distance between the plates by h and assume that the pressure in the entire space occupied by the gas is constant. It is experimentally shown that the gas sticks to both plates, where the gas velocity is zero directly near the lower plate, and coincides with the velocity v near the upper plate, and there is a linear distribution of velocities in the space between the plates. Speed v is proportional to the distance y from the lower plate and is expressed by the formula:

$$v(y) = \frac{y}{h} v \tag{9}$$

In order for such a state of motion to exist, tangential force must be applied to the gas from the side of the upper plate in the direction of motion, which balances the gas friction force τ per unit area. In other words, tangential stress is proportional to ratio $^{v}/_{h}$ instead of which $^{dv}/_{dy}$ can be taken. The proportionality factor between τ and $^{dv}/_{dy}$ is denoted by μ and it depends on the nature of the gas. Therefore, the law of friction (Newton's law) for gas has the form:

$$\tau = \mu \frac{dv}{dy}, \tag{10}$$

where the value μ is called the dynamic coefficient of gas viscosity.

The Navier-Stokes equation is a system of partial differential equations describing the motion and heat transfer of a non-ideal (viscous) incompressible fluid. They are one of the most important in aerodynamics and are used in mathematical modelling of many natural phenomena and technical problems. The system consists of two equations - the equation of motion and the continuity equation, the vector notation of which has the form:

$$\begin{cases} \frac{dv}{dt} = F - \frac{1}{\rho} grad\, p + v\Delta v, \\ div\, v = 0 \end{cases} \tag{11}$$

Where v – environment velocity, $\Delta v = \frac{\partial^2 v_x}{\partial x^2} + \frac{\partial^2 v_y}{\partial y^2} + \frac{\partial^2 v_z}{\partial z^2}$; ρ – density, F – mass density of external, in relation to the volume, mass forces, μ – dynamic viscosity coefficient, $v = \frac{\mu}{\rho}$ – kinematic viscosity coefficient, p – pressure. In this system of equations, the parameters μ and ρ are known and the mass forces F are given, so the system is closed and contains two unknown functions $v = (x, y, z, t)$ and $p = (x, y, z, t)$.

3.3 Bus body heat flow analysis

Currently we can move to the next step – investigate the heat flow inside the bus salon based on the temperature meaning on the vertical wall (marked area on fig.5c), which separates engine section with salon (fluctuates around 30-50 C). Let's emulate heat flow distribution in Ansys Fluent [14-15] thinking that the mentioned wall is equipped with some fan for getting the warm air masses from engine to heat the internal volume of salon (fig.6). Boundary conditions are:

Table 3. Formatting sections, subsections and subsubsections.

Parameter	Meaning
Inlet flow area	A on fig.6
Outlet flow area	B on fig.6
Gravity	-9.81 m/s^2 along Y axis
Velocity magnitude (inlet area)	0,25 m/s along X asis
Temperature (inlet area)	318 K
Temperature of internal volume and walls	293 K
Convection for static air	25 $W/m^2 K$

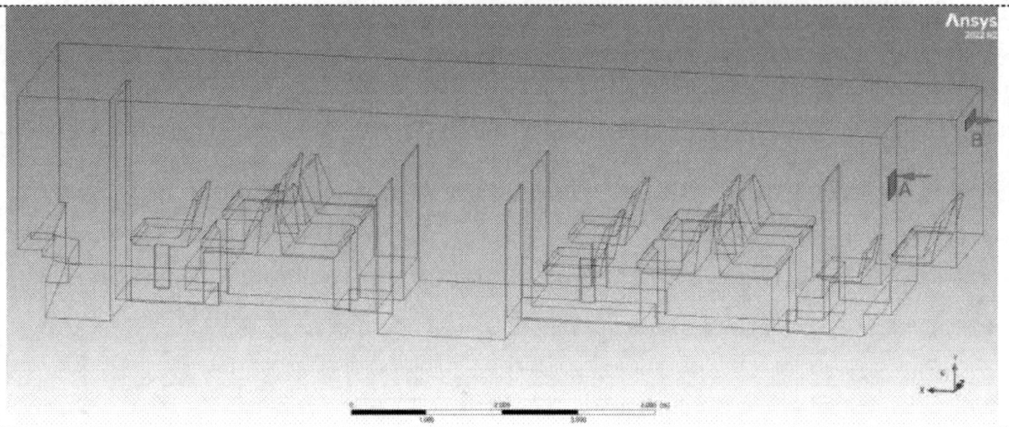

Figure 6. Bus salon volume boundary conditions for heat flow calculation - Ansys Steady State Thermal.

Let's get acquainted with the results based on velocity map (Fig.6): mostly laminar flow with around 0.15-0.22 m/s velocity is noticed it the top salon space are under the roof. Vertical wall behind the driver takes a hit on itself and the air flows splash, which cuts off some portion of heat from the back of driver. At the same time heat distribution is rather active in the frontal part of cabin (before the driver), so the air circulation is fully present here.

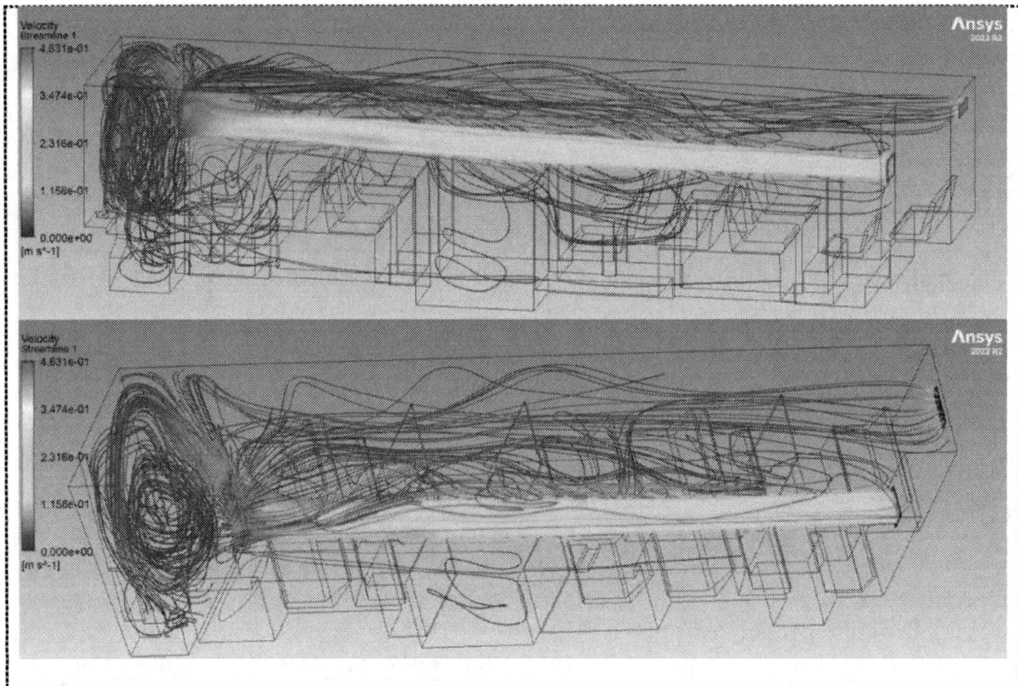

Figure 7. Velocity map of heat flow distribution along the salon space.

Let's check the temperature distribution along the selected plain (Fig.7) to analyze the situation on the driver's place – it fluctuates around 296-305 K, that is comfortable enough microclimate if we consider, that inlet temperature was taken rather high (45 C on the fan) and we have not calculated losses of heat due to the fact that the bus salon is not isolated and has gaps in the body, ventilation windows or hatches.

IOP Conf. Series: Materials Science and Engineering 1277 (2023) 012035 doi:10.1088/1757-899X/1277/1/012035

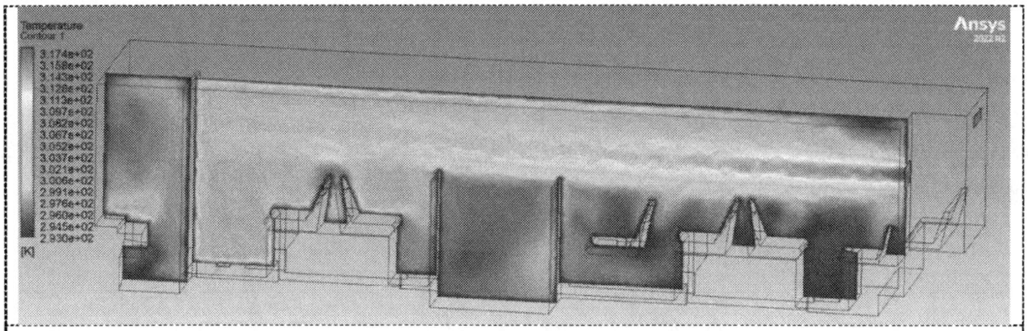

Figure 8. Temperature distribution along the defined plain.

For better visualization of volume heat distribution, it makes sense to present the next temperature maps as well – they allow to get enough understanding, how internal walls and height differences involve on the temperature spreading:

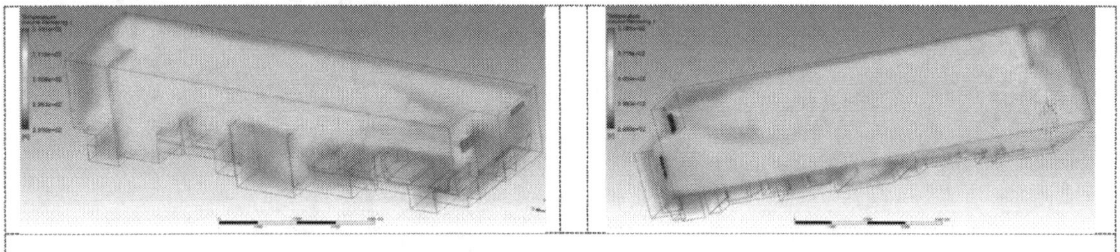

Figure 9. Temperature maps base on salon volume.

3.4 Microclimate parameters investigations

By having the temperature meaning in drivers place area the following parameters of the air are measured: temperature and air humidity (at 4 points: head, waist, floor and draft zone), air volume (Table 4).

During the measurements, the following experimental data were obtained (average values of the obtained parameters are given):

Table 4 Average microclimate parameters

Parameter	**Average value**
Temperature	23.2 ^{0}C
	23.7 ^{0}C
	25.0 ^{0}C
	23.4 ^{0}C
Humidity	82.0 %
	82.0 %
	71.0 %
	81.0 %
Amount of air	1 170 mL/s

The results of measurements in graphic format are shown in fig.10a,b,c. Graphs show the results of main microclimate parameters at work driver's seat Measurements were carried out during moderate bus movement in urban conditions, i.e. at low speeds. The graphs show the change of such parameters as air temperature, humidity and air volume.

Since the measurements were taken in the warm season, the external and internal temperatures are present the same 23-25 ^{0}C. As can be seen from the graph (Fig.10a), the highest temperature was in the floor area, dashboard, and accordingly, the humidity was the lowest in this area as well (Fig. 10b).

Measurements of the amount of air entering the bus cabin were carried out at the blow-off point air, i.e. near the open cabin. Analyzing the average results of the obtained results (Table 4), it is possible to draw conclusions, that the climatic conditions at the driver's workplace do not correspond to the values given in regulatory documents. Therefore, the driver's cabin, as well as the interior of the bus, need to have available an air conditioner and individual approach to each model of bus towards the heat distribution calculation, what was presented in the provided methodology.

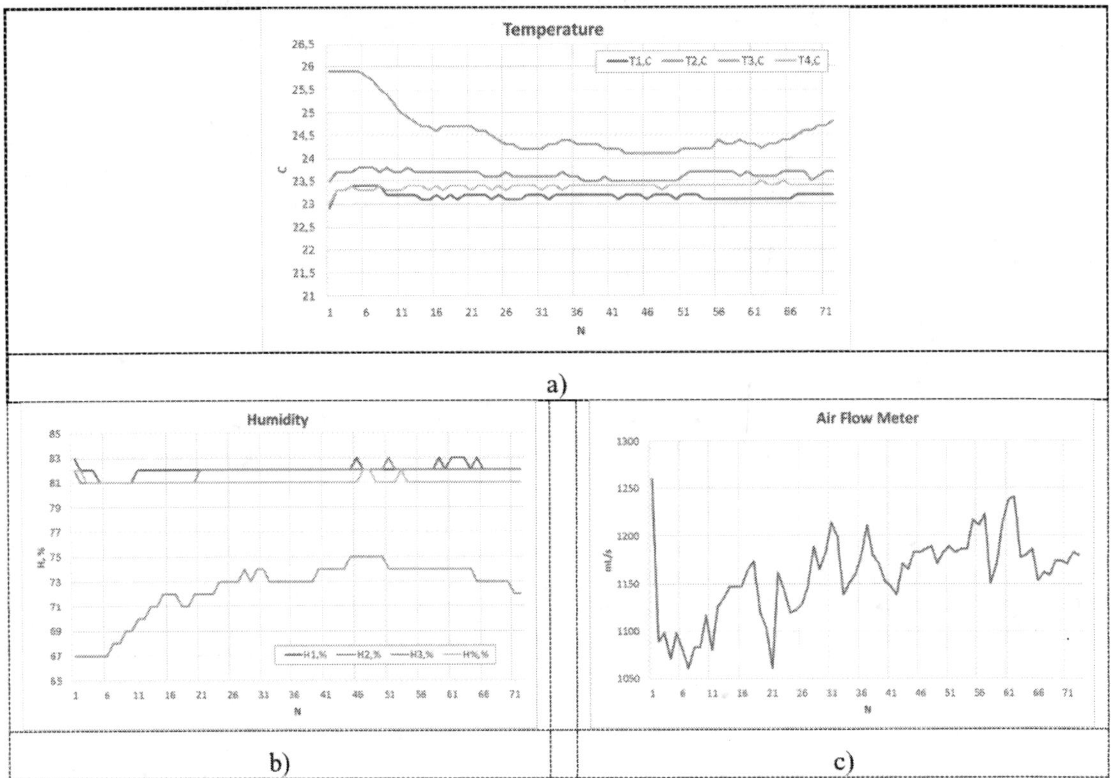

Figure 10. Graphical results of research: a) dependence of temperature on time; b) dependence of air humidity on time; c) dependence of the amount of air flow on time

4. Conclusions

Presented methodology of microclimate analysis with the step-by-step investigation starting from actual engine used on the bus though its heat radiation and further distribution along the bus salon space up to the driver seat has practical value for the engineers is described in the article. This could be actual first of all for small bus building companies, who are using existing powertrain components with the engine, gearbox, drive axles, etc. from the well-known manufacturers (MAN, Mercedes-Benz, Volvo, Scania, etc), but design and manufacture their own bus bodies. It's typical business model in sphere of bus building, so that's why our approach could be used effectively in this marketing branch. Additionally, the state of the microclimate in the interior of the vehicle is one of the priority topics during design phase, because comfortable microclimatic conditions form efficient transportation, correspondingly increasing the demand for passenger transportation.

When analyzing the thermal comfort of a bus driver's work, it is necessary to take into account not only external factors affecting the microclimate, but also internal ones (metabolic processes, heat transfer from the driver's body and his clothes). Provided results could be used for further researches as the background methodology for the bus bodies microclimate analysis.

References

[1] Regulation No 94 of the Economic Commission for Europe of the United Nations (UN/ECE) — Uniform provisions concerning the approval of vehicles with regard to the protection of the occupants in the event of a frontal collision
[2] Holenko K.E., Krainyk L.V. Osoblyvosti formuvannia nesivnykh struktur karkasu avtobusa typu low-entry u Visnyku NU “Kharkivskyi politekhnichnyi instytut” 2008 r.- 152s.
[3] K.J. Craig, University of Pretoria. Multidisciplinary Design Optimization of Automotive Crashworthiness And NVH Using LS-OPT/ 7th International LS-DYNA Users Conference, May 19-21, Dearborn, Michigan, 2002.
[4] K.E. Holenko, O.Z. Horbai, L.V. Krainyk “Kompiuterne modeliuvannia i analiz napruzheno-deformovanoho stanu karkasu kuzova avtobusa typu Low-entry” u Visnyku NU “Lvivska politekhnika”. “Dynamika, mitsnist ta proektuvannia mashyn i pryladiv.” 2008 r. № 614.
[5] K.E. Holenko, O.Z. Horbai, L.V. Krainyk Analiz vidpovidnosti Pravylam YeEK OON №66 karkasu kuzova avtobusa typu Low-entry – Silskohospodarski mashyny. Zb. sauk. st. Vyp. 17.- Lutsk: Red.-vyd. viddil LNTU, 2008. – 144 s.
[6] Voichyshyn Yu.I., Kruts T.I., Zinko R.V., Horbai O.Z. Doslidzhennia mikroklimatu salonu miskoho avtobusa. Suchasni tekhnolohii v mashynobuduvanni ta transporti, №1 (14), 2020. S. 49-57.
[7] L. Krainyk, Yu. Hai. Mikroklimat salonu avtobusa. Formuvannia normatyvnoi bazy. Avtobusobuduvannia ta pasazhyrski perevezennia v Ukraini: tezy dopovidei III-oi Vseukrainskoi naukovo-praktychnoi konferentsii, Lviv, 22-23 liutoho 2018 roku. Lviv: Vydavnytstvo Lvivskoi politekhniky, 2018. S. 14–15.
[8] ASHRAE Applications Handbook SI, American Society of Heating, Refrigerating and Air-Conditioning Engineers Inc., Atlanta, GA, USA, (2003).
[9] ASHRAE Fundamentals Handbook, American Society of Heating, Refrigerating and Air-Conditioning Engineers Inc., Atlanta, GA, USA, (2001).
[10] ASHRAE Standard 55, Thermal environmental conditions for human occupancy, American Society of Heating, Refrigerating and Air-Conditioning Engineers Inc., Atlanta, GA, USA, (2004).
[11] ASHRAE Standard 62, Ventilation for acceptable indoor air quality, American Society of Heating, Refrigerating and Air-Conditioning Engineers Inc., Atlanta, GA, USA, (2001).
[12] https://comfort.cbe.berkeley.edu.
[13] P.F. Zhuk, I.A. Yurchuk. Matematychni metody v aerodynamitsi: Navchalnyi posibnyk. – K.: NAU-druk, 2013. – 315 s.
[14] Sriram Chandrasekaran, Srinivas Garimella. CFD analysis of a solid pin-fueled small modular fluoride salt-cooled reactor. Nuclear Engineering and Design, 10.1016/j.nucengdes.2021.111612, 2022, Vol 387, pp. 111612
[15] Jhan Piero Rojas, Guillermo Valencia Ochoa, Jorge Duarte Forero. CFD Analysis of Swirl Effect in a Diesel Engine Using OpenFOAM. 10.15866/iremos.v13i1.18372, 2020, Vol 13 (1), pp. 8.

AUTHOR INDEX